ENVIRONMENTAL SOIL PROPERTIES AND BEHAVIOUR

ENVIRONMENTAL SOIL PROPERTIES AND BEHAVIOUR

Raymond N. Yong
Masashi Nakano
Roland Pusch

CRC Press
Taylor & Francis Group
Boca Raton   London   New York

CRC Press is an imprint of the
Taylor & Francis Group, an **informa** business

CRC Press
Taylor & Francis Group
6000 Broken Sound Parkway NW, Suite 300
Boca Raton, FL 33487-2742

© 2012 by Taylor & Francis Group, LLC
CRC Press is an imprint of Taylor & Francis Group, an Informa business

No claim to original U.S. Government works

Version Date: 20111221

International Standard Book Number: 978-1-4398-4529-5 (Hardback)

Library of Congress Cataloging-in-Publication Data

Yong, R. N. (Raymond Nen)
   Environmental soil properties and behaviour / Raymond N. Yong, Masashi Nakano, Roland Pusch.
      p. cm.
   Includes bibliographical references and index.
   ISBN 978-1-4398-4529-5 (hardback)
   1. Soil mechanics. I. Nakano, Masashi, 1937- II. Pusch, Roland. III. Title.

TA710.Y597 2012
624.1'5136--dc23   2011046919

Visit the Taylor & Francis Web site at
http://www.taylorandfrancis.com

and the CRC Press Web site at
http://www.crcpress.com

Contents

Preface

Soils are dynamic, living systems that constitute a vital part of the environment. The soil environment is the engine that provides the base or platform for human sustenance—food, shelter, and clothing. Food production, forestry, and mineral extraction are some of the life-support activities that depend on soils—in addition to their utility in constructed facilities. All the activities associated with soils require knowledge of their properties and their behaviour under various scenarios and requirements. Studies on soils and their characteristics, properties, and behaviour have been conducted in many different fields of science and engineering. Considerable progress has been made over the past 50 years in our understanding of soil behaviour, and especially in regard to the overriding physicochemical control of soil behaviour. Much of the progress has been due to (a) the concerted focussed research efforts of researchers, and the exchange and acceptance of ideas and information between different disciplines such as soil engineering, soil science, mineralogy, microbiology, engineering geology, and others, and (b) the transdisciplinary and multidisciplinary research studies mounted by these different disciplines. There has been heightened understanding of the significant roles of *geologic origin and regional controls* on the nature, properties, and response performance of soils.

It is currently recognized that the study of *soil properties and behaviour* is a maturing discipline founded on a base consisting of allied disciplines involved with soils, and that the tools and techniques for evaluation of interactions between soil particles and water have now attained a level of sophistication that permits us to further study the concepts and mechanisms involved in the development of soil properties and behaviour. Of the many challenges ahead, the more significant and pressing ones facing us today are environment-related issues. The subjects discussed in this present book focus on *environmental soil behaviour,* with particular attention to two main interrelated groups of soil environment issues: (1) the use of soil as an environmental tool for management and containment of toxic and hazardous waste materials, and (2) the impact of ageing and weathering processes, and of soil contamination on the properties and behaviour of soils. The key elements arising from these considerations are as follows: Soils represent our natural capital and our natural resource that will change as a result of impacts from both natural and anthropogenic stresses. Many of these changes are the result of long-term effects occurring as a result of biogeochemical processes. One needs to be aware of not only the nature of the stressors involved, but also of the kinds of soil-changing processes that are evoked in order to determine short- and long-term *soil quality* and *soil functionality.*

In the discussions in Chapter 1 on the origins and functions of soils, the basic soil functions are discussed, and the concepts of soil functionality are introduced. Chapter 2 provides the in-depth discussion on the nature of soils. Particular attention is given to soil composition and the basic physical attributes, since these are important factors in the interactions discussed in Chapter 3. The central focus of the subjects discussed in Chapter 3 is soil–water interactions and resultant relationships developed. These are basic issues in environmental considerations since they are the base upon which soil properties and behaviour are founded. Changes to the nature of soil–water interactions from external or internal stressors will impact directly on soil properties and behaviour. The properties and performance of swelling soils are considered as a separate set of discussions in Chapter 4 because of their interlayer attributes and also because they are most often used in the management and containment of hazardous and toxic wastes in ground containment devices.

Chapter 5 introduces the concept of soil environment stressors and the nature of their impact on soils. Distinction is made between natural stressor sources and stressor sources traceable to anthropogenic activities. The concepts of soil functionality indicators and soil functionality indices are discussed in detail as a means for determining the impact of stressors on a particular piece of soil mass.

The subjects included in Chapters 6 and 7 deal with the mechanical, thermal, and hydraulic properties of soils. The focus of the discussions of these subjects is toward their role in the development of soil functionality. The classical soil mechanics treatment of these subjects has been muted in favour of the aforementioned focus, since the many soil mechanics textbooks in print present complete details of shear strength, consolidation and hydraulic testing, properties and performance of soils, and their application in geotechnical engineering.

Chapters 8 and 9 are concerned with the exposure of soils to contaminants. In Chapter 8, the sorption properties and mechanisms involved in interactions between soils particles and various kinds of contaminants are considered, firstly in terms of the kinds of sorption mechanisms and later in terms of methods of assessment. These considerations are extended in Chapter 9 in regard to the transport and mobility of contaminants in soils. The intent of both of these chapters is to bring forward the role of soils in land disposal and/or containment of wastes and their products. The various processes involved in interactions between contaminants and soil particles are developed with a view to determination of the impact of chemical stressors on the various soil attributes.

The concerns addressed in the subjects covered in Chapter 10 are directed toward the effects of soil ageing and soil contamination on soil functionality, as seen from changes in soil properties and soil behaviour. For many obvious reasons, the roles of microorganisms and weathering processes are difficult to quantify. The preceding notwithstanding, it is well known that these factors (microorganisms and weathering processes) do impact considerably on the nature of soil–water interaction, with resultant effect on the soil properties and behaviour of the affected soil. Soil evolution, biogeochemical processes, seasonal freeze–thaw, and other time-related phenomena are many of the factors that impact directly on soil behaviour, rendering the study of *environmental soil behaviour* a challenging subject.

Whilst many of the discussions in the various chapters of this book address soil behaviour in terms of the nature of the various interactions between different soil constituents, much remains to be learnt about *environmental soil behaviour*. The authors have been made keenly aware of the importance and relevance of this subject (environmental soil behaviour), through their decades-long collaborative research studies on soil buffer–barrier properties and behaviour 100,000 years after construction. It is hoped that the material in this book will serve as a start of the much-needed transdisciplinary effort to develop a broader and deeper understanding of what happens to a soil under weathering and ageing processes, and especially under chemical stressors, and how we can determine and quantify the effect of biogeochemical processes on soil properties and behaviour. The answers to these and other related questions will contribute to the efforts to obtain a measure of sustainability of global environment and human sustenance. The authors are grateful to their colleagues for all their past discussions on many of the topics covered in this book. Their insights, comments, discussions, and suggestions have contributed much to the development of the many pieces of information developed in this book. We are only at the early stages of learning what environmental soil behaviour entails, and as mentioned here, much more needs to be learnt.

The Authors

Dr. Raymond N. Yong, Saanich, Canada, has 52 patents to his credit and has authored and coauthored 10 other textbooks and over 500 refereed papers in various journals in the field of geoenvironmental engineering. He is a fellow of the Royal Society (Canada) and a Chevalier de l'Orde National du Québec. He and his students were among the early researchers in geoenvironmental engineering engaged in research on the physicochemical properties and behaviour of clays, their use in managing the soil environment, and in buffer/barriers for HLW (high-level radioactive waste) and HSW (hazardous solid waste) containment and isolation. He is currently engaged in research on issues in geoenvironmental sustainability.

Dr. Masashi Nakano is emeritus professor of soil physics and soil hydrogeology at the University of Tokyo and director of the RISST (Research Institute of Soil Science and Technology), Japan. He has worked and educated many students as a professor of the University of Tokyo and has published many significant papers on mass transport in soils in the fields of soil physics and soil hydrology. He has served as a leader of land reclamation engineering for food production, a promoter of global environment research such as IGBP (The International Geosphere–Biosphere Programme) in Japan, and a counsellor for the research by JAEA (Japan Atomic Energy Agency) and its predecessors on clay barrier systems for radioactive wastes disposal since the inception of plans in Japan. He was recently a member of the Science Council of Japan and is now working on such issues in soil/clay science as adsorption/transport of chemicals on soils and mineral corrosion by microorganisms.

Dr. Roland Pusch is emeritus professor at Lund University, Sweden, and is presently guest professor at Luleå Technical University, Sweden, and honorary professor at East China Technological Institute. He has made significant contributions in research on the microstructure of clays and their impact on the properties and performance. He has been very active, nationally and internationally, in pioneering work on clay buffers and underground repository systems for HLW containment in association with Swedish Nuclear Fuel and Waste Management Company (SKB) and the European Commission. He has long been active in EU projects relating to HLW and HSW containment and isolation. He is currently the scientific head and managing director of Drawrite AB, Sweden, and is working on issues of long-term stability of clay buffers in HLW repositories and on design and performance of hazardous landfills.

1

Origin and Function of Soils

1.1 Introduction

We begin by establishing what is a soil. In the most general sense, a soil is a collection or accumulation of disintegrated rock fragments whose particle sizes can range from *boulders* measuring up to a few metres in dimension to much lesser sizes generally called *soil*, with particle sizes ranging somewhere close to 100 mm to sizes that cannot be seen by the naked eye—less than 0.0001 mm (in the micrometre range). Disintegration of rock produces fragments commonly referred to by the public as boulders, stones, gravel, sand, and clay—most often distinguished or characterized by particle (fragment) sizes, with the largest ones being stones and the smallest ones being clay. The forces and agents responsible for rock disintegration include mechanical, chemical, biologically mediated, and hydraulic.

Practical experience and reports in the popular literature have often shown differences in one's perception of what constitutes a soil, the reasons for which most generally lie in one's use and understanding of the soil material. Strictly speaking, soil materials could also include decomposed organic matter and other constituents such as evaporates. We use the term *soil material* in recognition of the fact that soil is a material consisting of various soil fractions identified as sand, silt, clay, organic matter, carbonates, oxides, and so forth. The term *soil* is now commonly used in place of *soil material* to mean the same thing. The nature and characteristics of the soil fractions constituting a soil will be discussed in greater detail in the next chapter.

1.1.1 Why Are We Concerned with Environmental Issues?

The very same forces and agents, which we will call *environmental forces/agents*, responsible for rock disintegration will persist after rock fragmentation. They will continue to act on the fragmented material (i.e., soil material), and in many instances, they will act in combination with forces resulting from actions associated with anthropogenic activities. One can expect that the nature of the various soil fractions constituting a soil material will change over time.

Changes in the nature of soil will be reflected in associated changes in the properties and characteristics of the affected soil. Because of this, it becomes important to recognize that if, for example, one designs and constructs an engineering facility that relies on various predetermined soil properties such as soil strength and hydraulic conductivity for its long-term survivability, any deterioration or degradation of these properties would threaten its survival. This means that unless changes in the measured soil properties used for design of an engineering facility are anticipated and factored into design considerations, the survival of the constructed facility could be jeopardized if negative changes in these properties occur. A very pertinent example of this is the engineered clay barrier used in containment of waste products. For secure long-term containment, one relies on the hydraulic transmission and chemical buffering properties to maintain their design capabilities. It goes without saying that deterioration of any of these properties will allow transport contaminants into the surrounding regions, thereby impacting the health of the various vegetative species and biotic receptors.

One needs therefore to (a) focus attention on the kinds of changes that will occur in soil due to the impact of environmental forces/agents and actions associated with anthropogenic activities, (b) provide an appreciation of what the environmental agents and anthropogenic activities are, and what kinds of forces/stresses are associated with these agents and activities, (c) discuss the impacts and various processes involved with these forces/stresses, and (d) establish the likely changes in the various soil properties and show how these impact on performance of the affected soils.

1.2 Soil Origin and Formation

There are several factors that control or influence the processes that are involved in producing the type of soil from parent rocks. Most obviously, the type of rock (composition and texture) leads the group of factors in terms of importance. This is closely followed by site conditions such as availability of water, climate, topography, and so forth. These factors have considerable influence on the kind of processes involved in breaking down rock to its various fragments, and also in the production of soil types from the broken rock fragments.

1.2.1 Parent Material

Rocks that when fragmented will eventually form soils are called *parent material*. They can also be referred to *source rocks*. These kinds of rocks fall into three general classes: igneous, sedimentary, and metamorphic. Igneous rocks are the product of magma. When they are the result of extruded lava, they are known as extrusive rocks. When the rocks are the result of unextruded magma, they are known as intrusive or plutonic rocks depending on the depth of the magma. The rate of cooling of the magma has a great influence on the size of mineral grains formed in the rocks. These minerals are important participants in the development of soil properties and characteristics. With higher rates of cooling in the extrusive rocks, one would expect the mineral grains to be fine, that is, very small. Slower rates of cooling will produce larger mineral grains. Since plutonic rocks are found at a greater depth than intrusive rocks, the mineral grains of the plutonic rocks will be larger than those found in intrusive rocks. Of the various kinds of igneous rocks such as andesite, felsites, durite, gabbro, basalt, and others basalts and granites are by far the greatest proportion of rocks that classify as igneous.

Sedimentary rocks are formed from the wind and water deposition of the material derived from chemical and physical breakdown of other source rocks, compressed under considerable geologic pressure and heat. The commonly found sedimentary rocks include sandstone, siltstone, dolomite, shale, and limestone. Metamorphic rocks are formed from metamorphosing of igneous or sedimentary rocks. Since considerable heat and pressure are required in the process of metamorphic transformation of the igneous and sedimentary rocks, there will be transformation of the original minerals found in the rocks. The common metamorphic rocks include quartzite, schist, slate, gneiss, and marble.

1.2.2 Weathering of Rock

1.2.2.1 Natural Processes

The natural processes involved in the transformation of source rocks are generally referred to as *weathering processes*. General usage will refer to this as *weathering*, with or without inclusion of the term *processes*. In the context of soil formation from rocks, weathering is a spontaneous reaction involving geologic material (rocks) and energy. Since it is a change in the direction of a decrease in free energy of the system, it is possible to predict the thermodynamic susceptibility of minerals to weathering. The assemblages of minerals in soils are reflective of the temporary equilibrium stages in the dynamic process of changes from source rocks to the final end product. The weathering sequence in development of clay minerals from source rocks may be represented in terms of a reaction series portraying soils as an equilibrium stage in a continuous progression from, for example, igneous rocks such as granite and basalt to a final end product such as laterite.

The intensity of weathering is highest where interfaces between the atmosphere, hydrosphere, biosphere, and lithosphere meet and overlap. A good example of this can be found in the upper soil zone or layer in temperate humid climate regions, and deeper soil zones in humid tropic regions.

Transformation of rock fragments to soil occurs primarily in the regolith: the region between solid rock and the soil.

The rate of weathering is influenced or controlled by several factors, not the least of which are the composition and size of the source rock (rock mass) or rock fragments, size and type of exposed surface areas of rocks, relative solubilities of rock mass, fragments and weathered material, permeability of the rock, source and availability of water (e.g., position of water table or hydrogeological setting), chemical composition of water, temperature, topography, oxygen content or availability, organic matter, cyclical freezing and thawing and wetting-drying cycles, climatic envelope, and others.

The primary types of actions responsible for the forces involved in fragmenting source rocks are physical and chemical in nature. The agents responsible for these soil-forming forces can be physical, chemical, or biological in nature. Strictly speaking, the actions of biological agents are actually biologically mediated chemical actions. In that sense, we see references in the literature to only two types of soil-forming actions or forces: physical and chemical. Under natural circumstances, that is, without any actions attributable directly or indirectly to anthropogenic input, these types of actions are often referred to as *chemical weathering* and *physical weathering*. In short, physical weathering and chemical weathering of rock are two of the *natural processes* of disintegration of rock that ultimately result in the production of soil. Figure 1.1 gives a summary illustration of the major weathering processes involved.

1.2.2.2 Physical Weathering

Physical weathering of rocks involves forces and processes that result in fragmenting of the rocks. Particles of much lesser sizes are obtained as a result of the breakdown of the various rock fragments or minerals. The agents producing these forces and processes could be (a) thermal in nature such as cyclic freezing and thawing, and heating and cooling, since rock components have different coefficients of expansion; (b) expansion of water in rock fissures as a result of freezing; (c) physical movement of glaciers atop the rocks, resulting in application of abrasive forces, and water on/against rocks and in rock fissures; and (d) physical pressures and loads from superposed glaciers and displacements due to plant growth and activities of animals. Note that in physical weathering, there is generally no alteration of the chemical composition of the rock material.

1.2.2.3 Chemical Weathering

Chemical weathering of rocks involves processes leading to transformation of the constituents and even loss of some constituents, addition of protons, and rearrangement of constituents into new materials. It should

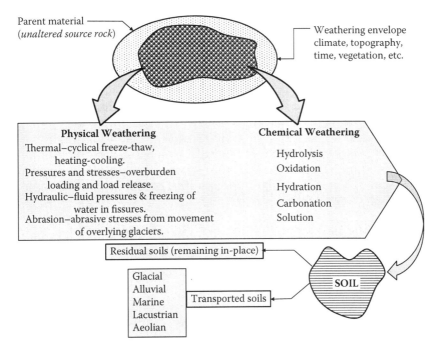

FIGURE 1.1
Soil formation beginning with weathering of unaltered source-rock mass (parent material) and ending with final resting place—that is, residual or transported.

be noted that chemical weathering will not occur without the presence of water; that is, water is an essential component in chemical weathering. The rates of chemical reactions in the weathering process are determined by temperature, moisture, composition, potential for leaching, and composition of the leaching solution. Chemical processes involved include hydrolysis, hydration, carbonation, oxidation, reduction, and solution. Protons in the leaching solution can be supplied from CO_2 in the air or from decaying vegetation or other acids added to the system.

Both the amount of water available and heat are big factors in the chemical weathering of the parent rock mass. Warm and humid climates are therefore instrumental in producing correspondingly high weathering of the minerals in the parent rock mass. The presence and amount of organic matter are also very important factors in the control and rate of weathering. If organic matter is allowed to remain, as for example in cool environments, high humidity would produce conditions that would successfully promote reactions between the organic acids and parent rock. However, if the environment is changed to a hot, humid environment, oxidation of the organic matter would preclude the formation of organic acids, resulting in little or no reactions between the organic matter and the parent rock.

1.2.3 Processes of Chemical Weathering

The major processes involved in formation of soil by chemical weathering agents include

- Hydrolysis: Hydrolysis is the reaction between the H^+ and OH^- ions of water and the elements or ions of a mineral in the source rock or other type of parent material. It is considered to be an important reaction for silicates. These are the primary rock-forming minerals. For hydrolysis reactions to cause weathering of rocks, it is necessary for the weathering products to be removed continuously. This means that fresh application of water is necessary throughout the weathering process. A generalized reaction of a silicate is given by Keller (1957) as

$$MSiAlO_n + H^+OH^- \leftrightarrow M^+OH^- + [Si(OH)_{0\text{-}4}]_n + Al(OH)_3$$

$$+ (M,H)Al^oSiAl^tO_n$$

  where the superscripts o and t refer to the octahedral and tetrahedral coordinations, respectively (as will be explained later in Chapter 2), and M and subscript n refer to metal cations and unspecified atomic ratios, respectively.

- Hydration: Hydration involves the chemical uptake of water. The minerals in the rocks form hydrous compounds with water uptake, resulting thereby in an increase in volume. The minerals involved with water uptake include the silicates, oxides of iron and aluminium, and sulphates. Because of the volume expansion of the rocks in water uptake, fragmentation of the rocks will occur. Whilst dehydration may occur in arid regions, it is not generally recognized as an important process in weathering.

- Oxidation: Oxidation occurs when an atom loses an electron e^-. When elemental oxygen combines with the weathering substance, elements are lost to the oxygen, which then becomes ionic. Technically speaking, oxygen need not necessarily be involved. For example, Fe may oxidize to FeS, FeS_2, $FeCl_3$. Oxidation of minerals by gaseous oxygen occurs by the intermediary action of water. Depending on the partial pressure of gaseous oxygen and acidity of the water, gaseous oxygen dissolved in water renders the resulting solution an oxidizing potential. Most rocks contain some iron in the form of sulphides or oxides. With this in mind, it is seen that interaction with this type of water promotes the process of oxidation, giving rise to weathering by oxidation of those rocks.

- Carbonation: Because rainfall passing through the atmosphere collects carbon dioxide (CO_2), rainwater will contain some amount of CO_2, resulting in the production of carbonic acid:

$$H_2O + CO_2 \rightarrow H_2CO_3 \rightarrow H^+ + CO_3^-$$

Carbonation, which is one of the dominant reactions, is the union of carbonic acid (H_2CO_3) with bases, the result of which is the formation of carbonates. This rainwater will attack minerals such as feldspar, hornblende, and olivine under normal temperatures and pressures. Carbonation in silicates is accompanied by the liberation of silica. This silica may remain as quartz or may be removed as colloidal silica. It is useful to note that present public concern regarding excessive discharge of CO_2 due to anthropogenic activities have often not given this aspect of acidification of rainwater as much attention as it deserves.

- Solution: The same carbonic acid responsible for carbonation will act as a solvent, attacking carbonate-type rocks such as limestones.

1.2.4 Parent Material and Other Influences

1.2.4.1 Parent Material

The composition and texture of the parent rock are important factors in the initial stages of weathering. Technically speaking, weathering processes are omnipresent in nature; that is, they are present all the time. The influence of parent material is relatively short-lived in regions with humid conditions and high temperatures. In contrast, in arid and arctic regions where the opposite conditions exist, the influence of parent material remains indefinitely.

Alkali and alkaline cation content of parent rocks are important factors in determining parent rock weathering products. Rocks containing no alkalis can only produce kaolinite or lateritic weathering products. Rocks containing alkalis and alkaline earth cations in addition to alumina, silica, and so forth, such as igneous rocks, shales, slates, schists, and argillaceous carbonates, will produce a variety of weathering products. Under different conditions such as climate, topography, and time, the same parent rock can produce both kaolinite and montmorillonite. Also, parent rocks with widely different compositions and texture can produce the same type of soil with a characteristic clay mineral composition.

1.2.4.2 Climate and Vegetation

Climate and vegetation often go hand in hand since climate affects vegetation and the products produced by decay of the vegetative (organic) material. The amount and kind of products resulting from the decay of vegetation are significant factors of weathering. Where vegetation is absent or rapidly oxidized, production of organic acids is minimal or absent, meaning that we will not have organic acids from this source as agents for weathering of parent material. Whilst hot and humid climates encourage oxidation destruction of vegetative organic material, cool humid climates allow for active organic acids from decaying organic material to react with the parent material, thereby contributing to the production of decay products.

The climatic factors of significant importance are temperature and rainfall. Warm and humid climates are favourable conditions for rapid weathering of the minerals in the source rocks. Continuously or persistent wet climates allow for percolation to remove decay products, thus allowing for weathering to proceed. At the other end of the spectrum, dry climates do not allow for removal of the decay products partly because of the much smaller amounts of decay material but mainly because of the absence of rainfall.

1.2.4.3 Topography and Time

We link topography and time together because of the importance of time of exposure of the parent material or weathering products to weathering agents. Percolation and infiltration of water through weathering products, in the natural environment context (i.e., without human intervention), are determined by topography. This influences the rate of erosion at the surface and hence the rate at which fresh parent material is brought close to the surface. In regions of high rainfall, low flat areas that are completely water saturated will produce conditions that will retard weathering.

The time factor is important also when weathering is moderately severe and the composition of the parent material permits production of various alteration products. Longer time periods allow for leaching and other weathering processes to proceed.

1.3 Soil Classification

The purpose of soil classification is to provide a means to convey information of the type of soil under consideration. Soil classification schemes may be simple descriptive schemes or may be detailed schemes that include grouping soils into various types or groups depending on certain features and characteristics. Whether or not the information on soil type would also convey some idea as to the general characteristics and properties of the soil depends on the protocols adopted in classifying the soil in question. There are several ways in which soils are classified, depending on the intended use of the soil. For example, geological classification of soil, which is mostly in terms of surficial deposits, is genetic but partly descriptive. This differs from those used in soil and geotechnical engineering, and from those in agriculture and soil science.

1.3.1 Subdivision of Surficial Deposits

The major subdivisions in the geologic classification of surficial deposits include

- *Residual soils*: Soils formed from weathering of source rock remain in place or are transported to other locations due to the actions of transport agents. Those soils that remain in place are called *residual* soils. Whether or not they retain most of the characteristics of the constituents in the source rock will depend on several factors related to further weathering processes of these soils. Those soils that retain the characteristics of the parent material are called saprolitic soils.

- *Transported soils*: The major transport agents responsible for transporting soils from one location to another include glaciers, wind, water, and humans. Discounting human activities, the soils formed from the actions of these transport agents include

  Glacial soils—From transport/deposition of rock grinding and erosion products. Soil formations following retreat of the glaciers are in the form of drifts, moraines, till, drumlins, and so forth.

  Lacustrian deposits—These soils are former lake sediments. The region where soil deposits are of considerable interest is the river discharge zone, that is, the zone where the river feeding the lake enters the lake. In a sense, this is a minor version of formation of deltas, which are characteristic of alluvial deposits.

  Fluvial deposits—These soils are the result of deposition of soil material transported by streams and rivers. It is not uncommon for the modern literature to include fluvial deposits under the umbrella of alluvial deposit.

  Alluvial deposits—These refer to alluvial fans and deltas and other water-borne deposits not associated with fluvial deposits.

  Marine soils—They are the result of sediments formed from deposition of suspended material in a marine environment, that is, sediment/soils obtained upon retreat of the ocean from a particular region. Typical kinds of soils are marine clays and silts.

  Aeolian deposits—Formed from wind transport. Soil materials carried by wind power range from fine particulates to sands, and soil formations are usually found as dunes, beaches, loess, and drifts.

1.3.2 Soil Horizons

The concept of soil horizons, part of the consideration of soil genesis, is one that is frequently used by pedologists and soil scientists working in agriculture. We define a soil profile as a succession of soil strata generally called soil horizons. These horizons are basically soil layers or strata constituting a mature soil profile, and are distinguished on the basis of distinct differences in texture and physical or chemical properties between neighbouring layers. In many instances, there will also be colour differences between these layers. These horizons are generally labelled O, A, B, and C—starting from the top O horizon and continuing downward to the bottom C horizon. In some disciplines, for example, soil science, there is also a R horizon that is also called a D horizon. Figure 1.2 shows the horizons and the main distinguishing features. We should emphasize that these horizons are not necessarily clearly separated; that is, the boundaries separating A from B and B from C are not well defined. In fact, it would be unusual to find clear separations between these horizons. The more usual situation would be transition zones between these horizons.

The degree or level of maturity of a soil profile plays an important role in how well the three horizons are discerned. This is because leaching is a major weathering agent in producing the different properties and characteristics that distinguish between the A and B horizons. In the A horizon, one finds soil material that has been altered by natural processes such as leaching and removal of leached products and other loose material by water or wind. This horizon is commonly called the leaching zone. The presence of concentrations of greenhouse gases in the atmosphere together with

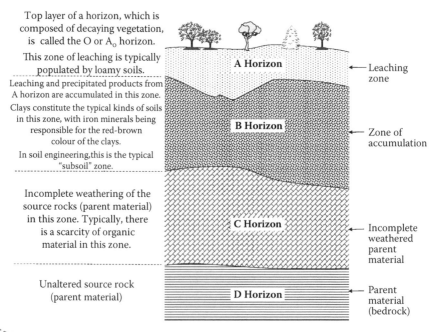

FIGURE 1.2

Soil horizons typical of a mature soil. Immature soils (soils that have not had much exposure to weathering and leaching phenomena) will not show distinct layering for zone classification.

decaying vegetation on the ground surface will contribute to the acidity of rainwater and snow, the primary leaching agents for this horizon. Note that when appropriate, the organic top surface of the A horizon is referred to as the O horizon, to identify or take into account the thin layer of organic material consisting of decaying vegetation, plant litter, and humus.

The B horizon is characterized by leaching products and other washed material deposited from the A horizon. The common name for this horizon is the *zone of accumulation*. Accordingly, one would expect to find various kinds of soil (clay) minerals, oxides, and organic matter in this stratum. This is probably the most common soil zone encountered in many soil engineering projects, and is generally referred to as the *subsoil*. Below this is the C horizon, normally considered to contain soil that is unaltered by weathering subsequent to deposition or formation. The top layer of the C horizon will contain partially weathered material, and the interface layer separating this horizon from unaltered solid rock is called the *regolith*. This is also sometimes referred to as the R or D layer.

Because of the slow reaction rates associated with weathering processes, development of mature soil profiles and soil horizons takes hundreds to thousands of years. As stated previously, the important factors involved in this time (development) period are source material, climate, vegetation, topography, and time. For example, warm and wet regions allow for more rapid chemical weathering of parent material, especially if the leaching agents contain organic acids. One could obtain a lesser time requirement for development of a mature soil profile with significant proportions of aluminium and iron in the zone of accumulation. Weathering of dark-coloured coarse-grained and fine-grained mafic rocks containing large amounts of iron and magnesium will produce soil profiles that will contain dark-coloured soils that would have high clay content. On the other hand, weathering of felsic rocks such as granites containing high proportions of iron and silica will produce soils that contain light coloured clays with significant proportions of quartz sands. The evidence shows that an understanding of how soil profiles are developed would be useful in the construction of classification schemes for soils.

1.3.3 Classification for Engineering Purposes

It is understood that whereas a full detailing of soil properties and behavioural patterns determined from measurements obtained from a comprehensive suite of laboratory tests might provide for an ideal soil classification scheme, this is neither feasible nor practical. This is because the engineering use of soils includes a large variety of applications and situations demanding specific knowledge of different types of soil properties and behavioural characteristics. The purpose of soil classification is to provide a general description of the soil in a terminology for classification that is understood by all concerned parties. To that end, it is apparent that soil classification schemes should be universal: that is, a large majority of stakeholders would be able to utilize the schemes in toto or as a starting point. With experience gained from associated laboratory tests, one could associate the identified soil with certain characteristic properties or behavioural patterns.

1.3.3.1 Particle-Size Differentiation and Textural Classification

The two most common types of soil classification for engineering purposes are *particle-size differentiation* and *textural*. Particle-size or grain-size differentiation of soils classifies or identifies soils on the basis of ranges of limiting sizes. In the left-hand bar chart shown in Figure 1.3, the distinction between gravel, sand, silt, and clay (called *separates*) is made on the basis of a limiting maximum or minimum size. Thus, for example, the limiting maximum size for clay, first proposed by Atterberg (1908, 1911), is 2 μm; or one could say that the limiting minimum size for silts is 2 μm. Sands and silts are also differentiated in terms of fine, medium, and coarse sizes. Fine sands are greater than 0.06 mm and less than 0.2 mm in particle size. Sand grains with sizes greater than 0.2 mm and less than 0.6 mm are classed as medium sands, and coarse sands are particles with sizes ranging from 0.6 to 2 mm. Particles greater than 2 mm are called *gravels*. We should point out that whilst the limiting

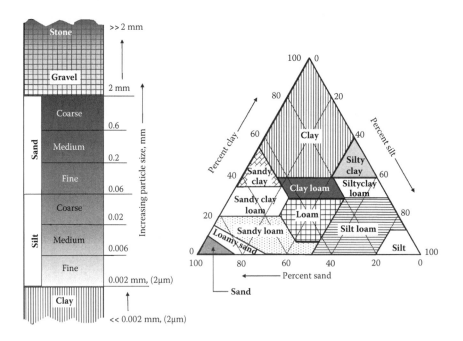

FIGURE 1.3

The left-hand bar chart shows the grain-size (particle-size) differentiation for classification of soils. The right-hand chart shows the textural classification of soils based on proportions of sand, silt, and clay in a soil.

size boundaries for these separates are considered by some as arbitrary, they have nevertheless been accepted in common practice.

The triangular chart shown in the right-hand side of Figure 1.3 is a textural chart that identifies soil type according to the proportions of sand, silt, and clay in the soil. As with the particle-size classification scheme shown on the left-hand side, textural charts classify soils based on particle sizes, meaning that these kinds of charts and classification schemes are more applicable for granular soils (gravels, sands, and silts). It is important to understand that the *clay* classification used in both the particle-size classification scheme and the textural chart refers to clay-sized particles. Some confusion may arise if one does not distinguish this from *clay minerals* as a particular type of soil. We will be discussing this in greater detail in the next chapter (Chapter 2).

1.3.3.2 Particle-Size Distribution Curves

A particle-size distribution curve is useful as an indicator of the kind of soil under consideration. This does not require an arbitrary division of particles into separate sizes (*separates*). Figure 1.4 shows some typical particle-size distribution curves in a semilog plot. The ordinate shows the weight, in terms of percentages, of particles finer than a particular size, and the abscissa shows the effective particle diameter.

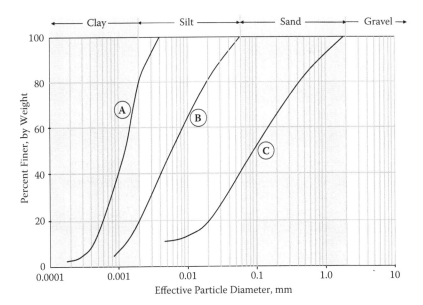

FIGURE 1.4
Particle-size distribution curves. Curve A represents a clay. Curve B is a clayey silt, and Curve C is a silty sand. Note that identification of soil types is based solely on proportions of *separates* in the soil sample and not on any particular behaviour characteristic.

To obtain a particle-size distribution curve, it is important to obtain a representative soil sample that is completely dispersed; that is, all the particles must be able to act individually and separately. Dispersion of a soil sample requires removal of cementing or bonding materials responsible for aggregating the individual particles, that is, deflocculating the soil sample. Typical cementing materials in surficial soil samples include carbonates, iron and aluminium oxides, and organic matter. Note that after removal of cementing material, the soil sample must be deflocculated or peptized. This is especially critical for clays since the forces of attraction that are always present between particles (see Chapter 2) will cause flocculation of the clay particles. Since the forces of attraction cannot be easily decreased, they need to be overcome by increasing the forces of repulsion between the clay particles (Chapter 2), generally accomplished by having a monovalent exchangeable ion on the clay and a low salt concentration in the solution. A common procedure is to add sodium ions (sodium metaphosphate) and washing the soil suspension free of soluble salt. The advantage of using sodium metaphosphate as the source of sodium ions is that the metaphosphate will form complexes with any remaining calcium or magnesium ions.

1.3.3.3 Atterberg Limits Classification

The initial work in agricultural science undertaken by Atterberg (1911) on classification of soils, designed to provide an assessment of the state of a soil in relation to its water content, serves as the basis for classification of soils in respect to its state of consistency. The consistency limits devised by Atterberg, shown in Figure 1.5, provide an indication of the plasticity or plastic behaviour of soils (mainly clays) in relation to this water content. These limits are most often referred to as the Atterberg limits, and the tests are part of the group of index tests that include determination of particle-size distribution.

The *shrinkage limit* (SL) shown in Figure 1.5 represents the water content boundary between a solid soil and a semisolid soil. Using the measured volume of the test sample as a guide, the shrinkage limit is the water content at which a further reduction in water content will not result in any observable decrease in volume of the test soil sample. In terms of a water content-volume change definition, the shrinkage limit is the only limit that can be rigorously defined. Using the relationship shown in Figure 1.5, the SL can be defined as the point where volume change is no longer linearly proportional to the change in water content. The SL is sometimes referred to as the lower plastic limit. The total shrinkage limit shown in Figure 1.5 is the point where the volume of the test soil sample is defined by its oven-dry volume.

The *plastic limit* (PL) is the water content at which any increase in water content beyond this limit would result in a plastic state for the test soil. A semisolid state of the test soil exists at water contents between the SL and PL. Increasing the water content of the test soil above the PL would produce a more plastic soil, to the point where the soil would exhibit liquid-like behaviour;

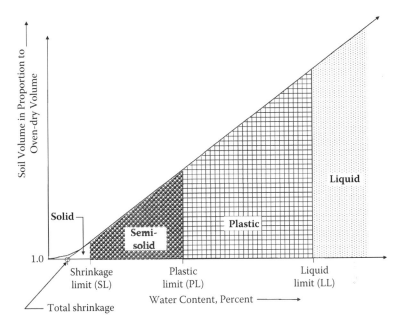

FIGURE 1.5
Pictorial representation of the Atterberg consistency limits. The ordinate represents the soil sample volume at any particular water content as a proportion of the oven-dry volume of the test sample.

that is, a liquid-like state of the soil would be obtained. At this point, the water content defines the *liquid limit* (*LL*) of the soil. The range of water contents between the plastic limit (*PL*) and the liquid limit (*LL*) is defined as the *plasticity index* (*PI*). Laboratory test procedures have been devised and standardized for determination of the various consistency limits.

1.3.3.4 Unified Soil Classification System

The *Unified Soil Classification System* (Waterways Experiment Station, 1953) is a soil classification scheme that uses field identification procedures and laboratory classification criteria to organize the soils under consideration into specific groups. Particle-size distribution information allows for initial grouping of soils into two major groups: coarse-grained and fine-grained soils—and a third group (highly organic soils) that is not differentiated by particle-size distribution. These highly organic soils are readily identified by colour, odour, spongy feel, and their fibrous nature.

For the two major groups, coarse-grained soils contain more than 50% of soil particles greater than the number 200 sieve size, and fine-grained soils contain more than 50% of particles passing the number 200 sieve size. The following notation is used for the soils in the various groups: gravels (G), sands (S), fine sands and inorganic silts (M), inorganic clays (C), organic silts, and organic clays (O). Since soils, as a rule, contain mixtures of different-sized particles, further refinement of the two broad groups into subgroups can be made with (a) information on proportions of fines in the coarse-grained group, using secondary notations of W for well graded and P for poorly graded and (b) the addition of Atterberg limits information for the fine-grained soils.

Laboratory classification criteria in the classification system use the coefficient of uniformity C_u (also called the Hazen coefficient) as an aid in grouping the coarse-grained soils into the W and P subgroups, where $C_u = D_{60}/D_{10}$, and D_{60} and D_{10} refer to grain sizes for which 60% is finer than, and for which 10% of the soil is finer than, respectively. For the other coarse-grained soils subgroups, the Atterberg limits are used in conjunction with a plasticity chart with the Casagrande A line (Figure 1.6). For fine-grained soils, the identification procedures focus on the soil particle sizes less than the number 40 sieve size using three particular indicators: (a) dry strength, indicative of crushing characteristics; (b) dilatancy, indicative of reaction to shaking; and (c) toughness, indicative of the consistency of the sample near the plastic limit. The types of soils classified under the *Unified Soil Classification System* are shown in the figure in relation to their plastic and liquid limits. Detailed descriptions of the various types of soils and the Casagrande A line can be found in Waterways Experiment Station (1953) and Casagrande (1937), respectively.

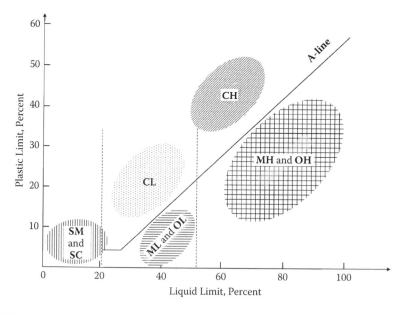

FIGURE 1.6
Plasticity chart used in conjunction with laboratory classification criteria in the Unified Soil Classification System. Detailed descriptions of the System and the Casagrande A line can be found in Waterways Experiment Station (1953) and Casagrande (1937), respectively.

For coarse-grained soils, the following subgroups are obtained:

- GW for well-graded gravels, gravel–sand mixtures, little or no fines
- GP for poorly graded gravels, gravel–sand mixtures, little or no fines
- GM for silty gravels, gravel–sand–silt mixtures
- GC for clayey gravels, gravel–sand–clay mixtures

The sands in the coarse-grained soils group are classified in the same manner:

- SW for well-graded sands, gravelly sands, little or no fines
- SP for poorly graded sands, gravelly sands, little or no fines
- SM for silty sands or mixtures of sands and silts
- SC for clayey sands or mixtures of sand and clay

The fine-grained soils (M and C) in this classification scheme are soils where more than 50% of the soil particles pass through the number 200 sieve size. For these soils, subgrouping is made on the basis of the water content at the liquid limit. For fine-grained soils with liquid limits less than 50%, the following subgroups are obtained:

- ML for inorganic silts and very fine sands, rock flour, silty or clayey fine sands, or clayey silts with slight plasticity
- CL for inorganic clays of low to medium plasticity, gravelly clays, sandy clays, silty clays, lean clays
- OL for organic silts and organic silty clays of low plasticity

For fine-grained soils with liquid limits greater than 50%, the following subgroups are obtained:

- MH for inorganic silts, micaceous or diatomaceous fine sandy or silty soils, elastic silts
- CH for inorganic clays of high plasticity, fat clays
- OH for organic clays of medium to high plasticity, organic silts

1.4 Basic Soil Functions

We define *basic soil functions* to mean the basic purposes or roles of soil in the ecosphere. In essence, a *functioning soil* is a soil that fulfils its natural or planned purpose, and *soil functionality* defines the capability of a soil to function in a role according to its nature, circumstance, and ecosystem constraints. For the purposes of the discussion on basic soil functions, we will consider *soil* in the context of *geoenvironment* as defined by the schematic shown in Figure 1.7. The geoenvironment, which is a specific compartment of the environment, includes a significant portion of the *geosphere* and portions of both the *hydrosphere* and *biosphere*. The *earth surface material* component from the *geosphere* shown in the figure is defined as *soil and other natural land surface materials* (organics, debris, etc.) from the geosphere: that is, soils in the A, B, and C horizons and the overlying surface material. From the hydrosphere and the biosphere, we include the receiving waters in the land environment and the living or life zone in the earth. Considering only the soil aspects, all of these contribute to the *soil ecosystem*.

There are a variety of roles for soil in the geoenvironment. For convenience, these can be grouped into three main categories:

- Natural in situ role of soil—The role of soil is in its natural state in all the settings (situations) found in the geoenvironment. This includes residual and naturally transported soils. Soils transported in support of anthropogenic activities are not included in this group.

- In situ role of soil with external intervention—The function of soil as a result of human intervention.

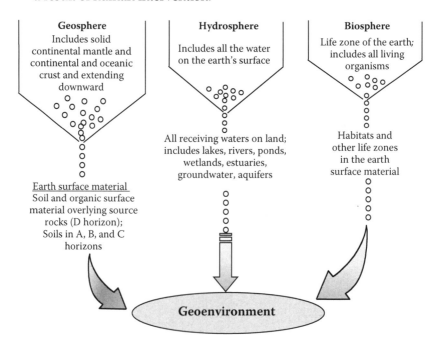

FIGURE 1.7
The various constituents of the ecosphere and their relationship to the geoenvironment. Note that the definition of *earth surface material* in the left-hand column of the figure applies to all references to the term *earth surface material*.

- Soil as a resource material—This includes soil used by itself as a resource material, for metal recovery, and for minerals extracted from the soil for use as ingredients for various uses.

In turn, these main groups will spawn numerous subgroups and sub-sub-groups. One should expect that there will be overlap between main and subgroups because of the variety of situations and requirements for soil functions. A number of these will be described in the following summary discussion.

1.4.1 Natural in Situ Role

As we have seen in Figure 1.7, the geoenvironment is a natural resource base that features soil as its primary component, and the natural resources in the geoenvironment can be considered as the natural capital of the geoenvironment. Figure 1.8 shows some of the principal ecosystems within the geoenvironment together with some of their major derivatives. The term *ecosystems* refers to a system where the various individual elements and organisms interact singly or collectively to the advantage or disadvantage of the whole. We consider the natural in situ role of soil to be one where the soil, which is in its natural state in the geoenvironment, functions without human or any

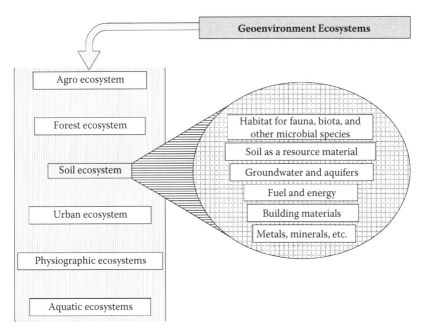

FIGURE 1.8
The soil ecosystem as part of some of the major ecosystems, resources, and features of the geoenvironment. Note that the physiographic ecosystems include, for example, coastal, alpine, desert, and arctic ecosystems.

other external intervention. In this main group, two main subgroups and perhaps some other lesser subgroups stand out prominently:

- Habitat for fauna, other biota, and microbial species. Loss of biodiversity is perhaps one of the best-known concerns in seeking active means for habitat preservation.

- Growing medium for plants, trees, and flora. Agricultural activities that involve the use of soil amendments or control of insecticides and pesticides are not directly included in this main group. These activities are grouped under the next major group: in situ *role of soil with external intervention.*

- Physical and chemical buffering system. Particular examples of these are the filtration capability, especially in transmission of groundwater to receiving waters, and contaminant partitioning or accumulative capacity of soils in respect to transport of contaminants.

- Groundwater storage and quality determination. Surface hydrology, hydrogeology, and the roles of aquifers and aquitards are particularly significant.

- CO_2 sink. Soil, organic matter, plants, trees, and others are important players in the carbon cycle.

- Heat sink and source for climate and air temperature control.

These groups and subgroups feature prominently in establishing ecosystem functioning. By this, we mean the *activities* of flora, fauna, plants, biota, microorganisms, animals, and so forth within the community defined by the ecosystem. In the broader sense, we can define a *functioning ecosystem* to include biological, chemical, and physicochemical activities that are characteristic of the type of ecosystem under consideration. The living organisms in a specific ecosystem represent the biodiversity in the ecosystem and are important in maintaining atmospheric composition and production of oxygen by photosynthesis and fixation of CO_2 and recycling of carbon, nitrogen, phosphorous, and other elements in the soil.

1.4.2 In Situ Role with Human Intervention

The subgroups describing the basic soil functions within this main group (and the next main group in Section 1.4.3) are by and large associated with the major activities and industries involved in the production of *food, shelter,* and *clothing*. These include

- Growing medium—The major activities and industries involved are agroindustries and other activities associated with production of raw material such as cotton and timber. What distinguishes this from the previous natural in situ role is the use of soil amendments and other technological aids designed to increase food and material production. In situ soil–water relationships, physical and mechanical properties, and fluid transmissivity, heat conductance, and transmission are some of the more important considerations.

- Foundation support and other ground engineering-associated systems—Major activities and products include foundation support for structures and infrastructure, tunnels and buried structures, buried pipelines, and transmission lines. The major sets of concerns revolve around the physical, fluid transmissivity, and physicochemical and mechanical properties of the in-place soil.
- Containment systems—We include waste landfills, mine tailings ponds, and repositories in the B, C, or D horizons in this subgroup. The major functional requirements revolve around natural soil buffering potential capacity, contaminant accumulation, and hydraulic permeability. The processes involved in situ are similar to those described in wastewater natural treatment systems (the next subgroup).
- Wastewater natural treatment systems—Slow-rate treatment, rapid infiltration, and natural wetlands are some of the natural treatment techniques used. The in situ processes involved include physical actions such as sedimentation; chemical reactions such as oxidation-reduction, ion exchange, and precipitation; and biological activities such as biological conversion and degradation.
- In situ groundwater harvesting, hydrocarbon resource extraction-exploitation, and related activities—These include the various in situ means for extraction of heavy oil, storage systems, and carbon sequestration.

1.4.3 Soil as a Resource Material

The subgroups in this main group are perhaps the most varied of the three main groups. As in the previous main group, these subgroups are best associated with activities and industries involved in the production of *food, shelter,* and *clothing.* Some examples of these include

- Engineering construction—This subgroup includes the use of soil as a construction material in varied forms, for example, sand and gravel for Portland cement concrete and bituminous concrete, embankments, sub-bases, and buffer-barrier systems.
- Mining industry (metal, mineral, aggregate, hydrocarbon resources, etc.)—Examples include extraction of oil such as tar sands, metals (bauxite industry), minerals such as quartz, calcite, and clay minerals such as kaolinite, montmorillonite, and illite.

- Paper, packaging, and paint industry—Paper coating, packaging, and paint pigment.
- Pharmaceutical-cosmetic industry—Laxatives and creams.
- Miscellaneous—Pottery, ceramics, oxidizing agents, catalysts, filter for UV emissions, thickeners, chalk, and so forth.

There are many more uses for soils other than the examples shown in this section. Suffice to say that the importance of soil, either in place of or as a resource material, cannot be underestimated. It is a living dynamic system and its properties and behaviour will inevitably change with time.

1.5 Concluding Remarks

We consider soil to be an important resource material that possesses a large variety of functions. Soil functionality is dependent on several factors and circumstances, some of which are controlled by natural forces and others by humans and their activities. In this chapter and in this book, we are concerned with the engineering properties and behaviour of soils. The brief discussion on soil formation (origin of soil) is designed to remind us about the importance of understanding where soil comes from because we need to appreciate that the properties and behaviour of a soil are intimately tied into its origin, and ultimately, its composition. Readers interested in learning more about soil genesis and formation should consult specialized textbooks dealing with these subjects.

We classify soils because we want to have a quick appreciation (idea) of a soil under consideration. The purpose or role of the soil will tell us what kinds of soil property and soil behaviour information we require. We recognize that many of the pieces of information require detailed and laborious laboratory tests. However, to sort out the various kinds of soils as candidates for the detailed tests, it is first necessary to choose appropriate candidates, and therein lies the dilemma: What kinds of simple tests and what kinds of information are needed? The discussion on soil classification has chosen to consider the role of soil in the context of soil engineering as the basis for classification. It is understood that detailed laboratory tests as befits the requirements of specific projects/plans must be undertaken to provide for a complete classification scheme.

1.5.1 Soil Functionality

The concept of *soil functionality* proposed herein for use in soil engineering is in essence similar in spirit to the concept of *soil quality* used in agriculture. *Soil quality* has been defined by the Soil Science Society of America (SSSA) as the capacity of a specific kind of soil to function, within natural or managed ecosystem boundaries, to sustain plant and animal productivity, maintain or enhance water and air quality, and support human health and habitation (Karlen et al., 1997). To ensure that the definition of soil functionality covers the various soil functions, we define *soil functionality* (SF) as "the capacity of a specific soil to function under designed circumstances to meet its planned intentions or requirement without any loss of its original functional capability." The use of the concept of soil functionality as a measure of the soil and its functional capability essentially establishes a base that allows one to measure changes in specific soil function capability, that is, specific *soil functionality indices* (SFI), to ensure that design or intended goals are met. This subject will be discussed in detail in Chapter 5.

To illustrate how one might use soil functionality and soil functionality indices in a soil engineering project, we use the example of the performance or role of a bottom engineered clay barrier (of a multibarrier system) of a nonhazardous waste landfill, as shown in Figure 1.9. The required properties of the clay barrier as placed in a multibarrier system is shown in the grey box located at the right of the figure. These are the properties required for the

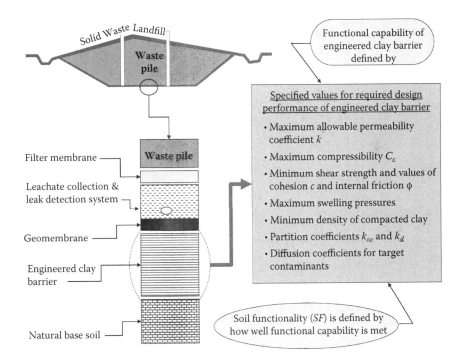

FIGURE 1.9

Typical landfill waste bottom engineered-barrier system for nonhazardous solid wastes. The functional capability of the clay used for the barrier is defined by the specified values of the various engineering properties required to fulfill design performance of the clay barrier. Soil functionality at any time after onset of service life is defined by how well functional capability of the clay barrier is met.

clay barrier to fulfil its design function. Any degradation of any of the properties will detract from the functionality of the clay. In essence, for this particular example, one uses these specified values as *indicators* to determine soil functionality indices, which in this case will be clay functionality indices.

References

Atterberg, A.M., 1908, *Studien auf dem Gebiet der Bodenkunde*, Landw, Versuchsanstalt, 69.

Atterberg, A.M., 1911, Über die physikalishe Bodenuntersuchung und über die Plastiszität der Tone, *Int. Mitt. für Bodenkunde*, 1:10–43.

Casagrande, A., 1937, Classification and identification of soils, *Proc. Amer. Soc. Civil Engrs.*, 73:783–810.

Karlen, D.L., Mausbach, M.J., Doran, J.W., Cline, R.G., Harris, R.F., and Schuman, G.E., 1997, Soil quality: A concept, definition, and framework for evaluation, *Soil Sci. Soc. Am. J.*, 61:4–10.

Keller, W.D., 1957, *The Principles of Chemical Weathering*, Lucas Brothers Publishers, Columbia, MO, 111 pp.

Waterways Experiment Station, 1953, Unified soil classification system, U.S. Corps of Engineers, Tech. Memo, 3-337, Vicksburg, MS.

2

Nature of Soils

2.1 Introduction

To understand and to evaluate or predict soil behaviour, one needs to have information on the properties of the soil under consideration. The properties of a particular soil are dependent on the nature of the soil, that is, its composition and how its various soil fractions interact with each other to provide the soil its intrinsic properties and attributes. Actions from environmental forces and human activities will impact the nature of a soil directly. These impacts will change the nature of a particular soil and can be anticipated with knowledge of the nature of the soil and the nature of the impacts. The various elements that combine to establish the nature of a soil and its properties are discussed in this chapter.

2.1.1 Soil Composition

We study soil composition because the nature of a soil is by and large determined by its composition. As we have seen from the discussion in Chapter 1, the composition of a soil is determined by the various pedogenic processes consistent with the particular region, climate, and anthropogenic activities. With the many different types of source materials, and the various factors and conditions governing weathering processes and regional controls, it follows that the composition of soils will vary from point to point and from location to location. To obtain a clearer picture of the nature of soils in general, it is expedient to consider them as complex systems consisting of solids, fluids, and gas. Common terminology refers to soils as *three-phase systems* consisting of a *solid phase,* a *fluid phase,* and a *gaseous phase.* In most cases, the gaseous phase is air, and the fluid phase is water with dissolved solutes. The solid phase consists of particles (soil solids) of various types such as carbonates, clay minerals, oxides, and so forth. The various types of soil solids are called *soil fractions.*

The various soil fractions, together with the fluid and gaseous phases in a soil, constitute the basic elements that define the composition of a soil. A key element in soil composition is the types of soil fractions. In large measure, this will influence the proportioning of each of the phases in a soil. In a typical soil, one would have such soil fractions as clay minerals, the various oxides and hydrous oxides, humic material or soil organic matter, and carbonates and primary minerals. How the various soil solids, fluid, and gas phases interact with each other will determine the characteristics and properties of the soil. In essence, soil composition is a fundamental feature of a soil that has considerable impact on the development of the structure (i.e., macrostructure) of the soil and the various physical and physicochemical properties of the soil. In turn, these properties will have a direct impact on the behaviour of a soil in response to a provocation. Take, for example, the mechanical behaviour of a soil in response to an external load applied to the soil. The response (mechanical) behaviour of a coarse-grained soil would be markedly different from that of a fine-grained soil. This is because of the differences in type of soil solids, soil structure, and interactions between particles, all of which are tied into the composition of the soil.

2.1.2 Coarse-Grained Soils

There are two groups of soil that are readily distinguishable according to their characteristic particle sizes. These are the coarse-grained soils consisting of gravels and sands, and fine-grained soils consisting primarily of soils whose particle sizes fall in the clay-sized range. Coarse-grained soils consist of soils that have particle sizes larger than 0.06 mm. For those who choose to include coarse silts in this category, the lower limit of coarse-grained particle size will be 0.02 mm. Because of their particle size, particle and soil mass behaviour are governed by gravitational forces. They are also called *cohesionless* soils because of the lack of cohesion between individual particles—generally known as soil grains because of their shape—and hence are unable to stand freely in a soil mass without the aid of side support in the dry granular state. The granular nature of these soils makes them good candidates for sturdy and stable soil masses when optimum packing of the soil grains is obtained. In terms of soil composition, the soil fractions consist primarily of inorganic crystalline material—mainly primary minerals such as quartz, feldspar, and primary mica.

2.1.3 Fine-Grained Soils

Most studies on the properties and behaviour of soils focus on fine-grained soils because of the predominance of these kinds of soils in projects dealing with the geoenvironment. The fine-grained soils are generally considered to have particle sizes finer than 0.02 mm, that is, ranging from and including medium silts and clays. Coarse silts, with particle sizes ranging from 0.02 to 0.06 mm, are sometimes included in this group. It is important to point out at this stage that the term *clays* used in most of the literature refers to soils with

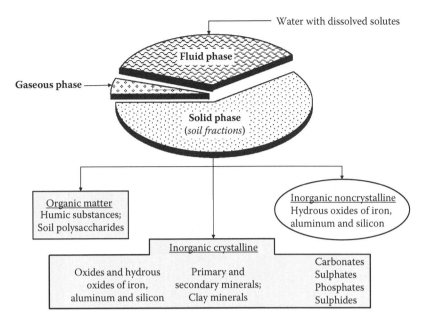

FIGURE 2.1
Pictorial diagram showing the three phases of a soil unit together with the various types of soil fractions in the solid phase.

particle sizes finer than 0.002 mm, that is, finer than 2 μm. Confusion arises when it is assumed to mean *clay minerals*, whose properties and behaviour could be considerably different because of the impact of molecular forces on the development of clay properties and response behaviour. These will be discussed later in this chapter.

For fine-grained soils, and especially the clays, in addition to inorganic crystalline material, the soil fractions also consist of inorganic noncrystalline material and organic matter. Figure 2.1 shows the major inorganic and organic soil fractions in a typical soil that we will call a *soil unit*. The three separate phases in a typical soil unit shown in the figure are gaseous, fluid, and solid phases. The gaseous phase generally consists of air and gases, and the fluid phase contains water and dissolved solutes. The solid phase soil fractions include

- *Inorganic crystalline material* consisting of primary minerals; clay minerals; oxides and hydrous oxides of iron, aluminium, and silicon; carbonates, sulphates, phosphates, and sulphides
- *Inorganic noncrystalline material* consisting of hydrous oxides of iron, aluminium, and silicon
- *Organic matter* consisting of humic substances such as humic and fulvic acids and humins, and polysaccharides

2.2 Clay Minerals

Clay minerals feature prominently in the clay-sized fractions of most soils. They are secondary minerals obtained from chemical weathering of primary minerals in metamorphic and sedimentary rocks. The processes involved include incongruent reactions such as alteration and transformation of primary minerals, and congruent reactions such as formation of new mineral structures from dissolved products. In incongruent reactions, the molar ratios of the chemical elements in solution will be different from those in the source rocks. The clay minerals formed from dissolution of primary minerals depend not only on the composition of the solution, but also on which ions are lost by leaching in the dissolution process and subsequent to the process. It is possible, for example, for a feldspar to weather first to montmorillonite and then to kaolinite, or directly to kaolinite depending on the conditions during weathering. Clay minerals such as montmorillonite or illite are formed from conditions where minimal or restricted leaching exists, that is, where cations tend to remain in solution. Excessive leaching of silica will produce iron and aluminium oxides.

The influential role of clay minerals in soil properties and soil behaviour is derived from their particle size and shape, their large specific surface area in contrast to the soil particles of larger dimensions, and their physicochemical activities. The types of electric charges, the source of charges, and the surface functional groups associated with the type of clay mineral are key to the type and level of physicochemical activities generated in a clay.

2.2.1 Unit Cell, Layer, and Mineral Structure

Clay minerals are phyllosilicates (layer silicates); that is, they are aluminosilicates with platy structural features consisting of sheets of silicon-oxygen tetrahedral and aluminium-oxygen octahedral units with small amounts of metal ions substituted within the crystal structure of the minerals. Figure 2.2 shows the aluminium-oxygen and silicon-oxygen combinations that form the basic molecular structural units. These units combine in various configurations to make up the different types of clay minerals found in clay soils. Textbooks dealing with clay mineralogy generally group phyllosilicates into five main groups according to their mineral structure. Of these groups, the uncharged 1:1 layer lattice minerals with no interlayer water such as kaolinites, and the 2:1 layer lattice minerals with interlayer hydroxide sheets or cations will be discussed in this book inasmuch as they are the most common and abundant types of clay minerals encountered in geoenvironmental projects.

The basic silica tetrahedral and the *Al-* or *Fe-* or *Mg*-octahedral structural units shown in Figure 2.2 are stacked together to form unit cells as shown in the example of a 1:1 unit cell (one tetrahedral unit and one octahedral unit) in the centre of the figure. The basic unit cells are linked laterally to form unit

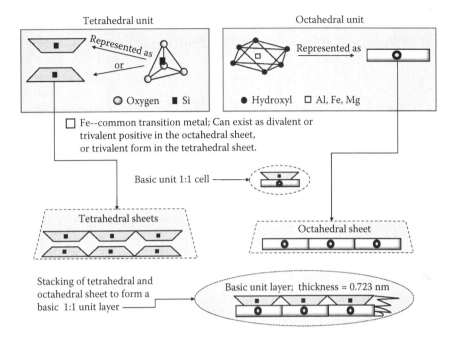

FIGURE 2.2
Basic molecular structural units in clay minerals.

layers, as shown, for example, in the bottom portion of the figure. The unit cell for the basic unit layer is called a *triclinic unit cell*. Another typical unit layer is composed of two tetrahedral sheets sandwiching one octahedral layer. This is called the 2:1 unit layer. Clay mineral particles are formed from a stack of vertically linked unit layers, and the type of linking together with the nature and distribution of the ions populating the tetrahedral and octahedral units will determine the attributes of the clay mineral obtained.

The planar surfaces and the edges of 1:1 or 2:1 layer lattice structures of clay minerals are called *reactive surfaces*, meaning that they react chemically with porewater and the solutes contained therein. Disruption of the unit layers in the course of mineral particle formation and other processes produces broken bonds at the layer surfaces and edges. These are crystal atoms for which valences are not completely "satisfied" or compensated. The planar bounding surfaces of 1:1 layer lattice mineral particles show siloxane and gibbsite surfaces on the opposing surfaces of the particles. If we picture a particle with the tetrahedral sheet at the top, we will have an exposed siloxane surface at the upper face and an exposed gibbsite surface at the lower face of the particle. The unit 1:1 layer shown in the bottom portion of Figure 2.1 best demonstrates this. The siloxane surface is defined by the basal plane of oxygen atoms that bound the tetrahedral silica sheet. For 2:1 layer lattice structures, we will have siloxane surfaces that bound both the upper and lower faces of the mineral particle.

2.2.2 Surface Functional Groups

We have previously indicated that surface functional groups associated with clay minerals are important features in the development of soil properties and resultant behaviour. In actual fact, they are also present in all of the soil fractions listed in the solid phase in Figure 2.1. In the context of soils, we can define *surface functional groups* as chemically reactive groups (molecular units) associated with the surfaces of the soil fractions. These functional groups make the surfaces of the soil fractions *reactive*. Figure 2.3 shows a clay mineral particle in respect to the properties of its upper and lower bounding surfaces. The lower bounding surface of the 1:1 layer lattice mineral particle is an octahedral sheet, as shown in the lower illustration in Figure 2.2. When two-thirds of the available positions in this sheet are filled with Al^{3+} ions, the mineral structure in this sheet is known as a *gibbsite structure* with the chemical formula $Al_2(OH)_6$. Layer lattice minerals with two-thirds of possible positions in the octahedral sheet filled are known as *dioctahedral minerals*. When all the positions in this sheet are filled with Mg instead of Al, the resultant structure is known as a *brucite structure* with chemical formula $Mg_3(OH)_6$, and the associated layer lattice mineral is known as a trioctahedral mineral.

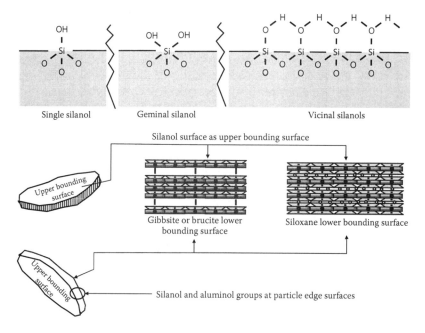

FIGURE 2.3

Properties of upper and lower bounding surfaces of clay mineral particles. The upper illustration shows some of the more common types of silanol groups that populate the siloxane upper bounding surface. The lower bounding surface is a hydroxide sheet, which can be a brucite layer or a gibbsite layer.

Because of the structural arrangement of the silica tetrahedra and the nature of the substitutions in the layers, siloxane-type surfaces are reactive surfaces. The structural arrangement of interlinked SiO_4 tetrahedra, with silicon ions underlying the surface oxygen ions, results in the formation of cavities bounded by six oxygen ions in ditrigonal formation. Isomorphous substitution in the octahedral and tetrahedral layers will result in development of resultant charges on the siloxane surface. By this process, the siloxane surface can be called a reactive surface. Figure 2.3 shows the various kinds of silanol groups of this type of surface.

The amount of silanol groups associated with the siloxane bounding surface is a function of the crystallinity of the interlinked SiO_4 tetrahedra. Silanol [$-SiOH$] and siloxane [$-Si-O-Si-$] functional groups can exist together on the surfaces of the silica tetrahedra. Of the various types of silanol groups, isolated (single silanols), geminal (silanediol), and vicinal groups are more common (Figure 2.3). Silanol groups may also be found within the structure of the particles. When surface silanol groups dominate, the surface will be hydrophilic.

Siloxane bridges are formed from the condensation of combined surface and internal silanol groups. The hydrophobicity of the siloxane surface is due to the presence of siloxane groups. Siloxanes tend to be unreactive because of the strong bonds established between the Si and O atoms and the partial π interactions.

2.2.3 Kaolin Group

2.2.3.1 Mineral Structure

The kaolin group of clay mineral is perhaps the most commonly found mineral in warm and moist climates, derived in large measure from acid leaching of the feldspars and micas in parent rocks. Other sources of kaolinite could be the resilication of aluminium-rich materials by hydrothermal alteration (Weaver and Pollard, 1973), and from precipitation of gels or solutions of silica and alumina (Dixon, 1977). The clay minerals in the kaolin group are 1:1 layer silicates, that is, one uncharged tetrahedral sheet linked to one uncharged octahedral sheet as shown in Figure 2.4. Although the name *kaolin* has been used historically to denote the mineral and also the mineral group in the parent rock, this name (*kaolin*) is now used to refer to the mineral group in the parent rock. The name *kaolinite* is used to mean the mineral itself. The kaolinite mineral belongs to the group of dioctahedral 1:1 layer silicates sometimes known as *kandite* that includes dickite and nacrite. Uncharged tetrahedral and octahedral sheets form the basic unit layer with a thickness of about 0.7 nm, and the kaolinite mineral is obtained from a stacking of these unit layers. The tetrahedral positions in the tetrahedral sheet are occupied by Si ions, and the octahedral sheet has two-thirds of the octahedral positions occupied by Al ions, and as stated previously, this is typical of a gibbsite structure.

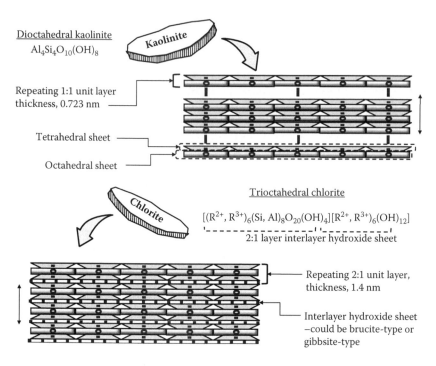

FIGURE 2.4

Basic repeating unit layers for 1:1 layer dioctahedral kaolinite shown in the upper portion of figure. The trioctahedral chlorite shown in the bottom of the figure has a 2:1 repeating unit layer with interlayer hydroxide sheets. The R^{2+} and R^{3+} in the ideal structural formula represent the sum of divalent and sum of trivalent cations, respectively. Note that the repeating layers are stacked vertically and stretched spatially to form the respective mineral particles. (Structural formula information from Newman, A.C.D., and Brown, G., 1987, *The Chemical Constitution of Clays*, Mineralogical Soc. Monograph No. 6, Longman Scientific and Technical, pp.1–128.)

Kaolinites are platy in morphology. The particle size, which is by and large dependent on the amount of disorder in the unit layers, shows an average equivalent diameter about 10 times that relative to its thickness. Dickites and nacrites are variants of the basic kaolinite structure, with dickite being more crystalline than kaolinite. They are large-sized crystalline minerals, and are relatively rare. Halloysite, which has the same basic unit layer as the kaolinite mineral, has often been included in the kaolin group. However, it should be noted that a monomolecular layer of water separates the contiguous unit layers, and that instead of the regular planar particle morphology, halloysites tend to assume tubular shapes.

2.2.3.2 Surface Functional Groups

Kaolinites and chlorites belong to the group of clay minerals with the 1:1 structure that show siloxane and gibbsite surfaces on opposite bounding surfaces of the particles (Figure 2.4). As stated previously, the amount of silanol groups on the siloxane bounding surface depends upon the crystallinity of the interlinked SiO_4 tetrahedra. The broken edges of the kaolinite particle expose hydrous oxide types of edge surfaces. These surfaces contain both silanol and aluminol groups. The broken octahedral sheets provide for Lewis acid sites [$Al(III)·H_2O$] that can bind OH groups in single coordination. Note that the gibbsite sheet, which acts as the lower bounding surface, will also have aluminol groups.

2.2.4 Chlorites

2.2.4.1 Mineral Structure

The lower schematic drawing in Figure 2.4 shows the basic structure of chlorite. This is basically a unit 2:1 layer structure that has a positively charged interlayer consisting of a coordinated octahedral hydroxide sheet that balances the unit 2:1 layer negative charge. This sheet differs from the octahedral sheet in the unit layer in that it does not have a plane of atoms that are shared with the adjacent tetrahedral sheet. The coordinated octahedral hydroxide sheets can be trioctahedral (brucite-type) or dioctahedral (gibbsite-type). It is not uncommon to find chlorite described as a mineral made up of mica layers held together by brucite or gibbsite sheets. Whilst cations such as *Fe, Mn, Cr,* and *Cu* are sometimes found in the hydroxy sheets, the more common hydroxy sheets are brucite and gibbsite sheets. The typical repeat spacing for the unit layers is 1.4 nm.

The chlorite group includes various types differentiated according to the kinds and amount of substitution in the tetrahedral and octahedral sheets (Grim, 1953). The substitutions in the tetrahedral sheet vary from Si_3Al to Si_3Al_2, and in the octahedral sheet the substitutions vary from Mg_3Al to Mg_4Al_2. Additionally, Mg^{2+} and Fe^{2+} are partially replaced by Fe^{2+} and Mn^{2+}, and Al^{3+} is replaced by either Fe^{3+} or Cr^{3+}. There are chlorites that have interlayer montmorillonite as part of their structure. These are sometimes called swelling chlorites.

2.2.4.2 Surface Functional Groups

Substitution of other cations for silicon and aluminium in chlorite results in development of a net negative charge in the mineral. Substitution of Al^{3+} for Si^{4+} in the tetrahedral layer results in a negative charge. Substitution of Al^{3+} by Mg^{2+} in the brucite layer gives a positive charge. The result of these substitution is a net positive charge that is the cation exchange capacity of the mineral.

2.2.5 Smectites

2.2.5.1 Mineral Structure

The hydrous aluminium silicate clay minerals classified as smectites include dioctahedral and trioctahedral structural configurations. The dioctahedral group of minerals includes montmorillonite, beidellite, and nontronite.

These are generally obtained from transformation and weathering processes of volcanic material and igneous rocks. The trioctahedral group of minerals includes saponite, sauconite, and hectorite obtained or inherited from the parent material. Dioctahedral smectites are 2:1 layer lattice minerals where the repeating unit layer consists of two sheets of SiO_4 tetrahedrons confining a central octahedral layer of hydroxyls with *Fe, Mg*, or *Li* ions (Figure 2.5). These unit layers are sometime called *lamellae* in the literature. Notations such as *lamellar* and *interlamellar space* are sometimes used in place of *layer* and *interlayer* space.

A characteristic feature of montmorillonites is the presence of exchangeable cations in the interlayer between unit 2:1 layers. These interlayer cations, their hydration characteristics, and the resulting swelling performance of montmorillonites, which are discussed in detail in Chapter 4, are significant factors in the use of smectites in engineered clay barriers and buffers. As opposed to the montmorillonites, beidelites are aluminium-rich smectites and nontronites are iron-rich smectites. The following list summarizes the substitution in the octahedral sheets.

Montmorillonite: Only *Si* in the tetrahedrons and *Al* in the octahedrons

Beidellite: *Si* and *Al* in the tetrahedrons and *Al* in the octahedrons

Nontronite: *Si* and *Al* in the tetrahedrons and *Fe* in the octahedrons

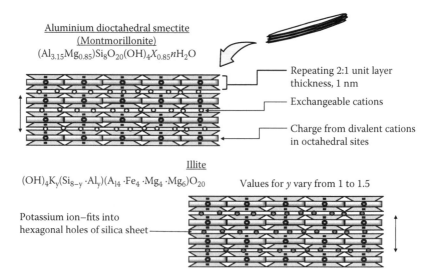

Aluminium dioctahedral smectite
(Montmorillonite)
$(Al_{3.15}Mg_{0.85})Si_8O_{20}(OH)_4X_{0.85}nH_2O$

Repeating 2:1 unit layer thickness, 1 nm

Exchangeable cations

Charge from divalent cations in octahedral sites

Illite
$(OH)_4K_y(Si_{8-y} \cdot Al_y)(A_{l4} \cdot Fe_4 \cdot Mg_4 \cdot Mg_6)O_{20}$ Values for *y* vary from 1 to 1.5

Potassium ion—fits into hexagonal holes of silica sheet

FIGURE 2.5
Basic repeating 2:1 unit layers for dioctahedral smectite (montmorillonite) and repeating 2:1 layer dioctahedral mica (illite) with their ideal structural formulas. Note that the repeating layers are stacked vertically and stretched spatially to form the respective mineral particles. Structural formula information for smectite is from Newman and Brown (1987) and from Grim (1953) for illite.

The trioctahedral minerals saponite, sauconite, and hectorite are not commonly found as clay fractions.

Hectorite: *Si* in the tetrahedrons and *Mg* and *Li* in the octahedrons

Saponite: *Si* and *Al* in the tetrahedrons and *Mg* in the octahedrons

Sauconite: *Zn* in the octahedral sheet

The term *bentonite* has sometimes been used in the literature to mean a swelling clay composed of a significant amount of montmorillonite. Bentonites are classified as dioctahedral smectites obtained as the alteration products of volcanic ash and are composed of primarily montmorillonite with measurable proportions of beidellite depending on the source of the bentonite.

2.2.5.2 Surface Functional Groups

Siloxane surfaces bound the opposite planar faces of each unit 2:1 layer of montmorillonites. The source of the charges is primarily from isomorphous substitution in the octahedral sheet of the unit layers, with magnesium or iron substituting for aluminium in this sheet. Table 2.1 gives a summary of the sources of charges for the minerals discussed in this section.

2.2.6 Micas, Illites, and Mixed-Layer Clays

The many minerals constituting the *mica group* are varied and complex because of the different types of substitution in their structures. The various types of mica minerals are distinguished by the nature of the isomorphic substitution of the octahedral sheet and the distribution of silicon ions in the tetrahedral sheet. For this book, we will limit our consideration to the clay micas; micas have a 2:1 layer structure for the basic unit layer, and cations in the interlayer separating the basic unit layers. These interlayer cations can be potassium, sodium, and calcium, with potassium being the more common interlayer cation. The lower illustration in Figure 2.5 shows the illite mineral.

Illites are platy in structure with variable thicknesses. They are hydrous clay micas that do not ordinarily expand from a 1.0 nm basal spacing (Grim, 1953), with a 2:1 layer structure as shown in Figure 2.5, with potassium occupying positions in the interlayer. These interlayer ions are not exchangeable. The potassium ions exist in 12-coordination, bonding six oxygens from one silica sheet to the adjacent six oxygens of the silica sheet of the next layer. The negative charge balancing the potassium cations comes from the substitution of aluminium for silicon in the silica sheet. Illites are nonswelling clays since they do not contain any expanding layer in their structure. There are illites, however, that are interlayered with montmorillonite. These are called

TABLE 2.1

Some Common Clay Minerals and Their Sources of Charges

Clay Mineral	Source of Charges	Isomorphous Substitution	Layer Charge[c]	Cation Exchange Capacity (CEC; mEq/100 g)	Specific Surface Area (m²/g)	Reciprocal of Charge Density (nm²/Charge)
Kaolinite 1:1	Surface silanol, edge silanol, and aluminol groups (ionization of hydroxyls and broken bonds)	Dioctahedral: 2/3 of positions filled with Al	<0.01	3–15	10–15	0.25
Illites 2:1	Isomorphous substitution, silanol groups, and some edge contribution	Usually octahedral substitution Al for Si	1.4–2.0	10–40	80–120	0.5
Clay Micas and Chlorites 2:1	Silanol groups, plus isomorphous substitution and some broken bonds at edges	Dioctahedral: Al for Si Trioctahedral or mixed Al for Mg	Variable	10–40	70–90	0.5
Vermiculites[a] 2:1	Primarily from isomorphous substitution, with very little edge contribution	Usually trioctahedral substitution Al for Si	1.2–1.8	100–150	300–500	1.0

| Montmorillonites[b] | Primarily from isomorphous substitution, with very little edge contribution | Dioctahedral, Mg for Al | 0.5–1.2 | 80–100 | 700–800 | 1.0 |

2:1

Note: Ratios of external to internal surface areas are highly approximate since surface area measurements are operationally defined; that is, they depend on the technique used to determine the measurement.

Ratios of external to internal surface areas are approximate since they are operationally defined; that is, they depend on techniques used to measure these areas.

[a] Surface area includes both external and internal surfaces. Ratio of external to internal surface area is approximately 1:120.

[b] Surface area includes both external and interlayer surfaces. The ratio of external particle surface area to internal surface area is approximately 5:80.

[c] Layer charge = number of moles of excess electron charge per chemical formula that is produced by isomorphous substitution (Sposito, 1989).

interlayered illites, and soils containing these kinds of illites are identified as *mixed-layer* clays.

2.2.7 Allophane

2.2.7.1 Mineral Structure

Allophane clays consist of aggregates of small hollow spherical particles in volcanic ash soils and podsols. Numerous studies of these clays, involving chemical composition analyses, x-ray analyses, infrared spectroscopy and high-resolution electron microscope observations, and others (Birrell and Fields, 1952), have shown that allophones are hydrous aluminium silicates. The spherical particles of allophane are considered to be poorly crystallized clay minerals consisting of a shell of *Al-Si* octahedral sheet with pores. The ideal chemical structure could be expressed as $1–2SiO_2 \cdot Al_2O_3\ 5–6H_2O$ according to Kitagawa (1971), and the hollow particles have diameters of 3.0–5.5 nm. The pore sizes are 0.3–0.5 nm. The mean specific gravity of allophones has been obtained as 1.91 Mg/m^3, and a specific surface area of 540–610 m^2/g has been observed using the glycerol and ethylene glycol monoethyle ether (EGME) adsorption method. Allophane can be formed with aluminium released from organic matter in weathering of volcanic ash. This could be transformed into halloysite, gibbsite, kaolinite, and some expandable 2:1 layer minerals with desilication in the final stage of weathering (Wada and Harward, 1974).

2.2.7.2 Surface Functional Groups

Allophone has *Si-OH* at both the inner and pore surfaces of the hollow particles. *Al-OH* is exposed at the outer surface (Wada, 1989). These hydroxyls act as surface functional groups depending on pH. The hydroxyls attached to *Si* together with the hydroxyls attached to *Al* endow the surface with negative charges. This results in adsorption of positive ions at high pH values. On the other hand, at low pH values, the hydroxyls of *Si* lose their charge and the hydroxyls of *Al* become significantly positive in charge, resulting in adsorption of anions. Accordingly, the charge of allophone is called a variable charge, similar to that of the edge surfaces of kaolinites and 2:1 layer silicates and also in the hydroxyls of organic acids.

2.3 Nonclay Minerals

The nonclay minerals most commonly found in clays include the oxides, hydrous oxides, carbonates, and sulphates. These have the capability of influencing the various soil transmission and strength properties, and the relationships established between particles of a soil, depending on their proportions and their distribution in the soil.

2.3.1 Oxides, Hydroxides, and Oxyhydroxides

2.3.1.1 Basic Structure

The various kinds of oxides present in soil are products of weathering of parent rock material. These include *oxides, hydroxides,* and *oxyhydroxides.* Weathering of iron-bearing silicates have iron bound in the silicates released through a combination of hydrolytic and oxidative reactions that will precipitate as iron oxide or iron hydroxide because of the low solubility of Fe^{3+} in the normal pH environment of soils (Schwertmann and Taylor, 1977).

The term *hydrous oxides* has been used in the literature to refer to both the noncrystalline and the crystalline form. Strictly speaking, the term *oxides* refers to the crystalline (mineral) form of the material. That being said, since the distribution of the hydrous oxides in a soil is very much dependent on whether it is in the mineral or amorphous form, it is important to distinguish between the two. Amorphous forms of the oxides have the ability to coat mineral particle surfaces because of the net electric charges on the surfaces of both the clay mineral particles and amorphous materials. Figure 2.6 illustrates this coating phenomenon for a microstructural unit (msu) composed of clay mineral particles (bottom schematic). Since the sign of the charges on the amorphous materials is pH dependent, one needs to pay attention to the amphoteric nature of these coated fractions when examining or studying interactions between coated soil fractions, as will be discussed in Chapter 3.

Oxides and hydrous oxide minerals are the main constituents of highly weathered tropical soils such as laterites and bauxites, for example, haematite, goethite, gibbsite, boehmite, anatase, and quartz. Whilst iron oxides are the most common form of oxides found in soils, the low solubilies of the oxides of aluminium and manganese, in addition to that of iron in the pH range found in most natural soils, mean that they are more common than the oxides of titanium and silicon. The surfaces of the oxides of iron, aluminium, manganese, titanium, and silicon essentially consist of broken bonds, a feature that distinguishes them from layer silicate minerals.

The main structural configuration of oxides is octahedral. Oxides of aluminium have octahedral sheets containing OH^- ions with two-thirds of the positions occupied by Al^{3+} ions. However, depending on how the OH^- ions in the octahedral sheets are distributed, one can obtain either gibbsite or bayerite. Gibbsite is obtained when the OH^- ions in each of the stacked octahedral sheets are directly opposite to each other and bonded by hydrogen bonds. Bayerite is obtained when the OH^- ions in each of the stacked octahedral sheets are positioned in the space formed by the OH^- ions in the opposing stacked sheet.

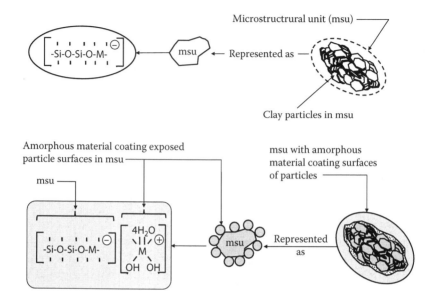

FIGURE 2.6

Schematic illustration of a natural clay with amorphous material coating the surfaces of particles in the clay microstructural unit (msu). The top schematic shows the packing of clay particles constituting the msu, and the bottom schematic shows the msu with amorphous material coating all the exposed surfaces of the particles in the msu. M is the coordinated metal.

Depending on the source of information, different terminologies are used to describe the oxides, for example:

$Al(OH)_3$ = aluminium trihydroxide = aluminium hydroxide = bayerite

$AlOOH$ = aluminium oxide-hydroxide = aluminium oxyhydroxide

γ-$Al(OH)_3$ = alumina trihydrate = γ-aluminium hydroxide = gibbsite

γ-$AlOOH$ = alumina monohydrate = γ-aluminium oxyhydroxide = boehmite

2.3.1.2 Surface Functional Groups

The discussion on surface functional groups for oxides is better viewed in terms of their surface characteristics, as illustrated in the bottom sketch in Figure 2.6. Representing an oxide crystal as MOH, where M is the coordinated metal, the charge on the surface of the oxide is dependent on the electrolyte concentration and pH of the porewater in contact with the oxide. Since oxide surfaces are partially hydrolyzed, their surface charges will be dependent on the pH of the contacting porewater (Bolt and Bruggenwert, 1978). This means that the charge developed on the surface of the oxide is dependent on the concentration of protons in the contacting porewater, the charge being developed in relation to the association or dissociation of protons. There is a point where the surface is uncharged, and when this occurs, this point is called the point of zero charge (*pzc*). The pH at which this occurs is identified as pH_o. Parks (1965) provides a relationship between cation valency and the *pzc* of its oxide as follows:

Monovalent ions giving an oxide M_2O	$pzc < pH\ 11.5$
Divalent ions giving an oxide MO	$8.5 < pzc < 12.5$
Trivalent ions giving and oxide M_2O_3	$6.5 < pzc < 10.4$
Quadrivalent ions giving and oxide MO_2	$0 < pzc < 7.5$

2.3.2 Carbonates and Sulphates

The relatively high solubilities of carbonates and sulphates compared to the clay minerals and oxides means that their presence in large amounts is more likely in arid and semiarid regions where limited leaching and high evaporation occur. Smaller amounts of the carbonate mineral can be found in other regions. These include calcite ($CaCO_3$), magnesite ($MgCO_3$), siderite ($FeCO_3$), dolomite ($CaMg(CO_3)_2$), trona ($Na_3CO_3HCO_3 \cdot H_2O$), nahcolite ($NaHCO_3$), and soda ($Na_2CO_3 \cdot 10H_2O$)—with calcite and dolomite as probably the most common. The influence of carbonates on the pH of the soil is significant since many of the chemical processes and reactions occurring in soils are pH sensitive.

Although one can list a few sulphate minerals such as gypsum ($CaSO_4 \cdot 2H_2O$), hemihydrate ($CaSO_4 \cdot 0.5H_2O$), thenardite (Na_2SO_4), and mirabilite ($Na_2SO_4 \cdot 10H_2O$), gypsum is the most common of the sulphate minerals found in soils, primarily in arid and semiarid region soils. Because of their high solubilities, the existence of the crystalline form of $MgSO_4$, Na_2SO_4, and other sulphate minerals is generally confined to the soil surface. By comparison, gypsum is at least 100 times less soluble than $MgSO_4$ and Na_2SO_4.

2.4 Soil Organic Matter

2.4.1 Characterization and Classification

By and large, the source or origin of natural organic matter in soils is from vegetation and animals—sometimes referred to as *debris*. Considering peat to be a completely separate case, the proportions of natural organic material in surface soils is small in comparison to the other soil constituents, in proportions as small as 0.5% to 5% by weight. Even though the proportions are small, the importance of organic matter in the development of soil properties and behaviour is highly significant—through formation and maintenance of a more robust soil structure and improved water retention characteristics. Classification of organic matter is generally undertaken on the basis of degradation of the organic matter in two broad groups as *unaltered organics* and *transformed organics*. As the name implies, unaltered organic matter is relatively fresh organic material that has yet to show the results of the various transformation processes.

Transformed organic materials by definition refer to those organic materials (debris) that exhibit neither (a) structural characteristics similar to the parent material nor (b) properties and attributes of the parent material. The processes leading to transformation of the parent material that are chemical and biologically mediated reactions include demethylation, and oxidative, reductive, and other electron transfer reactions catalyzed by enzymes (Flaig et al., 1975). There are two distinct subgroups that comprise the transformed organic materials group. One subgroup is generally identified as the *amorphous soil organics*. This subgroup, which consists of highly aromatic polymers that are rich with functional groups, includes humic acids, fulvic acids, and humins. Whilst some publications have included decayed material such as polysaccharides, lignins, and polypeptides in the *amorphous soil organics* subgroup, it is not uncommon to classify these under another subgroup listed as *various compounds*. It is not unusual to have polysaccharides extracted with the fulvic fraction, resulting in incorporation of these carbohydrates as a fulvic-related material. Polysaccharides are long molecules that are obtained as by-products of microbial metabolism, generally synthesized as a by-product from the breakdown of animal- or vegetal-derived organics. These polymeric carbohydrates have a definitive structure and thus fall into the subgroup of *various compounds*.

General techniques for characterization of these amorphous soil organics employ alkali treatment procedures on soil organics extracted from the parent soil, as shown in Figure 2.7. The organics dissolved in the supernatant are decanted, and the humic acid fraction is precipitated from solution with acid, leaving the soluble fulvic acid fraction in the supernatant. The constituents of the soil organic matter are as follows:

- *Humic acids*: Soluble in bases, but precipitate in acids
- *Fulvic acids*: Soluble in both bases and acids
- *Humins*: Insoluble in acids and bases

2.4.2 Structure and Surface Functional Groups

2.4.2.1 Structure

From the traditional point of view, the proposed basic structure of soil organic matter consists of carbon bonds combined with saturated or non-saturated rings (salycyclic or aromatic rings, respectively) as coiled macromolecules chains and gyration, with average molecular masses of 20,000 to

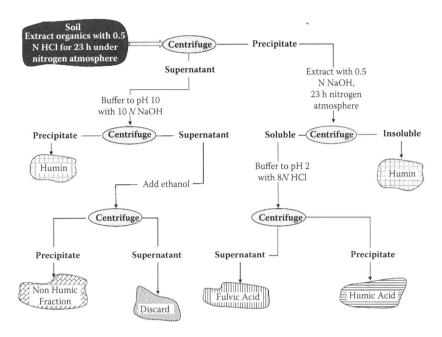

FIGURE 2.7
Extraction technique for determination of humic and fulvic acids, and humins.

50,000 Dalton units (Da), and radii of 4 to 10 nm (Piccolo, 2001). The results of multidimensional nuclear magnetic resonance spectroscopy (NMR) studies conducted in the first decade of this century have suggested that the structure of soil organic substances consists of supramolecular association or aggregates (less than 5,000 Da) comprising various organic molecules linked by metal bridging, H-bonds, and hydrophobic interactions, including polysaccharides, polypeptides, aliphatic chains, and aromatic lignin fragments as the major components based on the concept of micellar structure (Simpson et al., 2002). However, Asutton and Sposito (2005) have suggested that as the supramolecule association model does not elucidate and does not explain the changes in functional group composition occurring with separation of humic substances into different apparent mass or size fractions, "we need better information about the hydrophobic and H-bonding interactions that hold together humic aggregates of various sizes."

2.4.2.2 Surface Functional Groups

As opposed to clay minerals, which have hydroxylated surfaces with *OH* functional groups, soil organic matter exhibits a greater variety of surface functional groups. The sketch shown in Figure 2.8 illustrates some of the more common types of functional groups. The most common functional groups are hydroxyls, carboxyls, phenolics, and amines. The carboxyl group

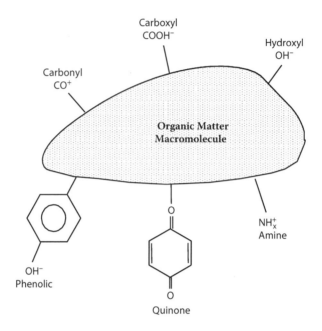

FIGURE 2.8
Some typical functional groups associated with soil organic matter. (Adapted from Yong, R.N., 2001, *Geoenvironmental Engineering: Contaminated Soils, Pollutant Fate, and Mitigation*, CRC Press, Boca Raton, FL, 307 pp.)

is the major contributor to the acidic properties of the soil organics. Carbon and nitrogen combine with oxygen or hydrogen to form the various types of surface functional groups shown in the diagram.

These functional groups control most of the properties of organic molecules that constitute the soil organic materials, and also their reactions with other soil fractions in a soil–water system. Depending on the pH of the soil and their respective pK_a or pK_b (their respective log acidity and log basicity constants), they can protonate or deprotonate (develop positive or negative charges). The decrease in aromaticity between humic acids, fulvic acids, and humins reflects the biodegradation sequence of humus, beginning with degradation of nonamorphous organics into humic acids, and continuing on to fulvic acids and finally humins.

Many of the different types of functional groups have been detected in the study of IR spectra information on tests on a soil organic reported by Yong and Mourato (1988), the details of which are shown in Table 2.2. These surface functional groups are classed as organic molecular units and cannot be diluted since they are part of the organic matter itself. Nonhumic materials of soil organics are generally composed of large numbers of aliphatic rings typical of polysaccharides such as those detected in the samples tested in the study. Because of the large variation or differences in the composition of soil organic

TABLE 2.2

IR Spectra of Organic Fractions Extracted from Soil Material Obtained from a Site

Adsorption Band (cm^{-1})	Organic Fraction	Description
3400	All fractions	OH stretching of free hydroxyls and hydration molecules
3000–2800	Fulvic acids, humins	Aliphatic C–H bonds
2200–2100	Humic acids, humins	COOH vibrations
1725	Humic acids, humins	C=O stretching of fluvic acids' COOH and ketones
1600 (large)	Humic acids, fulvic acids, humins	Aromatic bonds and some overlapping of strongly H-bonded C=O groups
1600 (small)	Nonhumic fractions	C–H deformations of aliphatic groups
1400 (large)	Nonhumic fractions	C–H deformations of aliphatic groups
1400 (small)	Humic acids, fulvic acids, humins	C–H deformations
1240	All fractions	C–H stretching and OH deformation of COOH
1140	All fractions	OH deformation of phenolic and alcoholic functional groups
1100–1000	Nonhumic fractions	Polymeric carbohydrates
950–450	Humins	Vibration of aluminium and silicon elements

Source: Adapted from Yong, R.N., and Mourato, D., 1988, *Can. Geotech. J.*, 25:599–607.

TABLE 2.3

Composition and Functional Groups for Fulvic Acid, Humic Acid, and Humin

	Fulvic Acid	Humic Acid	Humin
% Carbon content	40–50	50–60	50–60
% Oxygen	40–50	30–40	30–35
% Hydrogen	4–7	3–6	NA
Carbonyl, %	up to 5	up to about 4	NA
Carboxyl, %	1–6	3–10	NA
Quinone, %	2±	1–2	NA
Ketones, %	2±	1–4	NA
Alcoholic OH, %	2.5–4	up to 2	NA
Phenolic OH, %	2–6	up to about 4	NA

Source: Selection of data from Griffith, S.M., and Schnitzer, M., 1975, *Soil Sci. Soc. Amer. J.*, 39:861–869; Schnitzer, M. et al., 1973, *Soil Sci. Amer. Proc.* 7:229–326; Hatcher, P.G. et al., 1981, *Soil Sci. Soc. Amer. J.* 45:1089–1094.

matter, due to the differences in source material and decomposition processes, there will be wide ranges and values for the proportions of each of the kinds of functional groups shown in Figure 2.8. In addition, variations in the type of extraction and testing procedures will also contribute to the wide range of values reported, as seen for example in the values shown in Table 2.3.

Soil organic materials contain some sulphur in the range of 0% to 2% for humic acids and up to about 4% for fulvic acids. Since hydrogen in the oxygen-containing functional groups can be dissociated, the surface functional groups will provide the material with acidic properties. By and large, carboxyls and phenolic *OH* groups contribute significantly to the cation exchange capacity (*CEC*) of the soil organic material, and are considered to be the most important functional groups. They contribute significantly to the source of negative charge, which has been reported to range from 2 to 4 mEq/g (Greenland and Hayes, 1985). This compares with the charge range of 0.01 to 2 mEq/g for clay minerals.

2.5 Soil Particles

The soil particle properties of interest in the study of soil properties and behaviour are those that relate directly to particle interaction in the presence of porewater. These include (a) texture of soil particles, (b) surface functional groups associated with the electrical charges of soil particles, (c) cation exchange capacity, *CEC*, and (d) specific surface area, *SSA*. Although particle size and shape do not strictly qualify as surface properties, they will be included in this discussion because they relate directly to soil properties and behaviour such as specific surface area, soil density, and compaction of soils. The term *particles* is used here to include all the soil solids constituting the various soil fractions. There is a direct link between soil particle properties and many physical, mechanical, and physical-chemical properties, and environmentally associated evolution of soils. The links will become clearer when we discuss these properties and behaviour in the later chapters of this book.

2.5.1 Size, Shape, and Texture of Particles

2.5.1.1 Size and Shape

The size, shape, and texture of a soil particle are directly related to the morphology of soils, density, permeability, hardness, strength, and their hysteretic performance in response to external forces. The important items of note include the following:

- For particle sizes larger than clays and fine silts, size and shape go almost hand in hand; the composition of these particles, such as quartz, feldspar, chert, will generally result in the production of particles that are granular in shape.
- For granular-sized and granular-shaped particles, packing density together with particle strength are important factors in the

development of soil strength and resistance. Granularity and particle size distribution are also important in obtaining optimum density.

- As we have seen in the discussions in the previous sections, for clay particles such as the layer lattice minerals, the particles are platy in structure. The importance of size and shape, especially shape, can be seen in not only in (a) the packing and arrangement of soil particles (macro and microstructure) resulting in anisotropy of permeability, strength, and compaction, but also in (b) the amount of surface area per unit volume of soil (specific surface area, *SSA*) as it relates to the formation and properties of soil aggregates. These aspects will be discussed in greater detail when we discuss the subject of specific surface area of soils.

- Whether molecular forces will predominate over gravitational forces in interparticle action depends on how much control particle surface forces have on these actions. These will depend on the types of reactive surfaces presented by the clay particles, as discussed in the previous section. The greater the specific surface area is, the greater will be the potential for interparticle action to be dominated by molecular forces.

2.5.1.2 Texture

There are at least two important points of note in considering the role of surface texture of particles: (a) roughness (or smoothness) and (b) surface imperfections. In the case of roughness, this surface feature is more important in coarse-grained soils, where soil strength and integrity depends on packing and interparticle action involving surface resistance from frictional forces between particles. The rougher the particles, the greater the frictional resistance between adjacent particles in relative motion. In the case of surface imperfections, we note that these are important considerations in clay minerals since these are likely sources for electrical charges, due to unsatisfied valence charges. These will be discussed in the next section.

2.5.2 Electrical Charges

2.5.2.1 Sources of Electrical Charges

The second column in Table 2.1 gives the sources of charges associated with clay mineral particles. Substituting ions of lower positive valence for higher valence ions in the crystal lattice, shown for example in the third column in the table, will leave the clay layer lattice structure with a net negative charge. Substitution of aluminium for silicon in the silica (tetrahedral) sheet, and magnesium, iron, or lithium substituting for aluminium in the alumina (octahedral) sheet, account for the main sources of charges in the 1:1 and 2:1 layer lattice minerals—but only a minor part in the 1:1 kaolinite mineral. In this case, clay minerals have permanent charges regardless of the immediate surrounding pH. The amount of silanol groups on the siloxane bounding surfaces of these 1:1 and 2:1 layer lattice minerals depends on the crystallinity of the interlinked SiO_4 tetrahedra. The top illustration in Figure 2.3 shows three variations for the sources of charges associated with siloxane surfaces.

Breakage of edges of phyllosilicates will produce hydrous oxide-type of edge surfaces. Unsatisfied valence charges at the edges of particles due to the breakage constitute another source of electrical charges. These are generally referred to as broken-bond charges or pH-dependent variable charges. A significant source of charges for the 1:1 layer lattice kaolinites comes from these broken bonds; that is, the silanol and aluminol groups at the edges of the kaolinite particles. This is also the case for the various kinds of oxides in the oxide group.

These attract hydrogen or hydroxyl ions from the porewater, and the ease with which the hydrogen ion can be exchanged increases as the pH of the porewater increases, that is, as the hydrogen ion concentration in the porewater decreases. The Al^{3+} in the exposed edges of the octahedral sheets will complex with both H^+ and OH^- in the coordinated OH groups. On the other hand, the Si^{4+} will complex only with OH^-. Association of the surface hydroxyls with a proton occurs below the point of zero charge (*pzc*). This will produce a surface with a positive charge. The donation or loss of a proton by surface hydroxyls above the *pzc* will produce a negatively charged surface. In short, the charge due to broken bonds will increase as the pH of the porewater increases.

The surfaces of the hydrous oxides (of iron and aluminium for example) show coordination to hydroxyl groups that will protonate or deprotonate in accordance with the pH of the surrounding medium. Exposure of the Fe^{3+} and Al^{3+} on the surfaces provides development of Lewis acid sites when single coordination occurs between the Fe^{3+} with the associated H_2O; that is, $Fe(III) \cdot H_2O$ acts as a Lewis acid site. Charge reversal due to changing pH values is a significant characteristic of kaolinites and hydrous oxides, and charge reversal at the surfaces of the particles because of pH changes is the result of proton transfers at the surfaces. In the case of soil organic matter, the chemical structure of the macromolecule exerts a direct role on the nature of the surface functional groups and hence on the nature of the charge developed. The negative charges associated with soil organic matter are due to the ionization of hydrogen from carboxyl and phenolic hydroxyl groups.

2.5.2.2 Net Surface Charges and Surface Charge Density

The charge density for any clay mineral particle is the sum of all the charges acting on the total surface of the particle, that is, the sum of all the positive and negative charges. Strictly speaking, one should use the term *net surface charge densities* in referring to the sum of all charges acting on the total surface. However, since the terms *surface charge density* and *charge density* have been used in the literature to mean the net surface charge density, these terms will be used in this book. Without the presence of potential determining ions (*pdis*), the surface charge density of a soil particle σ_{ts} consists of σ_s, the permanent charge due to the structural characteristics of the clay particle (isomorphous substitution), and σ_h, the resultant surface charge density due to hydroxylation and ionization (net proton surface charge density). This can be expressed as

$$\sigma_{ts} = \sigma_s + F(\Gamma_H - \Gamma_{OH}) \tag{2.1}$$

where F refers to the Faraday constant; Γ is the surface excess concentration, that is, surface concentration in excess of the bulk concentration; Γ_H and Γ_{OH} refer to adsorption densities of H^+ and OH^- ions and their complexes, respectively; and $F(\Gamma_H - \Gamma_{OH})$ is the net proton surface charge density σ_h.

When $|\Gamma_H| = |\Gamma_{OH}|$, the *point of zero net proton charge* (*pznpc*) is reached, and the pH associated with this is designated as pH$_{pznpc}$. The point of zero net proton charge (*pznpc*) should not be confused with the *point of zero charge* (*pzc*) or the *isoelectric point* (*iep*). The pH$_{pzc}$, which represents the pH at which titration curves intersect, differs from the pH$_{iep}$, which represents the pH at which the zeta potential ς is zero. Whilst the separate pH values might be very close to each other (Figure 2.9), it is important to realize that the zeta

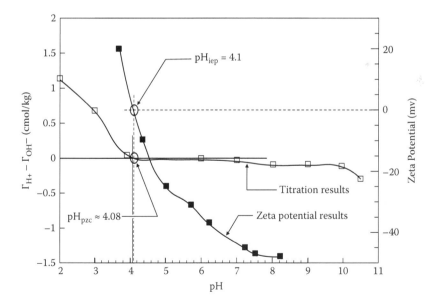

FIGURE 2.9
Titration test results and zeta potential measurements for a kaolinite. pH$_{iep}$ from zeta potential measurements is 4.1, and pH$_{pzc}$ is about 4.08 from titration test results. (Data reported by Yong, R.N., 2001, *Geoenvironmental Engineering: Contaminated Soils, Pollutant Fate, and Mitigation*, CRC Press, Boca Raton, FL, 307 pp.)

potential ç refers to the electric potential developed at the solid–liquid interface as a result of movement of colloidal particles in one direction and counterions in the opposite direction.

The distinction between pH_{pzc} and pH_{iep} is relevant not only because slightly differing definitions exist in the literature for the *pzc* and *iep*, but also because the possible presence of potential determining ions (*pdis*) would influence the respective pH values obtained. What this means is that these differences in the pH values are related to methods of determination of these particular charge density relationships and the influence of counterions in the inner and outer Helmholtz planes (discussed in Chapter 3) on these measurement. The methods for determining the influence of the charge densities and (influence) of the ions in the inner and outer Helmholtz planes are not the same. The zeta potential ç is computed from experimental measurements obtained with a zetameter using the Helmholtz–Smoluchowski relationship. One could reasonably argue that pH_{pzc} and pH_{iep} are operationally defined.

The *pzc* and *iep* can also be distinguished according to whether or not specifically adsorbed cations or anions are considered, as demonstrated in the graphical relationship shown in Figure 2.10. When H^+ and OH^- ions constitute the only potential-determining ions, the pH condition at which the adsorption densities of H^+ and OH^- ions and their complexes are equally balanced is characterized as the pH_{iep}. In terms of Γ, the surface excess concentration,

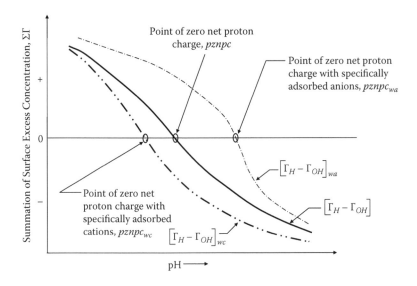

FIGURE 2.10
Net charge (σ_h) curves as determined by proton balance, with and without specifically adsorbed cations and anions. The solid curve represents the condition with only H^+ and OH^- ions as *pdis*. The subscripts *wc* and *wa* refer to *with specifically adsorbed cations* and *with specifically adsorbed anions*, respectively. (Adapted from Yong, R.N., 2001, *Geoenvironmental Engineering: Contaminated Soils, Pollutant Fate, and Mitigation*, CRC Press, Boca Raton, FL, 307 pp.).

this means that the pH_{iep} is obtained when $|\Gamma_H| = |\Gamma_{OH}|$. This distinguishes it from the situation where adsorbed ions from the porewater contribute to particle surface charges resulting in changes in the potential of the particle.

Specifically adsorbed cations will decrease the *pznpc*, whereas specifically adsorbed anions will increase the *pznpc*, as shown by the net proton surface charge density relationship σ_h in Figure 2.10 The solid curve in the figure represents the proton balance condition where H^+ and OH^- ions are the only *pdis*. Note that when the specifically adsorbed ions are cations (*wc*), a lower *pznpc* is obtained ($pznpc_{wc}$), and when the specifically adsorbed ions are anions (*wa*), a higher *pznpc* is obtained ($pznpc_{wa}$), as shown by the lower and upper dashed lines in the figure, respectively. Strictly speaking, since the point of zero charge is really the point of zero net charge, one should use pH_{pznc} in place of pH_{pzc}. However, we will follow common usage and continue with the use of pH_{pzc} to mean point of zero (net) charge.

2.5.3 Exchangeable Cations and Cation Exchange Capacity (CEC)

Exchangeable cations are the positively charged ions from salts in the porewater that are attracted to the surface of clay particles because of the net negative surface charge exhibited by the particles. They are termed *exchangeable* because one cation can be replaced by another of equal valence, or by two of one-half the valence of the original cation. By and large, the majority of exchangeable cations in clays are calcium and magnesium, with much smaller proportions of potassium and sodium.

Geological origin and environmental and regional controls (initial and subsequent leaching, etc.) are the major factors that determine which exchangeable ions will be present in a soil. Aluminium and hydrogen are the predominant exchangeable ions in acid soils. Clays derived from sedimentation-consolidation in seawater will show a predominance of magnesium and sodium. Whilst calcareous soils will contain mainly calcium, extensive leaching will remove the cations that form bases (calcium, sodium, etc.), leaving a clay with acidic cations, aluminium, and hydrogen.

As an example of the exchange process, we consider the case where a clay containing sodium as the exchangeable cation is washed with a solution of calcium chloride. Each calcium will replace two sodium ions, which are then removed from the soil through the washing process. This process, which is called *cation exchange*, or *base exchange*, is written as

$$Na_2\ Clay + CaCl_2 \leftrightarrow Ca\ Clay + 2NaCl$$

In short, cation exchange occurs when positively charged ions in the porewater are attracted to the surfaces of the clay fractions because of the net negative charge imbalance of the charged reactive surfaces of the clay particles. This stoichiometric process responds to the need to satisfy electroneutrality in the system, that is, replacing cations to satisfy the net negative charge imbalance of the charged reactive surfaces of the clay particles. The number of charged sites considered as exchange sites are determined by isomorphous substitution in the layer lattice structure of the clay minerals. The quantity of exchangeable cations held by a clay is called the *cation exchange capacity* (*CEC*) of the clay. The *CEC* is usually expressed as milliequivalents per 100 g of clay (mEq/100 g soil), as shown for example for the various clay minerals listed in Table 2.1.

For a clay containing solids whose net surface charges are pH dependent, the *CEC* of the clay is a function of the pH of the system. The clay solids that are included in this list are 1:1 and 2:1 layer-silicate minerals, natural organic matter, and the various oxides or amorphous materials. In kaolinites, for example, the values of *CEC* can vary by a factor of 3 between the *CEC* at pH 4 to pH 9 (Yong and Mulligan, 2004). A common technique used for measurement of *CEC* in clays is to use ammonium acetate (NH_4OAc) as the saturation fluid. In theory, since cation sorption should occur on all available sites, one is required to determine that it occurs without the presence of artefacts. Reactions between the saturating cation solution and clay fractions can produce erroneous results. The dissolution of $CaCO_3$ and gypsum in carbonate-rich clays when ammonium acetate (NH_4OAc) is used as a saturation fluid is an example of such an artefact. Because variations in measured *CEC* can occur due to experimental conditions, the results obtained are sometimes referred to as operationally defined values.

The relative energy with which different cations are held at the clay surface can be found by measuring their relative ease of replacement or exchange by a chosen cation at a chosen concentration. These measurements show that a small amount of calcium easily replaces exchangeable sodium, but the same amount of sodium does not replace much exchangeable calcium. The valence of the cation has a dominant influence on the ease of replacement. The higher the valence of the replacing ion, the greater its replacing power. Also, the higher the valence of the surface exchangeable ion, the more difficult it will be to replace the ion. For ions of the same valence, increasing ion size gives greater replacing power. The relationship used to determine replacement of exchangeable cations with the same positive charge and similar geometries as the replacing cations is given as follows: $M_s/N_s = M_o/N_o = 1$, where M and N represent the cation species, and the subscripts s and o represent the particle surface and the bulk solution.

Cations can be arranged in a series on the basis of the replacing power. The position in this series will depend on the kind of clay and on the ion being replaced, with the replacement positions being largely dependent on the size of the hydrated cation. The replacing power of some typical ions is shown as a lyotropic series as follows (Yong, 2001):

$$Na^+ < Li^+ < K^+ < Rb^+ < Cs^+ < Mg^{2+} < Ca^{2+} < Ba^{2+} < Cu^{2+} < Al^{3+} < Fe^{3+} < Th^{4+}$$

Changes in the relative positions of the lyotropic series depend on the kind of clay and the ion being replaced. The number of exchangeable cations replaced depends on the concentration of ions in the replacing solution. In heterovalent exchange, the selective preference for monovalent and divalent cations is dependent on the magnitude of the electric potential in the region where the greatest amount of cations is located. When the outside concentration varies, the proportion of each exchangeable cation to the total cation exchange capacity is determined by exchange-equilibrium equations. Perhaps the most commonly used relationship is the Gapon equation (Yong, 2001):

$$\frac{M_e^{+m}}{N_e^{+n}} = K \frac{\sqrt[m]{M_o^{+m}}}{\sqrt[n]{N_o^{+n}}} \tag{2.2}$$

where the superscripts m and n refer to the valence of the cations, the subscripts e and o refer to the exchangeable and bulk solution ions, and the constant K is a function of specific cation adsorption and nature of the clay surface. K decreases in value as the surface density of charges increases. Na^+ versus Ca^{2+} represents a particularly important case of exchange competition. Figure 2.11 shows the effect of the proportions of Na^+ and Ca^{2+} in three clays—kaolinite, illite, and montmorillonite—on the permeability of these clays. Note that the illustration shows a normalized format for the permeability ordinate.

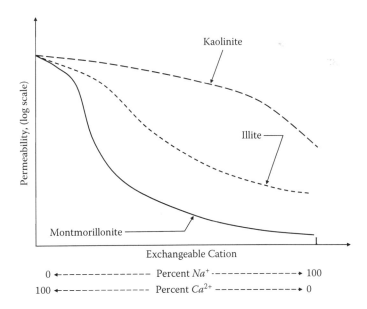

FIGURE 2.11
Schematic illustration showing the influence of exchangeable cations on the permeability of three different clays. Note that the ordinate shows a normalized permeability property.

When the amount of exchangeable calcium on the clay mineral is decreased, its release becomes more difficult. On the other hand, when the degree of saturation with sodium ions is decreased, they become easier to release. Potassium is an exception because its effective ionic diameter of 0.274 nm is about the same as the diameter of the cavity in the oxygen layer. This allows the potassium ion to just fit into one of these cavities, making it very difficult to replace. For other cations, it is the size of the hydrated ions rather than the size of the nonhydrated ones that controls their replaceability. For ions of equal valence, those that are least hydrated have the greatest energy of replacement and are the most difficult to displace. Although Li^+ is a very small ion, it is considered to be strongly hydrated and therefore to have a very large hydrated size. The low replacing power of Li^+ and its ready replaceability can be taken as a consequence of the large hydrated size, but there are in fact indications that Li^+ and Na^+ are only weakly hydrated in interlamellar (i.e., interlayer) positions.

2.5.4 Anion Sorption and Exchange

Anion attraction to clay solids is mainly associated with solids containing oxide surfaces. Clay mineral particles with negatively charged reactive surfaces will attract anions at the edges of the particles because of the presence of broken bonds, as has been discussed in Section 2.5.2. For example, $Al\cdot(OH)$ H_2O anion attraction is similar to those associated with anion attraction to oxide surfaces. The larger proportion of edge surfaces to planar surfaces in 1:1 layer silicates (kaolinites for example), in comparison to the 2:1 layer silicates (e.g., smectites), gives them a higher potential for anion attraction and sorption. Anion sorption capacity for illites appears to be attributable to their hydrous mineral characteristic. The three types of anion exchange in smectites and kaolinite include

- Replacement of OH^- ions of clay-mineral surfaces. The extent of the exchange depends on the accessibility of the OH^- ions; those within the lattices are naturally not involved.
- Anions that fit the geometry of the clay lattice, such as phosphate and arsenate, may be adsorbed by fitting onto the edges of the silica tetrahedral sheets and growing as extensions of these sheets (Pusch and Karnland, 1988; Pusch, 1993). Other anions, such as sulphate and chloride, do not fit that of the silica tetrahedral sheets because of their geometry and do not become adsorbed.
- Local charge deficiencies may form anion-exchange spots on basal plane surfaces.

The last mechanism contributes to the net anion exchange capacity of smectites. The other two may be important in kaolinite but are assumed to be relatively unimportant in smectite clays. The latter minerals commonly have an anion exchange capacity of 5–10 mEq/100 g but can be considerably higher for very fine-grained kaolinite and palygorskite.

Clay minerals such as allophane also attract anion owing to variable charge depending on pH. The surface positive charge increases with an increase in hydrogen ions in porewater because of the reaction of the hydroxyls attached to *Al* with hydrogen ions and, consequently, the amount of anion adsorption increases with a decrease in pH as shown in Figure 2.12.

2.5.5 Specific Surface Area (SSA)

Because of the many interparticle actions and interacting phenomena between clay particles and porewater involving the reactive surfaces of these particles—as will be discussed in the next chapter—the specific surface area (surface area per unit weight) in a clay is an important property. Many of the differences between clay minerals in respect to physical-chemical and mechanical properties such as water retention, plasticity, or cohesion can be explained by the different amounts of surface area between these minerals. This explains the characteristic high swelling performance of montmorillonites with their high specific surface areas. In general, one can obtain a good

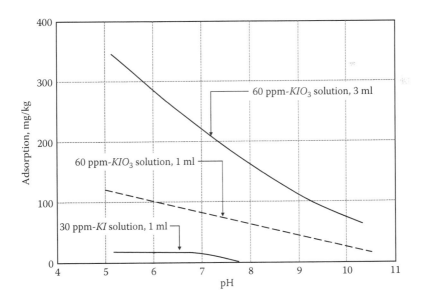

FIGURE 2.12
Adsorption of IO_3^- and I^- on allophane depending on pH owing to pH-dependant charge because of the reaction between hydroxyl of *Al* and the hydrogen ion when KIO_3 and KI solution of 1 or 3 mL are applied to volcanic ash soil. (Data from Nakano 2010—personal communication to Yong.).

appreciation of the nature of the surface properties of a clay from knowledge of the specific surface area (*SSA*) of the clay.

Clay mineral particles are plate shaped or tabular in shape because the layer-lattice structure results in strong bonding along two axes but weak bonding between unit layers. The thickness of a clay particle depends on the magnitude of the forces of attraction between the unit layers, and variations in specific surface area are due primarily to the difference in thicknesses of the particles, taking into account the participation of surfaces of all unit layers in those 2:1 minerals (i.e., interlayer surfaces).

Theoretically, we can calculate the surface area of a representative elementary volume (REV) of a clay if we have information on (a) the shapes and sizes of the individual clay particles, and (b) their distribution, as has been undertaken by Greenland and Mott (1985) using knowledge of the unit cell of a mineral to determine its representative surface area. The *a* and *b* dimensions of a unit cell for a dioctahedral 2:1 mineral such as a smectite were used together with Avogadro's number of such unit cells to arrive at a calculated specific surface area of 757 m²/g.

It is important to distinguish between surfaces available for interparticle action and interaction with water or other fluids. Figure 2.13 shows how the

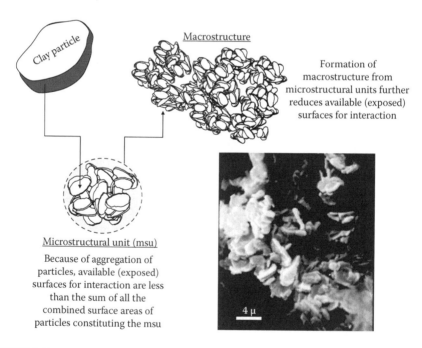

Clay particle

Macrostructure

Formation of macrostructure from microstructural units further reduces available (exposed) surfaces for interaction

Microstructural unit (msu)

Because of aggregation of particles, available (exposed) surfaces for interaction are less than the sum of all the combined surface areas of particles constituting the msu

4 µ

FIGURE 2.13

Surface areas of mineral particles and reduction of available (exposed) surface areas for interactions due to formation of microstructural units and aggregation into the macrostructure of a soil. Picture at bottom right is an SEM of kaolinite showing an aggregation of particles.

formation of microstructural units reduces the available or exposed surfaces and their surface areas.

Aggregation of packing of individual particles into packages identified as domains, peds, aggregate groups, floccs, clusters, and so forth creates the situation where the surfaces of individual particles exposed to fluids can be hidden or totally emasculated. As will be discussed in the later section and in the other chapters of this book, the formation and presence of these aggregated particle groups are factors that need full consideration in evaluation of soil properties and behaviour. Since different disciplines and interest groups use different terminologies to identify these aggregated particle groups, we will use the term *microstructural unit* (*msu*) for these aggregated particle groups. As shown in Figure 2.13, these microstructural units combine to form the macrostructure of a soil, thereby setting the stage for the many soil reactions evaluated as soil behaviour.

Knowledge of the type of clay mineral, the microstructural features of the clay, and also of the difference between exposed and nonexposed particle surfaces is most important. This is particularly relevant, for example, for smectitic clays, where the planar surfaces of the individual unit layers of montmorillonites (Figure 2.14) are active participants in particle interactions. The uncleaved state of the mineral particle exhibits an apparent planar shape. The surface area for this particle is defined by the two basal planes and the sides of the particle. Cleavage or separation of the dioctahedral 2:1 layer silicate mineral into n number of individual unit layers or particles with smaller numbers of unit layers will result in an increase in exposed particle surface areas (bottom portion of Figure 2.14).

In swelling clays, cleavage or separation of the layer silicate into individual unit layers will occur as a result of water uptake. The nature and amount of separation will depend on the chemistry of the water and also on the availability of the water. The footnotes in Table 2.1 highlight this phenomenon. The next chapter provides a detailed discussion of this important property.

Laboratory measurements are generally used to determine the specific surface area of clays since theoretical calculations are not only tedious, but also unrealistic if the clays contain different clay and nonclay minerals. In the procedure that is commonly used, a gas or liquid is used as the adsorbate for the clay solids. The choice of adsorbate is important since the amount of adsorbate that forms a monolayer coat on the surfaces of the clay solids (particles) needs to be determined. One needs to be sure that all the individual clay particles' surfaces are available for interaction with the adsorbate. This means that the clay particles must be in a totally dispersed state. The availability of clay particles in a totally dispersed state and the choice of adsorbate are the two most important factors in any laboratory measurement of the specific surface area (*SSA*) of a clay sample. Because of inherent uncertainties, laboratory determinations of the *SSA* of clay and other types of clays will produce operationally defined measurements of *SSA*.

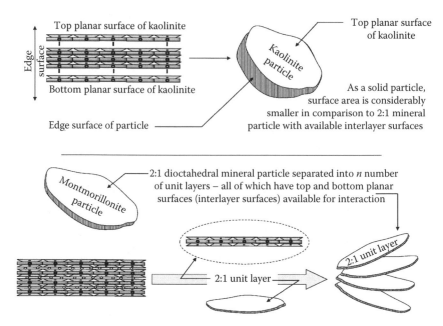

FIGURE 2.14
Surface areas of two clay mineral particles. Top sketch shows a well-crystallized kaolinite particle with edge surface and top and bottom planar surfaces contributing to the total surface area of the particle. In the bottom sketch, the 2:1 montmorillonite particle is shown with its single particle separated into *n* number of interlayers. The total surface area of the montmorillonite particle includes all the interlayer surfaces.

Historically, nitrogen gas, krypton, water vapour, and methyl alcohol were used as the adsorbates, with nitrogen gas being the more popular adsorbate of choice. One key consideration supporting this popular choice is that the number of molecules of nitrogen gas sorbed by the clay particles will be dependent on both the partial pressure of the gas and on the test temperature. Furthermore, since nitrogen gas is not able to migrate into the inner pores of aggregates and the interlayer of 2:1 layer silicates, this method is considered to provide observations pertinent to the outer surface area of particles, aggregates, and clay minerals in soils. This method allows one to measure a small SSA of soils. The relationship developed by Brunauer et al. (1938) is used to determine the amount (volume) of gas equivalent to a sorbed monomolecular layer of gas, since more than one layer of gas will be sorbed by the clay particles in the actual determination process. This relationship is generally known as the BET equation for multilayer sorption.

More recently, polar fluids have been used as the adsorbates. Because of their ability to penetrate and migrate into microscopic pores, the techniques described by Mortland and Kemper (1965) for ethylene glycol and Carter et al. (1986) for ethylene glycol-monoethyl ether (EGME) are used to observe the whole surface area, including the inner surface of aggregates and the interlayer of clay minerals. Common to all the techniques used to determine the SSA is the implicit requirement for the absorbate to only form a uniform monolayer coating on each individual clay particle. The SEM shown in Figure 2.13 will tell us why and how difficult this is. In short, the values of SSA determined are directly dependent on the procedure used and the analytical technique used to reduce the data obtained, leading one to conclude that the determination of SSA is operationally defined.

2.6 Soil Structure

2.6.1 Soil Fabric and Structure

We study soil fabric and soil structure because they constitute, in essence, the structural framework of a soil. We define *soil structure* as that property of a soil that establishes the integrity of the soil–water system. As a property, soil structure includes the distribution and arrangement of particles and the bonding mechanisms between particles, together with their mutual interactions and reactions with the porewater. The distribution and arrangement of soil particles can be viewed as the skeletal framework of the soil and is generally defined as the *soil fabric*. It is important to recognize this as a separate property or as a subproperty of soil structure. The scanning electron microscope (SEM) picture of the network of microstructural units of kaolinite shown in the bottom right of Figure 2.13 is a good example of a soil fabric picture.

The role and influence of soil structure on the properties and performance of soils can be studied from many different perspectives, depending upon end-purpose use or goals. The importance of the microstructural components and their contribution to the overall structure of a soil has long been recognized in the field of *soil science* because of their need to deal with such problems as soil moisture movement, soil tillage, water-holding capacity of soil, soil–water potentials, and water uptake of plants. In the field of *soil engineering*, attention has historically been directed more toward the physical and mechanical performance of soils. Early studies on clay structure in geotechnical engineering, for example, provided us with descriptions of flocculent, honeycomb, and "cardhouse" structures (Terzaghi and Peck, 1948). Studies such as those reported by Lambe (1953, 1958), Pusch (1966), and Yong and Warkentin (1966) called attention to the contributions made by the different clay minerals in combination with other clay fractions on the engineering properties and performance of clays.

Besides the significant contribution of soil structure to the strength and compaction properties of soils, it is abundantly clear that elements of soil structure such as packing of granular particles, aggregation of particles, and arrangement of clay minerals are central to the formation of the intrinsic pore structure for soils. The pore spaces in soils hold and allow flow of water and gas. In effect, soil structure plays a significant role in the water retention characteristics of soils, the permeability of soils to water and gas, and the rheological property of soils.

2.6.2 Granular Soil Packing

The packing of granular soil particles, that is, grains of soil, is very strongly dependent on the sizes, shapes, and distribution of sizes of the soil. For soil engineering purposes where stability and maximum resistance to shearing and shear displacement are required, optimum packing of the soil grains will provide the optimum density of the soil. This requires a specific range of sizes of the soil grains to fill the in-between voids created by larger particles in contact with each other, as shown in the left-side sketch in Figure 2.15 for idealized rounded soil grains. It is clear from the illustration shown in the left-hand sketch in the figure that it would be very unlikely that one would be able to find a distribution of soil grains that would completely fill all the voids in volume element depicted in the diagram. Idealized grain-size distributions of soil grains do not exist in natural soils. Note that the angularity—especially the sharp angularity of soil grains such as those obtained from crushed rock—will create more difficult conditions to obtain optimum density. In part, this will be due not only to the roughness of the surfaces of the particles, but also to length (*L*):width (*W*) ratios greater than one ($L/W > 1$).

The number of adjacent particles in contact with any particle is an important factor in obtaining optimum density and, hence, minimum porosity *n*.

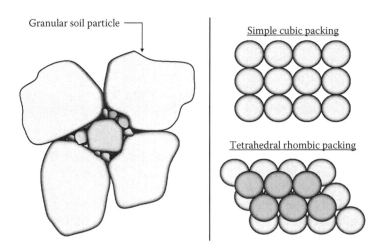

FIGURE 2.15
The left-hand drawing gives an example of an ideal particle-size distribution for optimum packing. The right-hand illustrations show a simple cubic packing of equal-sized spheres at the top and tetrahedral rhombic packing of the same kind of spheres at the bottom.

We define porosity n as the volume of voids in a representative elementary volume (REV) of soil divided by the total volume, that is, the REV. The number of particles in contact with any given particle is defined as the *coordination number N*. We can demonstrate the importance of obtaining the maximum coordination number to achieve optimum density by using the example of packing of spheres of equal radii R. The right-side sketches in Figure 2.15 show the maximum (top) and minimum (bottom) porosities obtained for the packing modes shown. There are five different packing modes that can be obtained with spheres with equal radii. These are

- *Simple cubic*: This packing mode is shown in the top sketch in the right-side illustration in Figure 2.15. The coordination number N is 6, meaning that any one sphere is in contact with 6 other spheres, and the porosity n obtained from this packing mode is $n = 47\%$.
- *Cubic tetrahedral*: The coordination number N for this mode of packing is 8 and the porosity obtained is $n = 39\%$.
- *Tetragonal sphenoidal*: For this mode of packing, $N = 10$ and $n = 30\%$.
- *Pyramidal*: For this packing mode, $N = 12$ and $n = 26\%$.
- *Tetrahedral rhombic*: This is shown in the bottom sketch in the right-side illustration in Figure 2.15. The coordination number N and the porosity n are identical to those obtained for pyramidal packing, that is, $N = 12$ and $n = 26\%$.

The comparison of N and n values obtained with the five different packing modes for the spheres of equal radii show that the simple cubic mode with the lowest N value has the highest porosity n. The highest numbers of N values obtained are with the pyramidal and tetrahedral rhombic packing modes, meaning that these two modes of packing gave one the optimum densities and minimum porosities for these kinds of spheres.

The study of packing of spheres with unequal radii, which is more representative of real soils, can be done mathematically. To do so, one needs to prescribe a limiting set of conditions, as demonstrated by Wise (1952) with one large sphere and all other spheres of smaller sizes placed such that they would all be placed on the surface of the large sphere. The first two spheres should not only touch each other but should also touch the surface of the large sphere, and each new subsequent sphere to be added must touch both the large sphere and at least two other spheres.

In natural sandy soils, the surface of soil grains is more or less coated or covered with organic matter and precipitated chemical compounds such as calcium carbonate and Al- and Fe-hydroxide and others. The soil grains of such soils contact each other or combine with clay minerals to form soil structures similar to aggregates because of the gluing action and cementation by the coating materials. To determine the distribution of particle sizes in sandy soils in experiments and laboratory tests, removal of gluing and cementing materials from soil particle surfaces and contact points between particles is necessary.

The aggregate size distribution is also an index of the structure of granular soil packing in natural soils. It is especially important to have information of water-bearing aggregates in relation to interaction with water since water-bearing aggregates affect the properties and behaviour of soils, for example, water retention, transport phenomena, and rheology.

2.6.3 Microstructure of Clays

The early studies of *soil behaviour*, as a formal discipline in the 1950s, provided the impetus for targeted research into the role of soil structure in the development of soil properties, behaviour, and response performance. It has long been understood that for many soil engineering applications, it was acceptable to consider soil as a continuum instead of a complex system of particulate media involving interacting soil solids and porewater. Accordingly, laboratory and field test procedures measured or determined the macroscopic performance of the soil, using testing-analytical tools that relied on data-reduction models developed for uniform and homogeneous media. Even today, the majority of analytical tools used for assessment of soil performance are still based on continuum mechanics principles.

Whilst deterministic considerations and analyses of soil as a continuum can be used successfully for many soil engineering applications, problems arise in predicting soil behaviour when the soils under consideration are (a) heterogeneous and structured and therefore require consideration of the structure of the medium such as the sensitive marine clays, and (2) such that the behaviour of the soils responds to both gravitational and molecular forces, and hence requires analyses that incorporate intermolecular forces. This is particularly true for clay soils with active clay minerals, that is, clay minerals with active and reactive particle surfaces (smectite is a good example of this type of clay).

The importance of soil structure in clay properties has long been recognized. Taylor (1948) classified soils into three types of fundamental structures: single-grained structure (granular soils), honeycombed structure, and flocculent structure. In addition, Taylor recognized the role of intermolecular forces, stating for example: "The finer the grains (of soil), the more noticeable becomes the effect of intermolecular forces" in reference to honeycomb-structured soils. In regard to flocculent-structured soils, he states that "in addition to the force of gravity, the molecular impact forces must be given consideration in the study of the action of these small particles." Present-day understanding of clays readily recognizes that clays are generally composed of soil–particle structural units grouped together in some coherent fashion to make up the macrostructure of the clay.

Figure 2.13 shows the basic elements that make up the microstructural units (*msu*) and their aggregation into the macrostructure of a soil. These microstructural units range from individual mineral particles to what is known in the various disciplines and interest groups as domains, flocs, clusters, peds, and aggregate groups, to cite a few of the names used in the literature. To determine the kinds of microstructural units, there are at least three broad groups of techniques available, depending on the sizes of the particles involved. The macroscopic technique that constitutes the first level of study of *msu* relies on recognition of *msu* with the naked eye. This is confined to granular particles. It is possible to see aggregate groups of particles generally called *peds*, or *crumbs*, consisting of a large number of indistinguishable (with the naked eye) clay particles. The two other groups of techniques for identifying microstructural units are microscopic and ultramicroscopic in nature. As the name implies, microscopic techniques rely on light microscopes for determination of *flocs*, and ultramicroscopic techniques use various forms of electron microscopic viewing to study *domains*, and so forth. Note that regardless of the technique used for determination of *msu*, the results obtained will be operationally defined since (a) the technique used will establish the nature of *msu* observed, and (b) the size of specimen studied is nothing more than a very miniscule portion of any laboratory or field soil sample and hence cannot be confidentially assumed to be totally representative of the soil under consideration.

We can also determine the presence or formation of *msu* by less direct methods, using the energies of interactions between particles. These will be more qualitative in nature than quantitative. Interactions between adjacent particles and interactions with water are by and large controlled by intermolecular forces since particle sizes are in the micron range, and since the specific surface areas of these particles and structural units are on the order of ten to hundreds of square metres per gram of soil. The total interaction energy between particles can be calculated from zeta potential test measurements relative to distance between surfaces for various particle arrangements. (The subject of particle interaction energies will be discussed in the next chapter.) Taking note that the zeta potential is directly related to the total interaction energy between particles, Yong and Sethi (1977) studied clay dispersibility using refiltration experiments as described by LaMer and Healy (1963a, 1963b). Measurements of refiltration rate relative to zeta potential provided information on the dispersibility of the clay studied by Yong and Sethi. The results showed that the higher the refiltration rate, the more flocculated the clay. The results obtained in conjunction with the corresponding zeta potential values are shown in Figure 2.16. The illustration shows that as the zeta potential ζ increases, the degree of dispersibility decreases, indicating that single particle interaction becomes less and less. At some point, flocculation of particles will occur, and interactions between the clay particles will be in relation to flocs (i.e., microstructural units). This technique allows one to obtain a mechanistic picture of not only the formation of microstructural units, but also their interactions.

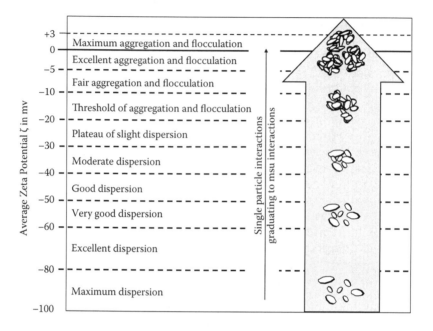

FIGURE 2.16

Interpretation of presence of microstructures from ranges of average zeta potential ζ measurements from dispersion stability study of clays. (Interpreted from data reported by Yong, R.N., and Sethi, A.J., 1977, Turbidity and zeta potential measurements of clay dispersibility, *ASTM-STP 623*, pp. 419–431.)

2.7 Interparticle Bonds

It is important to remember that when one speaks of soil structure and microstructural units, one is referring to material that exhibits inherent strength (resistance to shear displacement) and other properties such as transmissibility of fluid, heat, and so forth. In other words, microstructural units have physical, mechanical, physicochemical, and chemical properties. When such units are combined into a macrostructure that defines a clay, their integrated properties define the properties of that clay. It follows that the physical integrity of a clay is directly related to the nature, distribution of the microstructural units, and the bonding and interaction forces not only between the clay particles within the *msu*, but also between the various *msu*s themselves. There are two groups of bonds that exist between particles in microstructural units and between these units themselves, both of which are responsible for the development of the properties and characteristics of the *msu* and the macrostructure itself. The first group of forces and bonds deals directly with particle-to-particle interaction, that is, interparticle action. The second group of forces and bonds, which are those that are developed between particles with mediating forces and bonds from the solutes in water and the water molecules themselves, will be discussed in greater detail when we deal with clay–water interactions in Chapter 3.

2.7.1 Forces and Bonding

The forces and bonds associated with interparticle action consist of

- *Primary valence bonds between particles*: Except for cementation bonds, primary valence bonds are the strongest interparticle attraction forces; the activation energy for breakage normally exceeds 125.5 kJ/mol. They are dominant in heavily consolidated clays where the crystal lattices of adjacent particles are in contact.
- *London-van der Waals forces*: These forces are important because they operate in the range from 0.2 nm to more than 10 nm regardless of whether the particles are charged or uncharged. The bond strength is less than 4.184 kJ/mol.
- *Hydrogen bonds*: Hydrogen bonds are weak (4.184–12.55 kJ/mol) but their numbers can be large, and the net attraction force will thus be high.
- *Bonds by sorbed cations*: These are Coulombic bonds. The integrated attraction force between neighbouring but not contacting particles is significant. The approximate activation energy is intermediate to that of hydrogen bonds and primary valence bonds, that is, on the order of 41.84–62.76 kJ/mol.
- *Dipole-type attraction between particles with different charge*: Depending on the equilibrium or near-equilibrium pH of the immediate micro-environment, the edges of clay particles can be positively or negatively charged.
- *Bonds by organic matter*: The attraction forces are primarily due to hydrogen bonds, and purely physical coupling is obtained by embracing hyphae and flagellae. The bonds are flexible and can sustain large strain. However, their strength is very much dependent on the environment and can be short-lived.
- *Cementation bonds*: Precipitated matter binding particles together can develop bonds with strengths that can approach that of primary valence bonds (>125.5 kJ/mol). Their practical importance can be substantial, depending on the amount and nature of the precipitation as shown, for example, by the difference between plastic clay and claystone.

Short-range forces such as those developed as a function of ion–dipole interaction, dipole–dipole interaction, and dipole–particle site interaction are of considerable importance in arrangement of particles in a microstructure. A more complete discussion of these will be found in Chapter 3.

2.8 Microorganisms in Soils

Microorganisms in soils include bacteria, protozoa, fungi, algae, and viruses. They are smaller than plant and animal cells and are divided into two groups depending on their cell structures:

- *Prokaryotes (simple, single cells less than 5 μm)*: Prokaryotes have a nuclear region not encompassed in a membrane, with only a single strand of deoxyribonucleic acid (DNA).
- *Eukaryotes (single or more complex multicells that are greater than 20 μm)*: Eukaryotes have a nucleus surrounded by a membrane containing DNA molecules and are subdivided into unicellular organisms that have multipurpose cells and multicellular organisms (plants and animals) with special-purpose cells.

2.8.1 Types of Microorganisms

The various types of organisms and microorganisms in the soil environment fall within the Whittaker (1969) five-kingdom classification as shown in Figure 2.17. Whilst all of these organisms and microorganisms are important in the geoenvironment, our concern is with the effect of these bioorganisms on the soil engineering quality of a soil. In that respect, we consider bacteria and fungi to be the microorganisms of importance because of their role in (a) bioclogging, that is, changing the nature of pore spaces as a result of accumulation of living bodies and hyphae, resulting in a decrease in the infiltration rate of water, hydraulic conductivity, flow path of fluid in soils, and so forth. (Seki et al., 2006), and (b) alteration and decomposition of organic matter, transformation and bioweathering of clay minerals, resulting in changes in the properties and behaviour of the clay minerals. The discussion for bacteria will be conducted in the next subsection. Although viruses that are acellular are not included in the classification scheme shown in the figure, they are classified as microorganisms and are included in the short summaries of the kinds of organisms.

2.8.1.1 Protozoa

The protozoas, which include pseudopods, flagellates, amoebas, ciliates, and parasitic protozoa, have sizes that can vary from 1 to 2,000 mm. They

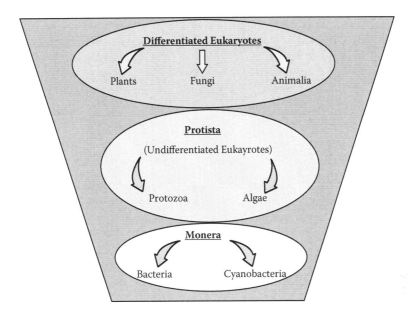

FIGURE 2.17
Organisms and microorganisms grouped according to the Whittaker 5-kingdom hierarchical system.

are aerobic, single-celled chemoheterotrophs, and are eukaryotes with no cell walls. They are divided into four main groups: (a) Mastigophora, which are flagellate protozoans; (b) Sarcodina, which are amoeboid; (c) Ciliophone, which are ciliated; and (d) Sporozoa, which are parasites of vertebrates and invertebrates. Protozoa scavenge particles such as bacteria, yeasts, fungal spores, and other protozoa. They are found in water and soil. Water is an absolute requirement for their survival.

2.8.1.2 Fungi

Fungi live mainly in the soil or on dead plants and are sometimes found in fresh water. Filamentous fungi (mould) are especially important because of their role in soil in clogging of pores from the production of hypha and mycelium, resulting in a decrease in hydraulic conductivity and gas diffusion coefficient in soils (Seki et al., 1998). Fungi are aerobic multicellular eukaryotes and chemoheterotrophs that require organic compounds for energy and carbon. Reproduction is by formation of asexual spores, and in comparison to bacteria, they (a) do not require as much nitrogen, (b) are more sensitive to changes in moisture levels, (c) are larger, (d) grow more slowly, and (e) can grow in a more acidic pH range (less than pH 5).

Under the general classification of fungi are slime moulds, yeasts, and mushrooms that form mycelium for adsorbing nutrients from a substrate.

Note that in the context of biological activities, the term *substrate* means food for microorganisms; that is, *substrate* serves as a nutrient source and a source of carbon, or energy for microorganisms. Yeasts are unicellular organisms that are larger than bacteria. They are shaped like eggs, spheres, or ellipsoids and reproduce by formation of buds or fission. The two main components of mushrooms are (a) the basidia, the fruiting body above the ground, and (b) the mycelium, which is below the ground.

2.8.1.3 Algae

Algae are single-celled and multicellular microorganisms that are green, greenish tan to golden brown, yellow to golden brown (marine), or red (marine). They grow in the soil and on trees or in fresh or salt water. Those that grow with fungi are called lichens. Seaweeds and kelps are examples of algae. Since they are photosynthetic, they can produce oxygen, new cells from carbon dioxide or bicarbonate (HCO_3^-), and dissolved nutrients including nitrogen and phosphorus. They use light of wavelengths between 300 and 700 nm. Red tides are indicative of excessive growth of dinoflagellates in the sea. The green colour in the body of lakes and rivers is due to the presence of algae that increase with eutrophication—attributable to the accumulation of nutrients such as fertilizers in the water.

2.8.1.4 Viruses

Although viruses are smaller than bacteria and require a living cell to reproduce, their relationship to other organisms is not clear. In order for them to replicate, they have to invade various kinds of cells. They consist of one strand of DNA and one strand of ribonucleic acid (RNA) within a protein coat. A virus can only attack a specific host. For example, those that attack bacteria are called bacteriophages.

2.8.2 Bacteria

There exist various species of bacteria in soils and sediments—alkalophilic and asidophilic bacteria in the ground and barophilic bacteria in sediments. They play an important role in soil since (a) they form microcolonies in large pores and voids and form biofilms on the walls of large cracks and gaps in soil, resulting in bioclogging of pores and changes in the mode of fluid flow in soils, and (b) they are thought to have the capability to transform clay minerals through mechanisms of weathering (bioweathering) and dissolution over long time periods. They are prokaryotes that reproduce by binary fission by dividing into two cells, in about 20 min. The time it takes for one cell to double, however, depends on the temperature and species. For example, the optimal doubling time for *Bacillus subtilis* (37°C) is 24 min and for *Nitrobacter agilis* (27°C) is 20 h.

The broad classification of bacteria is undertaken by the shapes of the bacteria: for example, rod-shaped bacillus that have diameters of 0.5 to 1 µm and lengths of 3 to 5 µm, spherical-shaped coccus that have spherical cell diameters of from 0.2 to 2 µm, and spiral-shaped spirillium that have cell diameters ranging from 0.3 to 5 µm and lengths ranging from 6 to 15 µm. The cells grow in clusters, chains, or single form and may or may not be motile. The substrate of the bacteria must be soluble. Some of the most common species of bacteria are *Pseudomonas, Arthrobacter, Bacillus, Acinetobacter, Micrococcus, Vibrio, Achromobacter, Brevibacterium, Flavobacterium,* and *Corynebacterium,* and within each species, there will be various strains, with each of them behaving differently. Some strains can survive in certain conditions that others cannot, and some species are dependent on other species for survival.

The living bodies of bacteria are composed of a variety of elements such as *C, H, O, N, P, S,* and trace elements, and by and large can be chemically expressed as $[A]CH_{1.7}O_{0.4}N_{0.2}$, where *A* designates *P, S, K, Mg, Mn, Ca, Fe, Co, Cu, Zn, Ni, Mo,* and so forth. Bacteria take these elements in vivo as nutrients, and these elements are used to form (a) the body of bacteria and (b) such metabolites as gas and extracellular polysaccharides as well as ATP (adenosine triphosphate) that supply the energy required for their growth and sustenance. Bacteria that utilize organic substrates for energy are called *chemoorganotrophs,* and those that use organics as a carbon source are called *heterotrophic bacteria.* Bacteria that utilize such inorganic compounds such as $H_2, NH^{4+}, NO^{2-}, S, H_2S, and Fe^{2+}$ as energy sources and utilize CO_2 as a carbon source (not organic matter) are called *autotrophic bacteria.*

Bacteria can also be classified in respect to their requirements for oxygen. Those that require molecular oxygen for respiration are aerobes (aerobic bacteria), and those that do not require molecular oxygen for respiration are anaerobes (anaerobic bacteria). Facultative bacteria grow in both environments, with and without oxygen. These organisms use molecular oxygen when it is present, and when low levels of oxygen are present they utilize nitrate or other oxides for their energy source. Energy is obtained from the respiration process by the release of hydrogen from a donor enzymatically, resulting in oxidation. When the hydrogen atom meets an acceptor, reduction occurs. Oxygen is the hydrogen acceptor during aerobic respiration. It is the best acceptor since the most free energy is released this way. Aerobic microbes thus grow faster than anaerobic strains. Water is produced as the end product. Carbonate, iron, nitrate, or sulphate ions serve as electron acceptors in anaerobic respiration. Methane, ammonia, hydrogen sulphide, or a reduced organic compound are the final products, depending on the hydrogen acceptor. They live in a wide range of temperatures, that is, 0°C to 120°C. The optimum temperature is a range of 25°C–50°C for a great number of bacteria, although there are some species (*thermophilic bacteria*) that will survive temperatures higher than 100°C, as for example, sulphate reducing bacteria. Depending on the species and circumstances, the life span of bacteria is different, that is, from a few days to about 300 days at the maximum.

By and large, bacteria live on the surface of solid substances. They form colonies and produce biofilms that are composed of the stacks of colonies on the surfaces of clay entities such as clay *msu* and clay masses. A common assumption for the mass of bacteria is about 10^{-12} g/cell, with an estimated size of bacteria to be about 0.25–2.0 μm. The maximum density obtained varies according to the species, for example, 4–6×10^{13} cells/kg-mineral for *T. ferrooxidans* and 2–4×10^{13} cells/kg-mineral for *T. thiooxidans* (Konishi et al., 1990, 1995). The mass of bacteria in colonies is assumed to be about 90 mg/cm³-colony, with the estimated size of colonies to be about 10–20 μm in diameter and 5–10 μm thick (Molz et al., 1986), with a maximum thickness of 10–50 μm.

The surface of bacteria has a positive charge at an extremely low pH and a negative charge at a pH higher than 4 (Marshall, 1976). It is important to note that when such biofilms are formed in soil, they will have the ability to change the apparent surface charge of the soil particles and *msu*, depending on the pH of the microenvironment. Bacterial growth rate is directly dependent on substrate concentration. The higher the substrate concentration, the higher the growth rate. The substrate can be a single compound or a mixture. When it is a mixture, the parameters such as biochemical oxygen demand (BOD), chemical oxygen demand (COD), total organic carbon (TOC), or total petroleum hydrocarbons (TPH) can be used on a mass or liquid basis to indicate the total substrate concentration.

2.8.3 Ecology in Soils

The existence and activities of microorganisms in natural soils and rocks, clay buffers, backfills, and host rocks in repository-type environments have been reported by many researchers (e.g., Huang and Schnitzer, 1986; Pedersen, 2000; Wang and Francis, 2005; Yong and Mulligan, 2004). The environmental factors that favour optimal activities of microorganisms include temperature, oxygen availability, presence of water, nutrients, and osmotic pressure. Some of the more important chemical factors impacting microbial activity include pH, toxicity, heavy metals, molecular structure, and co-metabolism. Whilst microorganisms prefer a pH range of between 5.6 and 9.0, there are some that can function at higher or lower pH environments than 5 as fungi. Sulphur-oxidizing bacteria can produce sulphuric acid, which will lower the pH to values below 1 (e.g., $H_2S + 2O_2 \rightarrow H_2SO_4$). At neutral pH, generation of carbon dioxide by the microorganisms helps to buffer the system and also helps to maintain neutral pH.

The three main temperature ranges wherein the various classes of microorganisms can grow optimally are as follows: (a) about 0°C to about 10°C for psychrophiles, (b) about 10°C to about 45°C for mesophiles, and (c) about 45°C to about 75°C and above for thermophiles. The lower and upper bounds of the temperature ranges are not absolute. It is commonly accepted that a 10°C increase in temperature will double the growth rate. Since normal ground temperatures in most habitable regions range from slightly below 0°C to about 45°C, psychrophiles and mesophiles are the most common types of microorganisms found in the ground. Capsule formation enables microorganisms to survive at reduced temperatures and grow as the temperature increases to more favourable conditions (Sims et al., 1990).

Water availability is essential for the survival of microorganisms, with moisture levels of 50% to 75% being considered to be ideal conditions. Water constitutes (a) the principal component of cell protoplasm and (b) the carrier for nutrient transport into the cell. An excess of water could result in oxygen limitations, because the low solubility of oxygen in water (9 mg/L at 20°C) makes it difficult for oxygen to be transported to bacteria in saturated soil pore spaces. Sufficient oxygen must be available for aerobic and facultative anaerobic bacteria, with a concentration of approximately 2.0 mg/L being considered to be a good value (Reynolds and Richards, 1996). This means that partly saturated soils are better conditions for optimal microorganism growth. Microbial activities in reduction of electron acceptors will affect pH levels through production of H^+.

Microbial flora in groundwater include (a) aerobic and microaerophilic heterotrophic microorganisms and (b) anaerobic iron-reducing and sulphate-reducing bacteria such as *Desulfovibrio* and *Desulfotomaculum*. Chemoautotrophs such as *T. thiooxidans* and *T. ferrooxidans* play significant roles in the subsurface environment, where organic matter is in short supply, because of their ability to (a) obtain energy required to sustain life by oxidation of inorganic matter, and (b) obtain carbon required for the formation of bodies by decomposing carbon dioxide. Microorganisms react with clay minerals by producing extracellular polysaccharides that coat the microorganisms through the various forms of interaction. Indigenous bacteria in soils, together with extraneous bacteria transported into soils by groundwater, will grow on the surfaces of soil particles in large pores or fractures with widths larger than 5 µm, provided that water and nutrients are available for growth in the microenvironment. Microbial flora will grow and induce biological dissolution/transformation of clays or metals following their adherence on the surfaces of clay particles, or following migration into fissures and cracks in clay buffer/barriers of more than 0.25 µm in width (the minimum size of bacteria).

2.8.4 Bacterial Growth Kinetics

For bacterial growth, the maximum rate of growth and metabolism occurs during the log growth phase, in which the log of the number of cells versus time is linear according to Monod (1949). As the substrate oxygen or nutrient becomes limiting or is depleted, pH shifts result, or toxic

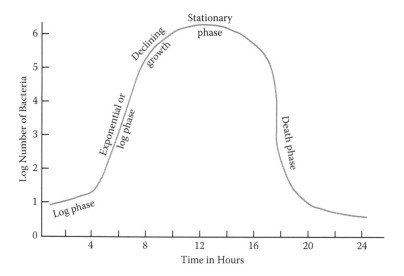

FIGURE 2.18
Bacterial growth as a function of time. (Adapted from Yong, R.N., and Mulligan, C.N., 2004, *Natural Attenuation of Contaminants in Soils*, CRC Press LLC, Boca Raton, FL, 319 pp.)

components start to accumulate as the declining growth phase occurs (Figure 2.18). The rate of growth decreases, and microorganisms begin to die. The stationary phase occurs as the numbers of cells produced and dying are equal. This occurs usually over a 12- to 36-h period. The death phase occurs when the number of microorganisms dying exceeds the amount growing, and the cells may either be inactivated or may start to degrade and break apart.

The growth rate of living bacteria can be expressed as (Monod, 1949)

$$\frac{d\rho_m}{dt} = (\mu - \kappa)\rho_m \tag{2.3}$$

where ρ_m is the biomass density, μ is the intrinsic growth rate, and κ is the intrinsic decay rate coefficient. Nakano and Kawamura (2010) have suggested that in view of the ageing and lifespan of bacteria, the intrinsic growth and decay rate coefficients can be considered as functions of time as follows:

$$\mu = \mu_{max}\left(1 - a_\mu \cdot \exp(-b_\mu t^2)\right) \tag{2.4}$$

$$\kappa = \kappa_{max}\left(1 - a_\kappa \cdot \exp(-b_\kappa t^2)\right) \tag{2.5}$$

where μ_{max} and κ_{max} are the maximum values of μ and κ, respectively; t is time; and a and b are constants.

We can observe the microbial growth kinetics under conditions that substrate and oxygen concentration control an increase in biomass, as for example: (a) under conditions where the microenvironment is isolated from the outside, such as those in simple laboratory culture experiments, or (b) where the inflow of substrate and oxygen into the system is smaller than the demand by microbes in natural soils. In such cases, we describe bacteria growth kinetics using the equation that explicitly includes the factors relating to the concentration of substrate and oxygen. The two types of equation describing this event are (a) those taking into consideration the effects on the intrinsic growth rate only (Molz et al., 1986), and (b) those multiplying the intrinsic growth rate by the inhibition term (Thullner, 2003).

$$\frac{\partial \rho_m}{\partial t} = \left(\mu_{max}\frac{c_s}{K_s + c_s}\frac{c_o}{K_o + c_o} - \kappa\right)\rho_m \tag{2.6}$$

$$\frac{\partial \rho_m}{\partial t} = \mu_{max}\frac{c_s}{K_s + c_s}\frac{c_o}{K_o + c_o}\frac{k_{inh}}{k_{inh} + c_{inh}}\rho_m \tag{2.7}$$

where c is the concentration, K is the half-saturation constant in solution or gas, subscripts s and o refer to substrate and oxygen respectively, c_{inh} is the concentration of substances inhibiting the growth, and k_{inh} is the inhibition constant.

In the environment, following the expiration of a group of bacteria due to certain inhibition factors, a new group of bacteria may arise after an interval from bacteria originating from the outside or from indigenous bacteria. For the long term, they would repeat the growth and decay cycle as in process A shown in Figure 2.19. As another case, we can consider an event where the decay process overlaps with the next growth process that occurs in the new group of bacteria, resulting in a steady population of bacteria at any time as shown in process B in Figure 2.19. The changes in population of bacteria in soils will be influenced by the local environmental conditions in the pores of soil.

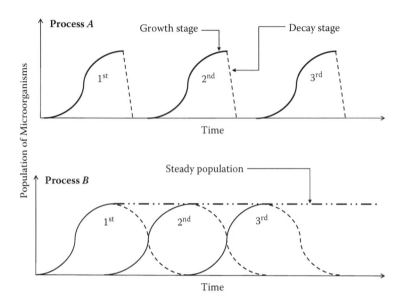

FIGURE 2.19

Schematic diagram of the survival mode of bacteria with reproduction for the long term. Process *A*: an event reproduced at an interval. Process *B*: an event with reproduction overlapping with decay. The solid lines represent the growth stage, and the dashed lines represent the decay stage.

2.9 Laboratory Determinations

2.9.1 Scope of Tests

In this section, we will be concerned with what should normally be conducted in laboratory tests in respect to determination of soil type and its physical properties for a particular soil being considered for soil and geoenvironmental engineering applications. The scope of laboratory tests (extent and detailed scrutiny) for type of soil and physical properties is dependent on at least two main groups of factors. Group I is the most obvious group of factors, whilst the second (Group II) is perhaps somewhat hard to fully grasp. To a large extent, this is because the two factors in Group II are intertwined. Defining the various impacts and their consequences will depend on the kinds of expertise "brought to the table."

Group I:

- Type of engineering application
- Economic considerations
- Time constraints

Group II

- Impact of "failure to perform design functions" on safety, health, and economic welfare of the public and environment
- Expertise of parties—owners, regulators, contractors, consultants, and others involved.

2.9.2 Soil Type and Mineral Analysis

2.9.2.1 Soil Type

Some of the tests conducted by experienced soil engineers and practitioners for determination of soil type do not classify as rigorous or quantitative tests. Upon first encounter with a sample of a particular soil for consideration for an engineering application, the experienced practitioner uses *sight* (*colour*), *smell*, *feel*, and in some instances, *taste*. With these preliminary sight-smell-feel-taste *observations*, one can deduce whether (a) the particular soil is granular or clayey or somewhere in between, (b) some organic material is present in the soil, and (c) the soil is plastic or nonplastic. Whilst these preliminary observations are essentially qualitative in nature, they are nevertheless very useful since they provide one with an initial appraisal of the soil under consideration. With the basic laboratory tests for soil type, this initial appraisal can be tested and confirmed or denied.

The basic laboratory tests generally used to identify *soil type* include

- Sieve analyses and hydrometer tests for determination of grain-size distribution and soil texture, and to classify soil type. The important point to be made here is to ensure that the grab samples used for these tests are representative of the soil under consideration, especially if reduction of sample size is necessary. Pulverization of aggregate groups, quartering procedures, and preparation and allocation of samples are some of the procedures that are partly qualitative in nature and also sensitive to operator technique.

  The top portion of the protocols shown in Figure 2.20 contains some of the principal elements pertaining to sieve analyses for particle-size distribution analyses and also for separation of the grab samples into sand, silt, and clay. Sieve separation into sand, silt, and clay is necessary if one needs information on the types of minerals in the soil under consideration.

- Consistency tests to determine liquid limit, plastic limit, shrinkage limit, and plasticity index. It is important to note that since sample preparation and operator technique are significant factors in laboratory testing for consistency limits, the results obtained should be considered as operationally defined.

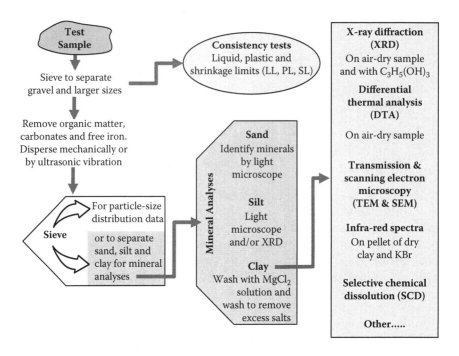

FIGURE 2.20
Test techniques and procedures for initial identification of soil type, consistency, and minerals. For detailed information on clay mineralogy (right-hand panel), one begins with the top technique (XRD) and proceeds downward with other supplementing techniques for more clarity in mineral identification.

- Tests to determine the physical properties such as density, water content, degree of (water) saturation, porosity, void ratio, and specific gravity of solids. Other tests to determine such properties as soil rheology, compactibility, transmissivity, and others do not classify physical properties. These will be discussed in the next few chapters.

2.9.2.2 Mineral Analysis

The various types of tests for mineral analysis are shown in Figure 2.20. There is no single method that can be considered to be fully satisfactory for identification of the kinds of minerals in soils. In part, this is because (a) soil is composed of a variety of minerals and (b) there will always be a range in composition of the crystal structure of clay minerals from different sources. Techniques for determination of a single mineral will be complicated because of interference between the various different minerals and the less than pure crystal structures of the minerals. Accordingly, several methods are required to seek identification of the different minerals in soils, especially if quantification of these minerals is required.

X-ray diffraction (XRD) is the most useful method of identification. This is especially true if the samples are treated. Glycerol ($C_3H_5(OH)_3$) treatment for detection of montmorillonite and K-saturation and heating to collapse vermiculite are the most common treatment methods. A comparison of the x-ray diffraction spacings d obtained from the [001] planes of Mg-saturated air-dried, Mg-saturated glycerol-solvated, K-saturated air-dried, and K-saturated heated at 500°C samples of montmorillonite, vermiculite, and chlorite shows:

Mg-saturated, air-dried:

  Montmorillonite, vermiculite, chlorite $d = 1.4–1.5$ nm

Mg-saturated, glycerol solvated:

  Montmorillonite $d = 1.77–1.8$ nm

  Vermiculite, chlorite $d = 1.4–1.5$ nm

K-saturated, air-dried

  Chlorite, vermiculite (with interlayer aluminium) $d = 1.4–1.5$ nm

  Montmorillonite $d = 1.24–1.28$ nm

K-saturated, heated (500°C)

  Chlorite $d = 1.4$ nm

  Vermiculite (contracted), montmorillonite
  (contracted) $d = 0.99–1.01$ nm

X-ray diffractograms combine with differential thermal analysis (DTA) and measurements of infrared absorption spectrum (IR spectra) to provide a useful grouping of tools for mineral identification. Clay minerals have absorption bands in the infrared region of the energy spectrum in the range of molecular bond vibration frequencies. Many of these bonds are not specific to one mineral because they are due to interatomic bonds common to many minerals. However, this should not limit the use of information from IR spectra since other supporting techniques such as the use of DTA can assist in sorting mineral identification.

Clay minerals lose water or undergo phase changes that give off or require heat at specific temperatures. The loss of water molecules causes an endothermic reaction in which heat is taken up by the sample. In contrast, an exothermic reaction is the result of heat given off by the sample, occurring because of a phase change in the structure of the mineral. The temperatures at which these reactions occur are characteristic of the mineral. The use of differential thermal analysis as a tool utilizes these thermal change characteristics. In essence, DTA provides information on the temperature at which changes occurs in a mineral when it is heated continuously to a high temperature. The intensity of the change is directly proportional to the amount of the mineral in the sample being tested.

2.9.2.3 Soil Fabric and Microstructural Features

The traditional methods for viewing soil fabric include light microscopy and electron microscopy. For a successful and accurate presentation of soil fabric, it is necessary to obtain a representative sample of the soil in a manner that preserves the natural particle arrangement. This can pose severe problems since one requires thin sections and flat smooth surfaces. For light microscopy, one generally uses specimens with a thickness of 10–30 μm obtained by microtomes equipped with special steel or diamond knives. Light microscopy using special wavelengths and optics can be used for identifying soil fabric details with a size down to about 3 μm.

For electron microscopy techniques, the thickness of the section from which micrograph pictures are taken is of fundamental importance. The ideal thin section is a two-dimensional plane. The thicker the section, the smaller the possibility of identifying voids included in the clay matrix. The question of what thickness one can accept depends on the required resolution and the need for discerning microstructural details. Specimens used in transmission electron microscopy (TEM) are generally 20–100 nm thick. It is expected that objects with sizes of 1 nm are discernible with this technique. A detailed discussion of quantitative analysis of microstructural features can be found in Pusch and Yong (2006).

For sample preparation, impregnation of clay samples with a suitable well-hardening substance is used to replace the porewater. Impregnation of a sample without prior drying can be made by repeated impregnation of special water-soluble substances such as Durcupan and Carbowax. This procedure removes the porewater and replaces it with a substance that becomes sufficiently hard for microtomy. In the case of monomers used for impregnation, replacement of the porewater by the monomer without disturbing the network during the replacement phase or during polymerization of the monomer is a problem that needs to be addressed.

2.10 Concluding Remarks

A recurring question in soil characterization or soil identification is "How detailed should we be in conducting tests for soil characterization/identification?" There is no simple answer to this simple question, the reason being that this depends on several factors as identified in Section 2.9.1. The presence of various kinds of clay fractions and their interactions are important factors in the evaluation of the surface properties of clays.

Because of the pH dependency of the surface properties of such fractions as organic matter, amorphous materials, and even some clay minerals, interactions occurring between various clay fractions will change the characteristic *SSA* and *CEC* of the clays. We see from the discussion in the early part of this chapter that the surface properties of the mineral particles are more or less dominated by the hydroxyl surface functional groups, whereas organic matter possesses a greater variety of surface functional groups. We will learn in the next few chapters about the importance of surface properties of soils in relation to soil integrity and in regard to interactions with contaminants.

The importance of the role of microstructure in the control of the hydraulic conductivity and other transmission characteristics of clay, and in the development of clay integrity and rheological properties, cannot be overstated. We need to emphasize that the microstructural units and groups of microstructural units are never static, in the sense that restructuring of the microstructural units and groups of such units will always occur in response to internal and external provocative gradients.

The physical properties of soils discussed in this chapter relate directly to the status of a soil. In soil and geoenvironmental engineering applications, one is generally interested in the initial or natural state of the soil under consideration. These include organic matter content, soil density, water content, degree of water saturation, and consistency limits. There are some that would argue that physical properties should also include inherent soil strength and hydraulic conductivity of the soil. We believe otherwise since these fall under classifications of mechanical and transmission properties, to be discussed in the next few chapters.

References

Asutton, R., and Sposito, G., 2005, Molecular structure in soil humic substances: The new view, *Envr. Sci. Tecnnol.*, 39:9009–9015.

Birrell, K.S., and Fields, M., 1952, Allophane in volcanic ash soils, *J. Soil Sci.*, 3:156–166.

Bolt, G.H., and Bruggenwert, M.G.M., 1978, *Composition of soil*, in G.H. Bolt and M.G.M. Bruggenwert (Eds.), *Soil Chemistry: Basic Elements*, Elsevier Scientific Publishers, Amsterdam, 281 pp.

Brunauer, S., Emmett, P.H., and Teller, E., 1938, Adsorption of gases in multimolecular layers, *J. Am. Chem. Soc.*, 60:309–319.

Carter, D.L., Mortland, M.M., and Kemper, W.D., 1986, Specific surface, in A. Klute (Ed.), *Methods of Soil Analysis, Part 1: Physical and Mineralogical Methods*, Monograph 9. Am. Soc. Agron., pp. 413–423.

Dixon, J.B., 1977, Kaolinite and serpentine group minerals, in J.B. Dixon and S.B. Weed (Eds.), *Minerals in Soil Environment*, Soil Sci. Soc. Amer. J., 41:357–403. Madison, Wisconsin

Flaig, W., Beutelspacher, H., and Reitz, E. 1975. Chemical composition and physical properties of humic substances, *Soil Composition,* Geisking, J.E., (Ed), Springer-Verlag, Berlin, 1:1–219.

Greenland, D.J., and Hayes, M.H.B. (Eds.) 1985, *The Chemistry of Soil Constituents,* John Wiley and Sons, Chichester, 469 pp.

Greenland, D.J., and Mott, C.J.B., 1985, Surfaces of soil particles, in *The Chemistry of Soil Constituents,* 1985. D.J. Greenland, and M.H.B. Hayes. (Eds.), John Wiley & Sons, Chichester, pp. 321–354.

Griffith, S.M., and Schnitzer, M., 1975, Analytical characteristics of humic and fulvic acids extracted form tropical soils, *Soil Sci. Soc. Amer. J.,* 39:861–869.

Grim, R.E., 1953, *Clay Mineralogy,* McGraw Hill, New York, 384 pp.

Huang, P.M., and Schnitzer, M., 1986, *Interactions of Soil Minerals with Natural Organics and Microbes,* 3rd. printing, SSSA Special Publ. 17. Madison, Wisconsin.

Hatcher, P.G., Schnitzer, M., Dennis, L.W., and Maciel, G.E., 1981, Aromaticity of humic substances in soils, *Soil Sci. Soc. Amer. J.* 45:1089–1094.

Kitagawa, Y., 1971, The "unit particle" of allophone, *Amer. Mineral.,* 56:465–475.

Konishi, Y., Asai, S., and Katoh, H., 1990, Bacterial dissolution of pyrite by *Thiobacillus ferooxidans, Bioprocess Eng.,* 5:231–237.

Konishi, Y., Asai, S., and Yoshida, N., 1995. Growth kinetics of *Thiobacillus thiooxidans* on the surface of elemental sulfur. *Am. Soc. Microbiology,* 61:3617–3622.

LaMer, V.K., and Healy, T.W., 1963a, Adsorption–flocculation reactions of macromolecules at the solid–liquid interface, *Rev. Pure Appl. Chem.,* 13:112–133.

LaMer, V.K., and Healy, T.W., 1963b, The role of filtration in investigating flocculation and redispersion of colloidal dispersions, *J. Phys. Chem.,* 67:2417.

Lambe, T.W., 1953, The structure of inorganic soil, *Proc. ASCE,* No. 315.

Lambe, T.W., 1958, The structure of compacted clay, *J. Soil Mech.* and Found. Div. ASCE, 84: SM2, p.34.

Marshall, K.C., 1976, *Interfaces in Microbial Ecology,* Harvard University Press, Cambridge, MA.

Molz, F.J., Widdowson, M.A., Benefield, L.D., 1986. Simulation of microbial growth dynamics coupled to nutrient and oxygen transport in porous media, *Water Resour. Res.,* 22:1207–1216.

Monod, J., 1949, The growth of bacterial culture, *Annu. Rev. Microbiol.,* 3:371.

Mortland, M.M., and Kemper, W.D., 1965, Specific surface, in C.A. Black (Ed.), *Methods of Soil Analysis: Part 1,* Amer. Soc. Agron., pp. 532–544.

Nakano, M., 2010. Personal communication to R.N. Yong on studies by Hamamoto, S., Yamaguchi, N., and Nakano, M. on pH influence on adsorption of IO_3^- and I^- on allophone.

Nakano, M., and Kawamura, K., 2008, Estimating the corrosion of compacted bentonite by a conceptual model based on microbial growth dynamics, *Appl. Clay Science,* 47:43–50.

Newman, A.C.D., and Brown, G., 1987, *The Chemical Constitution of Clays,* Mineralogical Soc. Monograph No. 6, Longman Scientific and Technical, pp. 1–128.

Parks, G.A., 1965, The isoelectric points of solid oxides solid hydroxides and aqueous hydroxo complex systems, *Chem. Rev.* (London) 65:177–198.

Pedersen, K., 2000, Microbial processes in radioactive waste disposal, SKB Technical Report TR-00-04, 97 pp.

Piccolo, A., 2001, The supramolecular structure of humic substances, *Soil Sci.,* 166:810–832.

Pusch, R., 1966, Quick clay microstructure, *J. Eng. Geol.*, 3:433–443.

Pusch, R., and Karnland, O., 1988, Hydrothermal effects on montmorillonite: A preliminary study, SKB Technical Report TR88-15, SKB, Stockholm.

Pusch, R., 1993, Evolution of models for conversion of smectite to non-expandable minerals, SKB Technical Report TR93-33, SKB, Stockholm.

Pusch, R., and Yong, R.N., 2006, *Microstructure of Smectite Clays and Engineering Performance*, Taylor & Francis, Spon Press, London, 328 pp.

Reynolds, T.A, and Richards, P.A. 1996, *Unit operations and processes in environmental engineering* PWS Publishing Co., Boston, 798 p.

Schnitzer, M., Ortiz de Serra, M. I., and Ivarson, K., 1973, The chemistry of fungal acid-like polymers and of soil humic acids, *Soil Sci. Amer. Proc.* 7:229–326.

Schwertmann, U., and Taylor, R.M., 1977, Iron oxides, in J.B. Dixon and S.B. Weed (Eds.), *Minerals in the Soil Environment*, Soil Science Society of America, Madison, WI, pp. 145–180.

Seki, K., Miyazaki, T., and Nakano, M., 1998, Effects of microorganisms on hydraulic conductivity decrease in infiltration, *Eur. J. Soil Sci.*, 49:231–236.

Seki, K., Thullner, M., Hanada, J., and Miyazaki, T., 2006, Moderate bioclogging leading to preferential flow paths in biobarriers., *Ground Water Monit. Remediation*, 26(3):68–76.

Simpson, A.J., Kingery, W.L., Hays, M.H.B., Spraul, M., Humpfer, E., Dvortsak, P., Kerssebaum, R., Godejohann, M., and Hofmann, M., 2002, Molecular structures and associations of humic substances in the terrestrial environment, *Naturwissenschaften*, 89:84–88.

Sims, J.L., Sims, R.C., and Mathews, J.E., 1990, Approach to bioremediation of contaminated soil, *J. Hazardous Waste Hazardous Mater*, 4:117–149.

Sinclair, J.L., Kampbell, D.H., Cook, L., and Wilson, J.T., 1993, Protozoa in subsurface sediments from sites contaminated with aviation gasoline or jet fuel, *Appl. Environ. Microbiol.*, 59(2):46–47.

Sposito, G., 1989, *The Chemistry of Soils*, Oxford University Press, New York, 277 pp.

Taylor, D.W., 1948, *Fundamentals of Soil Mechanics*, John Wiley & Sons, New York, 700 pp.

Terzaghi, K., and Peck, R.B., 1948, *Soil Mechanics in Engineering Practice*, John Wiley & Sons, New York, 566 pp.

Thullner, M., Van Cappellen, P., and Regnier, P., 2005, Modeling the impact of microbial activity on redox dynamics in porous media, *Geochim. et Cosmochim. Acta*, 69:5005–5019.

Wada, K., and Harward, M.E., 1974, Amorphous clay constituents of soils, *Adv. Agr.* 26:211–260.

Wada, K., 1989, *Minerals in Soil Environments*, 2nd ed., Soil Sci. Am. Soc., Madison, New York.

Wang, Y., and Francis, A.J., 2005, Evaluation of microbial activities for long-term performance assessments of deep geologic nuclear waste repositories, *J. Nuclear Radiochemical Science*, 6:43–50.

Weaver, C.E., and Pollard, L.D., 1973, *The Chemistry of Clay Minerals*, Elsevier, Amsterdam, 213 pp.

Whittaker, R.H., 1969, New concepts of kingdoms or organisms: Evolutionary relations are better represented by new classifications than by the traditional two kingdoms, *Science* 163(863):150–160.

Wise, M.E., 1952, Dense random packing of unequal spheres, *Phillips Res. Rep.*, 7:321–343.

Yong, R.N., and Warkentin, B.P., 1966, *Introduction to Soil Behaviour*, MacMillan, New York, 451 pp.

Yong, R.N., and Sethi, A.J., 1977, Turbidity and zeta potential measurements of clay dispersibility, *ASTM-STP 623*:419–431.

Yong, R.N., and Mourato, D., 1988, Extraction and characterization of organics from two Champlain Sea subsurface soils, *Can. Geotech. J.*, 25:599–607.

Yong, R.N., 2001, *Geoenvironmental Engineering: Contaminated Soils, Pollutant Fate, and Mitigation*, CRC Press, Boca Raton, FL, 307 pp.

Yong, R.N., and Mulligan, C.N., 2004, *Natural Attenuation of Contaminants in Soils*, CRC Press LLC, Boca Raton, FL, 319 pp.

3

Soil–Water Systems

3.1 Introduction

In Chapter 2, we described soils as being three-phase systems: gaseous, liquid, and solid. The water content of a soil is a significant determinant of its behaviour. Properties of soils, such as strength, compressibility and consolidation, thermal and hydraulic conductivity, plasticity and consistency, change measurably with corresponding changes in water content. Changes in water content occur all the time in the field. These changes can be large in surficial soils because of evaporation and also because of infiltration from rainfall. Changes in water content in subsurface soils occur because of water movement due to external and internal forces.

How water is retained (held) in soils is a very important feature of soils, not only because it can be considerably different between various kinds of soils, but mainly because it has a great influence on several soil properties. For example, the water-holding mechanisms involved differ sharply between cohesionless and cohesive soils, that is, between granular soils and clays. In this chapter, we will be discussing the nature of forces or mechanisms of water retention in soils, the energy of water retention, and how these relate to soil engineering applications.

3.2 Water Retention

The two ways in which the water content of a soil is expressed are in terms of a mass or a volumetric basis. In terms of a mass basis, the water content ω is defined as the water contained in the sample per mass of oven-dry soil: $\omega = W_w / W_s$, where W_w and W_s represent the mass of water and mass of soil solids, respectively. It is not uncommon in soil engineering practice to express ω in terms of a percentage. Water content expressed in terms of a volumetric basis for a moist soil sample is defined as θ and is determined as $\theta = V_w / V$, where V_w and V represent the volume of water and volume of moist soil, respectively. The relationship between ω and θ is $\theta = \omega\,\gamma_d/\gamma_w$, where γ_d and γ_w represent soil dry density and density of water, respectively. The mass basis for expression of water content is generally used in soil engineering applications, whilst the volumetric water content format is most often used in applications dealing with water uptake and transport.

3.2.1 Granular (Cohesionless) Soils

Water retention in granular (cohesionless) soils is mainly by capillarity, that is, the adhesive forces to the surfaces of the granular particles are stronger than the cohesive forces of the water molecules. Figure 3.1 shows a granular soil idealized as being composed of a bunch of capillary tubes. The sketch on the left-hand side of the figure reminds one of the mechanism of water rise in a capillary tube and the simple set of calculations to determine the height of water rise. In the case of the cohesionless soil sample shown at the right-hand side of the figure, if one assumes an equivalent capillary diameter of 0.01 mm, the calculated height of water uptake in the soil using the idealized capillary tube concept would be about 300 cm for $\alpha = 0$, and $T = 72.5 \times 10^{-3}$ Nm^{-1} at room temperature.

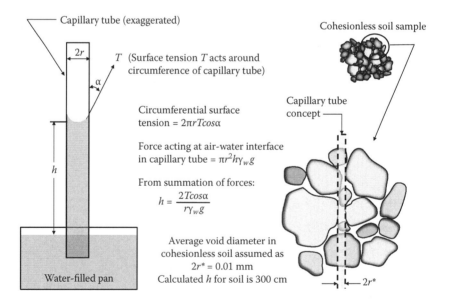

FIGURE 3.1
Height of rise in a capillary tube (left), and application of capillary tube concept to a cohesionless soil. The diameter of the capillary tube is $2r$, and the diameter of the average voids in the cohesionless soil is assumed to be $2r^* = 0.01$ mm. α refers to the contact angle, and g refers to the acceleration due to gravity.

3.2.1.1 Wetting and Drying

It is understood that whilst the capillary tube calculation procedure for water uptake of cohesionless soils is too simplistic, the view of water uptake by capillary activity in such soils is appropriate. Since the void spaces in granular cohesionless soils exhibit unequal or nonuniform capillary diameters, determination of water retention should recognize that water retention values are conditioned by whether measurements are made on the wetting or drying process of the soil. Consider the left-hand sketches shown in Figure 3.2. The top sketch shows the uneven capillary in a drying (desorption) process from a totally saturated condition, including full saturation of the $2r_2$ diameter bulb. If $P_1 < 2\pi r_1 T \cos\alpha$, water will be retained in the top $2r_1$ diameter portion of the tube.

Consider the case of water uptake by the cohesionless soil from an initial dry state, as shown in the bottom left-hand sketch in Figure 3.2. The water will reach the top of the smaller diameter ($2r_1$) capillary and will not proceed further. What is lost in the wetting-drying phenomenon is the amount of water contained in the $2r_2$ bulb—for the same P_1 pressure condition. The effect of uneven capillary bores (diameters) is best demonstrated by viewing

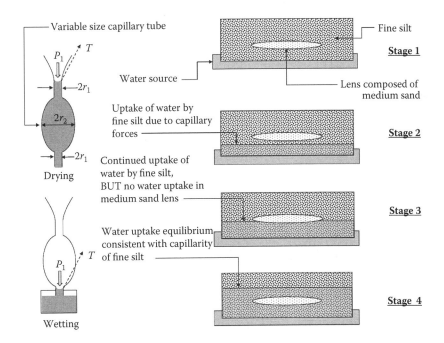

FIGURE 3.2
Schematic illustration of hysteric performance in water retention as a result of wetting-drying phenomenon (left), and uptake of water (wetting) in a bulk fine silt soil with an included medium sand lens.

the wetting of a mass of very fine silt shown in the right-hand set of sketches in Figure 3.2. The example shows an included medium sand lens in the fine silt mass (Stage 1). We assume that the equivalent average capillary diameter in the fine silt is very much smaller than the equivalent average capillary diameter of the sand. Stage 2 shows the beginning of water uptake by the fine silt, continuing onward to Stage 3. Since the equivalent average diameter of the sand is considerably larger than that of the fine silt, no water uptake occurs in the medium sand. This is the same phenomenon demonstrated in the bottom left-hand sketch of the figure. Note that water uptake continues in the fine silt, to an extent consistent with equilibrium requirements (Stage 4).

3.2.2 Cohesive Soils

Clays constitute the main constituent of cohesive soils. Because the structure of clays includes both macro- and micropores (Figure 3.3), water retention is by both capillary-type phenomena and molecular activity, some of which have been described previously in Section 2.5 of Chapter 2 in relation to the surface properties of the various clay fractions. This subject is addressed in the next subsection and in greater detail in the latter part of this chapter when we discuss soil–water interaction and energy relationships.

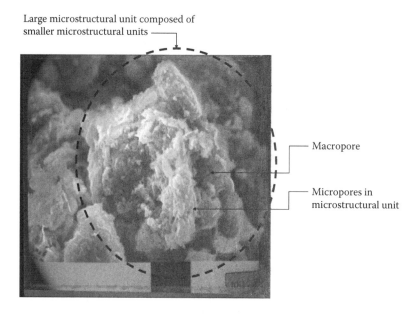

FIGURE 3.3
Scanning electron microscope (SEM) picture showing a large microstructural unit composed of smaller microstructural units. The width of the black band at the bottom (near the centre) represents 10μ.

For the discussion in this section, we can consider the retention of water in macropores of clays as equivalent to that demonstrated by sand/fine silt capillaries for the macropores, with the exception of the first two to three layers of water next to the surfaces of clay particles where intermolecular activity may be dominant. In the case of micropores, water retention can be considered to range from retention by micro capillarity phenomena at the one end of the spectrum to retention by intermolecular activity involving the surface functional groups of clay particles. Some might consider the latter as being mechanisms of hydration.

3.2.2.1 Hydration of Clays

The forces holding water molecules to clay particle surfaces arise from both the water and the clay mineral particles. Water is a dipolar molecule with a separation of centres of the two hydrogen atoms into a V-shape, with the oxygen atom forming the bottom of the V (Figure 3.4). The angle of the V-shape that separates the two hydrogen atoms is 104.5°, and the length of the arms of the V at which end the hydrogen atoms reside are 0.0957 nm. The diameter of a water molecule is nearly equal to 0.3 nm—using van der Waals' radius. Water molecules associate in tetrahedral arrangements with each oxygen surrounded by four others, held by hydrogen bonding.

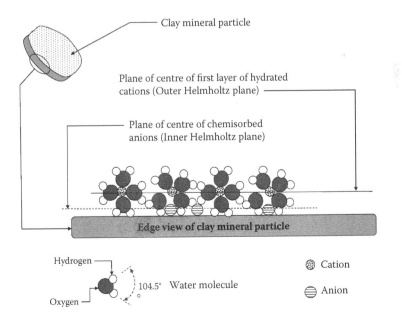

FIGURE 3.4
Schematic representation water attachment to clay mineral particle surface. Anions are specifically sorbed on the clay mineral particle surface.

The hydrogen ions of water give rise to hydrogen bonding of water molecules to the exposed oxygen atoms of clay mineral particle surfaces. The particle contributes both the negative charge and the oxygen or hydroxyl surface to attract molecules. Since cations in water are always hydrated, the exchangeable cations held near the negatively charged particle surface will hold some of the water at the surface of the particle as water of hydrated ions, as shown in Figure 3.4. Note that cations are hydrated to different degrees and that the hydration of the clay mineral particle surface will depend upon the cation present.

The main bonding force holding the first layer of water to the surface of a clay mineral particle is the hydrogen bond. The second (next) layer is held to the first, again by hydrogen bonding, but with a weaker force because of the distance of orienting influence of the particle surface on the water molecule. It follows that each successive layer will be held less strongly. Whilst there is a lack of total agreement as to the thickness of the hydration layer, there appears to be a three-layer (i.e., three water layers) consensus. Much depends on the type of clay mineral, the nature of the cations and anions in the water, and their distribution. There is no distinct demarcation between the water of hydration and free water.

3.3 Clay–Water Interactions

3.3.1 Electrified Interface and Interactions

The distribution of ions in the porewater of a clay–water system is a function of the electrostatic and chemical reactions between the ions and the charged surfaces of the clay particles. Of the various kinds of electrochemical models proposed to explain and determine the nature and distribution of ions adjacent to the clay particle surfaces (Stern, 1924; Grahame, 1947; Kruyt, 1952; Sposito, 1984; Ritchie and Sposito, 2002), there is common agreement on the diffuse double-layer concept. The model that exemplifies this concept is generally identified as the diffuse double-layer (DDL) model, as shown in the schematic illustration in Figure 3.5. The partly hydrated cations and anions in the inner Helmholtz plane shown in the figure are potential determining ions (*pdis*) which are bonded to the reactive surfaces by ionic and covalent bonds. They contribute directly to the net electric charge and potential on the surfaces of reactive particles. The structured water adjacent to the surface of the particle is due to the specifically adsorbed ions at the interface. Values of up to 1.14 g/cm^3 for water density have been measured for the first layer of water molecules, decreasing as the layers are added to 0.917 g/cm^3 at about four water layers before increasing to 1.0 g/cm^3 for free water. The viscosity of the hydration layer as measured by diffusion of ions near the surface is

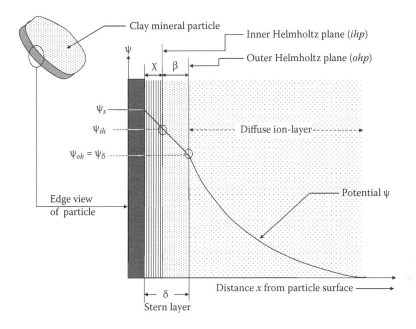

FIGURE 3.5
Generalized DDL model showing electrified interface with aqueous solution containing dissolved solutes.

much greater than that of free water—about 2.5×10^{-3} Pa s—and the dielectric constant decreases as one approaches the surface of the particle.

The complexes that are formed between the surface functional groups and the ions at the electrified interface are inner sphere complexes, and they have direct contact between them without interruption from any water molecule (Ritchie and Sposito, 2002). When a layer of water molecules interrupts contact between these, the complexes are classified as outer sphere complexes. The inner Helmholtz plane (*ihp*) and outer Helmholtz plane (*ohp*) shown in Figure 3.5 identify the positions of the inner sphere and outer sphere complexes with corresponding distances of χ and β, respectively. The thickness of the Stern layer δ, which includes the *ihp* and the *ohp*, according to the Grahame (1947) model, is obtained by summing up the distances as $\delta = \chi + \beta$. According to the DDL model, the diffuse ion layer consists of the counterions beyond the Stern layer needed to satisfy the net negative charge of the reactive particles.

3.3.1.1 Diffuse Double Layers

The interactions of diffuse double layers between adjacent and proximal clay particles are responsible, to a large extent, for the properties of swelling, plasticity, water retention, and other processes involving aqueous solution interaction with the clay particles. This makes it important for one to have a better appreciation of what constitutes the diffuse double layers and how these double layers feature in the development of the various properties of soils.

The exchangeable cations (counterions) in the diffuse double layer shown in Figure 3.5 are located at some distance from the surface of the clay particle. Whilst the electrical forces between the net negatively charge surface and the positively charged ions serve to attract the cations to the particle surface, the thermal energy forces them to diffuse away from the surface. The balance of Coulomb electrical attraction and thermal diffusion leads to a diffuse layer of cations, with decreasing concentration as one moves further away from the particle surface. The theoretical distribution of cations in the diffuse double layer can be calculated using some simplifying assumptions, namely: (a) clay particles can be considered as simple charged plates for which the electric field is described by the Poisson equation, and (b) the distribution of ions is described by the Boltzmann equation. Yong and Warkentin (1975) show that with these assumptions, the relationship for n_+, the number of cations per unit volume at a distance x from the clay particle surface, is obtained as

$$n_+ = n_i \left(\coth \frac{x}{2} \sqrt{\frac{8\pi z_i^2 e^2 n_i}{\varepsilon \kappa T}} \right)^2 \tag{3.1}$$

where n_i and z_i are the concentration and valence of the ith species of ion in the bulk solution; and ε, e, κ, and T represent the dielectric constant, electronic charge, Boltzmann constant, and temperature, respectively. Calculations using Equation (3.1) for n_+ for a 0.001 M salt with monovalent ions (Yong and Warkentin, 1975) give a value of n_+ at $x = 5$ nm as 0.016 M. This shows that the concentration of cations at this distance of 5 nm from the particle surface is at least 16 times that of the concentration of cations in the bulk solution.

The extent of the diffuse double layer is related to the concentration and valence of the cations in the bulk solution. The lower the concentration and valence, the larger the ratio of n_+ / n_i. Monovalent ions at low concentrations provide the largest diffuse double layers. Increasing either the valence or the salt concentration in the porewater will reduce the extent of the diffuse double layers (Figure 3.6). For a divalent ion at the same concentration of 0.001 M used in the previous calculation example, the value of n_+ at $x = 5$ nm is 0.004 M, showing that the concentration of cations at this distance is one-fourth that of the monovalent ions as shown in Figure 3.6.

3.3.1.2 Electric Potential ψ

The electric potential ψ_s at the surface of the clay particle, which is defined as the surface potential and which varies with electrolyte concentration and the nature of the charge of the clay particle, decreases to a potential of ψ_δ at the

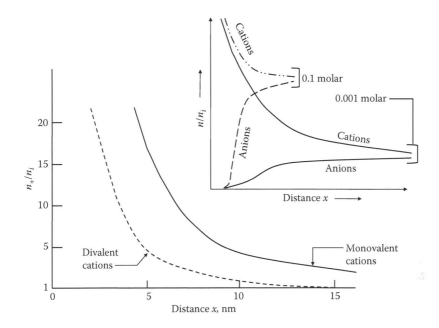

FIGURE 3.6

The bottom-left diagram shows the theoretical distribution of ions at a charged clay particle surface in relation to the influence of valence on the thickness of the diffuse double layer. Top right shows the influence of salt concentration on the theoretical distribution of cations and anions (n refers to the number of cations or anions).

outer Helmholtz plane (*ohp*). Although this is shown as a linear drop from ψ_s to ψ_δ in Figure 3.5, there is considerable discussion as to the exact nature of this potential drop. Electrokinetically, ψ_δ is considered to be equal (or almost equal) to the zeta potential ζ – the electric potential in the slipping plane between the fixed and flowing liquid.

At distances beyond the *ohp*, the electric potential ψ is described by the Gouy Chapman diffuse double-layer (DDL) model. With this model, one can compute the electric potential ψ as a function of the distance x from the charged particle surface. Because of chemical bonding processes and complexation in the δ region shown in Figure 3.5, simple electrostatic interaction calculations do not apply in this region. To determine ψ in relation to x, the following assumptions and conditions are prescribed:

- Coulombic interaction occurs between the charged ions (cations and anions) in solution and the charged clay particle surfaces.
- The interactions between ions in solution and clay particles are determined in terms of the electric potential ψ and are described by the Poisson relationship in respect to variation of ψ with distance x away

from the particle surface, as shown in Figure 3.5 as $d^2\psi / dx^2 = -4\pi\rho /\epsilon$, where ρ is the space charge density and ϵ is the dielectric constant.

- The ions in solution are considered to be point-like in nature, that is, zero-volume condition.
- The space charge density ρ of the ions that contribute to the interactions can be described by the Boltzmann distribution: $\rho = \sum n_i z_i e \exp(-z_i e\psi / \kappa T)$.

With these conditions, the Poisson–Boltzmann relationship for ψ can be obtained as follows:

$$\frac{d^2\psi}{dx^2} = -\frac{4\pi}{\epsilon} \sum_i n_i z_i e \exp\left(\frac{-z_i e\psi}{\kappa T}\right) \tag{3.2}$$

The relationship for ψ can be obtained (e.g., Kruyt, 1952; van Olphen, 1977; Yong, 2001) as follows:

$$\psi = -\frac{2\kappa T}{e} \ln \coth\left(\frac{x}{2}\sqrt{\frac{8\pi e^2 z_i^2 n_i}{\epsilon \kappa T}}\right) \tag{3.3}$$

The relationship between the surface charge σ_s and surface potential ψ_s for constant surface charge minerals is

$$\sigma_s = \left(\frac{2n_i \epsilon \kappa T}{\pi}\right)^{\frac{1}{2}} \sinh \frac{z_i e}{2\kappa T}\psi_s \tag{3.4}$$

For constant surface charge minerals, the surface charge density is constant. Accordingly, if n_i and z_i are increased, the electric potential ψ_s at the surface of the clay particle will be reduced, and the thickness of the diffuse double-layer will be correspondingly reduced. For pH-dependent surface charge clay particles, we need to account for the dependence of the surface potential ψ_s on the potential determining ions (*pdis*). The Nernst equation for the surface potential ψ_s is given as $\psi_s = 2.303\ \kappa T/\epsilon\ (pH_o - pH)$, where pH_o is the pH at which the surface potential $\psi_s = 0$. Accordingly, the relationship between the surface charge density σ_s and surface potential ψ_s for pH-dependent surface charge clay particles is

$$\sigma_s = \left(\frac{2n_i \epsilon \kappa T}{\pi}\right)^{\frac{1}{2}} \sinh 1.15 \frac{z_i e}{\epsilon}\left(pH_o - pH\right) \tag{3.5}$$

3.3.2 Interaction Energies

3.3.2.1 Energies of Interaction in the Stern Layer

The energy of interaction in the inner Helmholtz plane (IHP), E_{ihp}, is obtained directly from a knowledge of the interaction between the ions in solution and forces emanating from the surface of a clay particle. From a consideration of Coulomb forces, we obtain (Yong, 2001):

$$E_{ihp} = E_c = \frac{z_i z_j e^2}{4\pi\varepsilon r} \tag{3.6}$$

where E_c is the Coulombic interaction energy, z_i and z_j are the valences of the i and j species of ion in the region under consideration, and r is the distance between the centre of the ith ion and jth ion.

The interaction energy in the outer Helmholtz plane (OHP), E_{ohp}, includes interaction with the Coulomb forces and also the short-range forces due to (a) ion–dipole interaction, (b) dipole–dipole interaction, and (c) dipole–site interaction. The Coulomb force is the strongest, and its effect is felt at long distances, quite often exceeding some chemical bonding forces.

Taking r as the distance between the centre of a dipole and the corresponding ion, the ion–dipole interaction E_{id} is given as follows:

- Ion–dipole interaction E_{id} given as $E_{id} = -\mu z e \cos\theta / 4\pi\varepsilon r^2$
- Dipole–dipole interaction E_{dd} is given as follows: $E_{dd} = -\frac{\mu_1 \mu_1}{4\pi\varepsilon r^3 D}$
  where μ is the dipole moment, D is a function of angles of the dipoles, and the subscripts 1 and 2 refer to the respective dipoles.
- Dipole–site interaction energy E_{ds} is similar in almost all respects to the ion–dipole interaction energy, with the exception that the distance r is now taken to be the distance between the centre of a dipole and the charge site on the particle.

3.3.2.2 London–van der Waals Energy and Total Intermolecular Pair Potential

The *London–van der Waals force*, which is the force between ions and between molecules, is sometimes called *London dispersion forces* or simply *van der Waals forces*. This force, which operates between all molecules regardless of the electrostatic conditions of molecules, exceeds the dipole-dependent forces. The van der Waals force is operative from 0.2 nm to more than 10 nm in distance and is a repulsive or attractive force depending on the distance between molecules. The van der Waals interaction energy E_{vdw} is given by $E_{vdw} = -c_i c_j / r^6$, where r is the distance between molecules, and c_i, c_j are the London dispersion force constants of molecules i and j. These consist mainly of the ionization potential, dielectric constant, and polarizability.

The potential due to the repulsive force between a pair of molecules can be determined or expressed in two different ways: (a) using an empirical form of hard-sphere potential or power-law potential and (b) using the exponential potential established on the basis of quantum mechanics. The power law potential E_{rp} is described by $E_{rp} = A/r^n$, where r is the distance between the molecules, A is a coefficient, and a commonly adopted value is $n = 12$. The exponential potential is given as follows: $E_{rp} = b \exp(-cr)$, where b and c are coefficients.

The total intermolecular pair potential E_{tip} used for uncharged colloidal particles, which is called the Lennard–Jones form or 12-6 potential, is commonly described as follows:

$$E_{tip} = \frac{A}{r^{12}} - \frac{c_i c_j}{r^6} \tag{3.7}$$

In clay–water systems with charged ions and charged clay minerals, the total intermolecular pair potential is given as

$$E_{tip} = \frac{z_i z_j e^2}{4\pi\varepsilon r} - \frac{c_i c_j}{r^6} + b \exp(-cr) \tag{3.8}$$

This intermolecular pair potential will describe the mode of ion desorption on clay minerals, and in addition, the properties of soil water such as density, viscosity, and self-diffusion coefficient of water molecules, with proper constants.

3.3.2.3 DLVO Model and Particle Interaction Energies

The DLVO (Derjaguin, Landau, Verwey, and Overbeek) particle interaction energy model, to which the diffuse double-layer (DDL) concept is applied, is useful for calculations of interparticle or interaggregate interaction. In this instance, calculations account for the nature of the charged clay particle surfaces, chemical composition of the clay-water system, and particle configuration. Figure 3.7 shows the three principal configurations for particle interaction used in most calculations. The energy interaction calculations consider van der Waals' attractive force and the DDL forces as the primary forces responsible for the energies of interaction between the particles.

The particle interaction models reported by Flegmann et al. (1969) form the basis for calculation of the maximum energies of interparticle action for similar charged surfaces in face-to-face and edge-to-edge particle arrangement, and the relationship reported by Hogg et al. (1966) is used for dissimilar charged surfaces (edge-to-face). In this instance, the surface potentials are defined by the zeta potential at the edges and at the faces of the particles. The zeta potential at the edge of the particle is assumed to be one-fifth that on the face of the particle, consistent with the observation reported by Ferris and Jepson (1975).

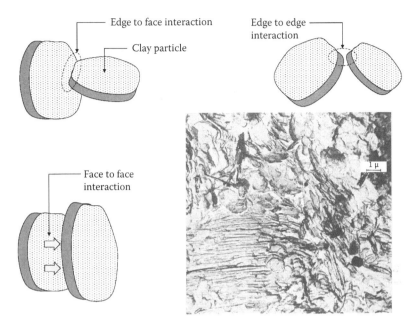

FIGURE 3.7
Sketches show the principal modes of interaction between particles used for interparticle energy calculations. Transmission electron microscopic (TEM) picture shows an example of the three modes of interparticle and interaggregate interactions for a cleavage surface in a sheared kaolinite sample.

Assuming constant surface potential and potential-determining ion influence on surface potentials (i.e., pH dependent), the energy of repulsion between interacting parallel-faced particles, E_r^{ff}, is obtained as follows:

$$E_r^{ff} = \frac{4n_i \kappa T (yz_i)^2 \exp\left(\dfrac{-D_H x}{2}\right)}{D_H \left[1 + \exp(-D_H x)\right]} \tag{3.9}$$

where $y = -(e\psi_{oh}/\kappa T)$ is a dimensionless potential; e, ψ_{oh}, κ, and T have been previously defined as electronic charge, potential at the *ohp* (see Figure 3.5), Boltzmann constant, and temperature, respectively; z_i and n_i are valence and bulk concentration of counterions, respectively; x is the interparticle distance; and D_H is the Debye–Hueckle reciprocal length. The distance x can be taken as the distance between particle surfaces minus the distance of closest approach determined by the size of the adsorbed hydrated ion. For hydrated Na ions for example, this would be 1.58 nm.

The energy of attraction between interacting parallel-faced particles, E_a^{ff} utilizes the London–van der Waals attraction energy for two similar flat plates, and is obtained as follows:

$$E_a^{ff} = \frac{A}{12\pi x^2} \qquad (3.10)$$

where A is the Hamaker constant. The total interaction energy E_T^{ff} for the face-to-face configuration of particles is the sum of the E_r^{ff} and E_a^{ff} energies. The relationship for this total energy as affected by *NaCl* concentration is shown in Figure 3.8, together with the results for the face-to-edge and edge-to-edge total energies.

To calculate the face-to-edge repulsion and attraction energies, two interacting spheres are adopted as the model, with one sphere having a very large radius a_f with potential ψ_f in comparison to the other with a much smaller radius a_e with its corresponding potential ψ_e. The interaction of the very large radius a_f sphere in comparison to the other very small radius sphere is meant to portray a flat particle interaction with a particle edge. The energies of face-to-edge repulsion E_r^{fe} and attraction E_a^{fe} are given as follows

$$E_r^{fe} = \frac{\varepsilon a_f a_e \left(\psi_f^2 + \psi_e^2 \right)}{4(a_f + a_e)} \left[\frac{2\psi_f \psi_e}{\psi_f^2 + \psi_e^2} \ln \frac{1 + \exp(-D_H x)}{1 - \exp(-D_H x)} + \ln \left[1 - \exp(-2D_H x) \right] \right] \qquad (3.11)$$

$$E_a^{fe} = -\frac{A}{12} \left(\frac{r_a}{x_a^2 + x_a r_a + x_a} + \frac{r_a}{x_a^2 + x_a r_a + r_a} + 2 \ln \frac{x_a^2 + x_a r_a + x_a}{x_a^2 + x_a r_a + x_a + r_a} \right) \qquad (3.12)$$

where $x_a = x/(2a_f)$, and $r_a = a_e/a_f$.

The calculation of the edge-to-edge repulsion and attraction energies assumes small interacting spheres with identical radii of a_e. The energies of repulsion E_r^{ee} and attraction E_a^{ee} are obtained as follows

$$E_r^{ee} = \frac{\varepsilon a_e \psi_e^2}{2} \ln \left[1 + \exp(-D_H x) \right] \qquad (3.13)$$

$$E_a^{ee} = -\frac{A}{12} \left(\frac{1}{x_a^2 + 2x_a} + \frac{1}{x_a^2 + 2x_a + 1} + 2 \ln \frac{x_a^2 + 2x_a}{x_a^2 + 2x_a + 1} \right) \qquad (3.14)$$

The calculated results shown in Figure 3.8 (Yong et al., 1979) used actual zeta potential measurements from experiments using a zetametre. The assumption of the edge zeta potential being one-fifth of that for the surface of the kaolinite particle needs to be verified or determined for other types of clay minerals, since the magnitude of the edge potential affects the magnitude

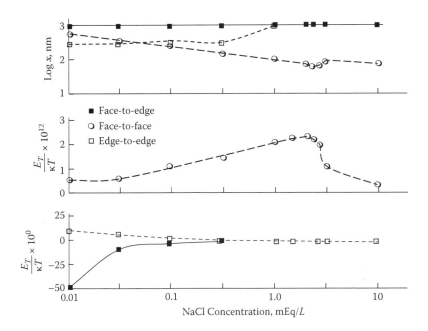

FIGURE 3.8

Maximum interaction energies ($E_T/\kappa T$) and their corresponding interparticle distances (x) at various *NaCl* concentrations for the three different particle configurations. The total interaction energy E_T is the sum of the repulsive and attractive energies for the respective particle configurations.

of the net energies of interaction. For more precise calculations using the DLVO model relationships, different values of Hamakaer's constant for the faces and edges of particles should be used. Calculations based on the DLVO model show good accord for ideal systems where particle separation distance is above 3 nm. At lower particle separation distances, the repulsive energy is overwhelmed by the van der Waals forces, as will be seen in the later discussion relating to swelling pressures of clays.

3.3.2.4 Energies in Particle Interaction and in Interlayer Space

When two clay particles are less than 1.5 nm apart, the exchangeable ions are uniformly distributed in the interparticle space and do not separate into two diffuse double layers: one associated with each particle. Under these conditions, there is a net attraction between particles. For clay minerals, one should be mindful of interactions in interlayer space, that is, the space between repeating layers of clay minerals, in addition to interactions between clay particles. When interparticle and interlayer distances exceed 1.5 nm, diffuse double layers are formed, and net repulsion results. The concentration of ions is higher in the plane midway between the particles and repeating layers. Figure 3.9 provides a schematic illustration of the interactions in interlayer space and development of the electronic potential, using the sketch of the montmorillonite mineral shown in Figure 2.5 in Chapter 2.

The Poisson–Boltzmann relationship between two charged particle surfaces (see Equation 3.2) is rewritten as

$$\frac{d^2y}{dx^2} = K^2 \frac{\exp(y)}{2} \qquad (3.15)$$

where $y = -z_i e\psi/\kappa T$ is a dimensionless potential, and $K = \sqrt{8\pi e^2 z_i^2 n_i / \varepsilon \kappa T}$. To determine the midplane potential ψ_m which tells one about the energy of interaction between the two particles or interlayers (Figure 3.9), the solution given by Langmuir (1938) is used. This solution, which specifies (a) the boundary condition that $y = y_m$ at the midplane between the particles for the first integration, and (b) the boundary condition that $y \to \infty$ at $x = 0$ for the second integration for Equation (3.15), is given as

$$y_m = 2\ln\frac{\pi}{Kx} \qquad (3.16)$$

The solution for the concentration of ions at the plane midway between the particles is given as

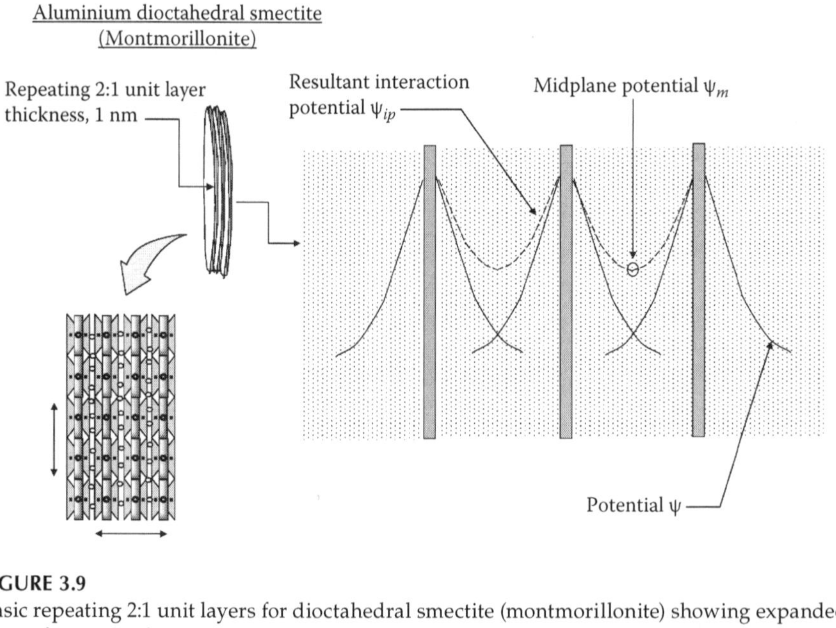

FIGURE 3.9
Basic repeating 2:1 unit layers for dioctahedral smectite (montmorillonite) showing expanded view of repeating layers and the interaction of their electric potential ψ and development of the resultant interaction potential ψ_{ip}.

$$c_c = \frac{\pi^2 c_o}{K^2 x^2} \qquad (3.17)$$

The Langmuir solution neglects anions, a condition that is acceptable at small distances between particles; that is, at low water contents, the anions are mostly excluded from the space between clay particles.

3.3.2.5 Calculated Swelling Pressures from Interaction Models

Calculations of swelling pressures in clays are useful in the assessment of potential behaviour of swelling clays used for engineered barriers and buffers. In most instances, the theoretical diffuse double-layer (DDL) models used for these calculations assume interactions between parallel particles or unit layers. This simplifying assumption provides a basis for comparison with experimental swelling pressure information where samples are prepared with parallel particle orientation for swelling pressure tests. There have been tests conducted on clay samples without control on initial particle orientation, that is, on samples with indeterminate soil fabric and soil structure. Whilst swelling clay samples can be prepared without particle orientation control, it is difficult for comparisons of measured swelling pressure to be made with results obtained from theoretical models, since these models will have to assign effective particle spacings to various particle orientation configurations other than parallel. Whilst one could argue that the DLVO model might be used for such instances, one needs to remember that whilst the DLVO model permits one to make calculations for interactions of particles with face-to-face, face-to-edge, and edge-to-edge configurations, it is difficult to assign the proportions of any of these configurations to the overall soil fabric and soil structure.

Common methods for obtaining parallel particle orientation for swelling pressure tests include controlled slow evaporation of shallow pans of very dilute montmorillonite suspensions (Yong and Warkentin, 1959) and high-pressure consolidation (Alammawi, 1988). Figure 3.10 shows the results from the high-pressure consolidation tests on a Na-montmorillonite saturated with a 10^{-3} MNaCl reported by Alammawi, together with calculations from DLVO theory and from a modified Gouy Chapman DDL model (G-C DDL model). The modifications for the G-C DDL model reported by Yong and Mohamed (1992) were required to account for the energies in the Stern layer (i.e., energies in the inner Helmholtz and outer Helmholtz planes). In the unmodified G-C model (without inclusion of the Stern energies), the charge on the surface of the particle σ_s must be balanced by the total space charge σ_p in the soil solution to preserve electroneutrality. To account for the interactions in the Stern layer, the surface charge at the surface is balanced by the additional contributions from the charges due to the ions in the inner Helmholtz and outer Helmholtz planes, σ_{ihp} and σ_{ohp}, respectively.

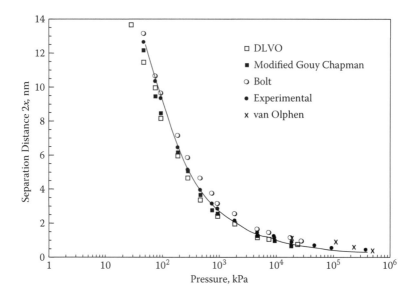

FIGURE 3.10
Comparison of experimental swelling pressure results with calculated results from theoretical models. Experimental results are from tests reported by Alammawi (1988).

The results shown as "Bolt" in Figure 3.10 are from the experimental results reported by Bolt (1956) also for *Na*-montmorillonite at the same salt concentration. The "van Olphen" results added to the previously reported results of Yong and Mohamed (1992) are interpreted from calculations reported by van Olphen (1963) for pressures required to remove the first four water layers next to a typical montmorillonite unit. The comparison of calculated values and experimental results shown in the graph indicate acceptable agreement between calculated and measured swelling pressure results, especially at the higher particle separation distances (2*x*). We will discuss these results in greater detail when we address the topic of water uptake and movement in partly and fully saturated clays.

3.4 Soil–Water Energy Characteristics

3.4.1 Concept of Soil–Water Potential

Of all the physical properties of soil, the energy relationship of soil for water is perhaps the most important. It describes the energy with which water is held in a soil mass in relation to its water content. Many different terms have been used to describe the energy with which water is held in soils. This is because terms were required, both in research where thermodynamics terminology was usually used, and in practical soil–water work and engineering, where descriptive terms were used. These descriptive terms included such terms as *soil–water tension* and *soil suction*, indicating that the water in soil is in equilibrium with a pressure less than atmosphere (soil–water tension), and that the soil exerts a force to take in water (soil suction).

The energy with which water is held to a soil at any water content has been specified as the soil–water potential. For example, Buckingham (1907) defined the capillary potential of a soil as the work required per unit weight of water to pull water away from the mass of soil, having in mind that this potential is due to the capillary forces holding water in soil. This potential decreases as the water content increases, and vice versa. To demonstrate the macroscopic relationship between energy by which water is held in soil and its water content, we consider a simple laboratory desorption test using a Buchner-type apparatus as shown in the top left-hand sketch in Figure 3.11.

This type of laboratory test measures the amount of energy required to push water out from the soil sample contained in the apparatus, using air pressure acting on the top of the soil sample to drive water out from the sample. In this demonstration exercise, one applies air pressure to the soil sample contained in the apparatus as illustrated schematically in the figure.

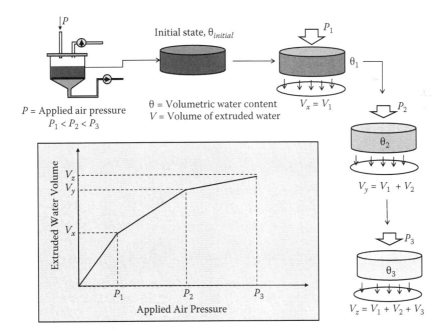

FIGURE 3.11
Schematic illustration of desorption-type experiment using the Buchner-type pressure apparatus. The right-hand sketches illustrate the steps in the test procedure, and the bottom left-hand graph shows the results obtained.

After equilibrium is reached at each pressure increment, the amount of water extruded or discharged is determined.

A relationship can be developed between water content in the soil and pressure applied, taking note of the initial water content (before pressure application), the amount of water discharged under the pressures applied, and making the necessary calculations to track the equilibrium water content remaining in the soil. This is the water-holding capacity of the soil at that particular applied pressure. This sequence of pressure application is shown in the right-hand sketches in Figure 3.11 and the results plotted in the graph shown at the bottom left-hand portion of the figure. The graphical results portrayed in Figure 3.11 are presented in terms of water content and applied pressure. Water retention or water-holding capacity is viewed from the soil–particle frame of reference in terms of suction. This is the opposite of the frame of reference which considers water-holding capacity in terms of the work required to move water into or out of the sample: the soil–water potential. In that sense, whilst both measurements should give equal results in terms of magnitude of effort required, suction measurements are expressed positively, whereas potential measurements have the opposite (negative) sign. Figure 3.12 shows the results for three typical kinds of soils using the pressure membrane as the desorption-provoking device. In this case, the soil–water potential is used in place of the applied pressure as the abscissa. The discussion relating to

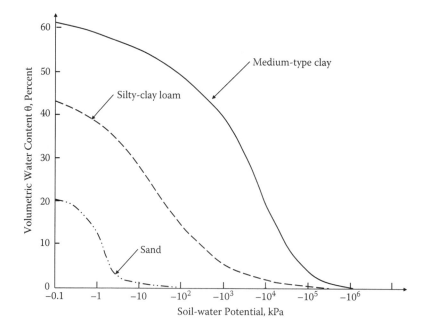

FIGURE 3.12
Typical soil–water potential relationships for sand, silty-clay loam, and a medium-type clay. The results are obtained as desorption curves using the pressure membrane apparatus.

the pressure membrane technique is given in Section 3.4.2 when methods for determining soil–water potential are addressed.

As we have learnt from the discussions in the previous section, different types of forces are involved in respect to how water is held to soil particles, between soil particles, and within the soil mass itself. This means that water in soils is held within the soil with different mechanisms depending on whether one is concerned with water surrounding individual soil particles, interlayer water, interparticle water, or "bulk" water in soils. For soil engineering purposes, all of these kinds of "waters" are lumped together as *porewater* (in the macroscopic sense). However, for geoenvironmental engineering purposes, where water uptake, unsaturated flow, solute transport, heat flow, and so forth are important considerations, there is a need to distinguish between the mechanisms by which water is held in the soil. Many of these issues will be discussed in the later sections of this chapter and in the other chapters at the appropriate juncture.

3.4.1.1 Components of Soil–water Potential

Figure 3.1 in Section 3.2 showed the simple capillary experiment using a glass column containing clean sand, demonstrating that water is held in the sand by capillary action and that the height h of the water drawn up into the sample is $h = 2T\cos\alpha/r\gamma_w g$. This glass column is shown schematically in the left-hand illustration in Figure 3.13, where the height of capillary rise in the sand column is designated as h_s. The sketch on the right-hand side of the figure shows a similar glass column containing a dry compact inorganic clay also placed in a shallow pan of water. In this column, we will see that the height of water rise (water uptake) H in the column, which is the sum of h_c and Δh_c, far exceeds h_s of the sand column in the equilibrium state. The height of the capillary rise in the sand column h_s in Figure 3.13 is determined by the factors described in Figure 3.1 in the derivation of the relationship for h. In the case of the clay column shown on the right side of the figure, the height of capillary rise h_c is a computed value based on the average pore radius of the clay sample. In reality, this is not distinguishable from the total height of rise in the clay column, as will be discussed later when we address the mechanisms involved in water uptake and movement in soils.

In geotechnical and soil engineering, the height of capillary rise h_s in the sand column is generally described as being due to *capillary suction*, meaning that capillary forces are responsible for suction of the water in the sand column. From a thermodynamic point of view, one could define a capillary potential ψ_c that represents a measure of the energy by which water is held by the sand particles by *capillary forces*. Buckingham (1907) defined it as the potential due to capillary forces at the air–water interfaces in the sand pores holding water in the sand.

The requirements of analyses of the soil–water system as a whole are better satisfied if the height of water uptake in a soil column is determined in

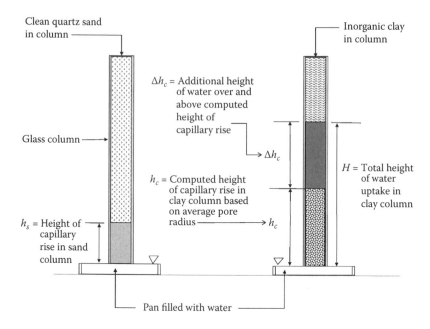

FIGURE 3.13
Capillary rise in quartz sand column (left) and water uptake in an inorganic clay (right). Note that fractioning of H into h_c and Δh_c is performed theoretically; that is, the capillary rise in the clay column is a calculated value based on the average pore radius in the sample. In reality, it is not possible to partition H.

terms of the work required to move water up to its equilibrium height H. The total work required to move water into (and out of) a soil, which is defined as the *soil–water potential* ψ, describes the water-holding capability of soil; that is, it describes the energy by which water is held to (or attracted to) the various soil solids comprising the soil under consideration. In soil–water systems where the soil solids possess reactive surfaces such as the clay column shown in the right-hand sketch in Figure 3.13, the interaction of the reactive surfaces of the soil solids (e.g., clay particles) with water in combination with the microstructure of the clay are responsible for the height H of water uptake in the clay column. It is particularly useful in providing a simple picture of the kinds of internal forces that will contribute to water movement and retention in soils.

In general, the work associated with actions on soil water can be quantitatively represented by Gibbs free energy if one uses thermodynamic concepts. Specifically, one can consider the energy with which water is held in a soil mass as the energy of the water in a soil. Gibbs free energy per unit mass of soil water is termed the *chemical potential*. Analyses of water retention and movement in a soil–water system are better satisfied using chemical potentials. Since the energy state of soil water (porewater) in partly saturated soils is determined by such factors and forces such as (a) capillary action, (b) solute concentration, (c) interactions with the reactive particle surfaces, and (d) external forces such as gravity, overburden load, surcharge, and static water pressure, it follows that the chemical potential of soil water is the summation of the free energies associated with all these factors and forces. It is less than that of free water under one atmospheric pressure at the same elevation and temperature of the test sample. This chemical potential, which is defined as the total soil–water potential ψ (Iwata et al., 1995), is a negative quantity.

The components of the total soil–water potential ψ are described as follows:

- *Total soil–water potential ψ* is defined as the chemical potential derived from the sum of the various forces, and is equivalent to the work required to move a unit quantity of water from the reference pool to the point under consideration in the soil. It is a negative quantity.

- *Matric potential ψ_m* is a reflection of the property of the soil matrix. This is specified as $-(2\sigma/r)\nu_w$, where σ is the surface tension of water, r is the pore radius, and ν_w is the specific volume of water ($= 1 \text{ cm}^3/\text{g}$). This is equivalent to the Buckingham capillary potential, where the sorption forces in the sand column experiment are due to capillary phenomena. In the case of soils containing particles with reactive surfaces, there are other types of forces that need to be taken into account. A good example of this is the case of clays, and especially the case of swelling clays. Complications surrounding analysis of interlayer swelling and microstructural effects and influences do not permit easy resolution in terms of capillary "forces." According to Sposito (1981), the matric potential ψ_m includes the effects of dissolved components of the soil–water system on the chemical potential μ_w.

- *Gravitational potential $\psi_g = -gh$*; where h is the height of the clay above the free water surface, and g is the gravitational constant. If the point in the clay under consideration is below the surface, h is a negative quantity, and hence the relationship becomes positive.

- *Osmotic potential ψ_π* is specified as $-\dfrac{RT}{1000}\sum \pi_i n_i = \Pi \nu_w$ for nonideal solutions, where π_i is the osmotic coefficient (1 for ideal solutions), n_i is the molality of solute i, and Π is the osmotic pressure and is equivalent to the work required to transfer water from a reference pool of pure water to a pool of soil solution at the same elevation, temperature, and so forth. The term *osmotic potential* is generally used in conjunction with swelling clays. For nonswelling soils, the general term *solute potential ψ_s* is more appropriate. To avoid confusion, the more general term *solute potential* is favoured. It is not unusual for the literature to report on the use of ψ_s as the osmotic potential ψ_π, that is, $\psi_\pi = \psi_s = nRTc$, where n is the number of molecules per mole of salt, c is the concentration of solutes, R is the universal gas constant, and T is the absolute temperature.

- *Pressure potential* ψ_p arises from externally applied pressure transmitted through the fluid phase of the soil–water system.
- *Pneumatic (air) potential* ψ_a arises from pressures in the air phase.
- *Overburden potential* $\psi_o = (P_0 - P_{ex})v_w$ due to overlying surcharge, where P_0 is the reference pressure and P_{ex} is the surcharge pressure.
- *Soil water potential* ψ_w, defined as $\psi_w = \psi_m + \psi_\pi$, is the sum of the matric and osmotic potentials. This should not be confused with the total soil–water potential ψ.

3.4.2 Measurements of Soil–Water Potentials

Soil–water potential measurements are made for at least two purposes: (a) in the laboratory to define the equilibrium relationship between potential and water content for a candidate soil, and (b) in situ to measure soil suction for irrigation, water movement, water availability, and soil engineering purposes. The techniques and equipment used for determination of soil–water potential as the energy status of soil porewater are to a large extent dependent on whether the determination is undertaken as a laboratory procedure or an in situ assessment. There are at least three types of systems used to determine the soil–water potential of a candidate soil. Not all of these are suitable for use in the laboratory. The more common ones include (a) tensiometers, (b) pressure plates or pressure-membrane systems, and (c) thermocouple psychrometry.

3.4.2.1 Tensiometer for in Situ Measurements

The tensiometer basically measures the *water tension* in a clay and is only useful for measurements of clay-water tensions where pressure differences are less than one atmosphere. Richards and his co-workers have contributed much of the initial work with tensiometers (Richards, 1928; Richards and Gardner, 1936). Figure 3.14 is a schematic illustration that shows the elements constituting a type of tensiometer used for in situ measurements. Water in the tensiometer is in contact with soil water through the ceramic tip shown at the bottom of the tensiometer.

As the soil dries, a tension is exerted on the water in the tensiometer, and ordinarily, this is measured with a manometer or a Bourdon gauge. More recently, sensing devices such as electronic pressure transducers have been used. The tensiometer will function until a negative pressure is reached, at which air migrates through the ceramic cup. At this time, one can conclude that this determines the air-entry value of the ceramic cup. Generally, this point is about −85 kPa.

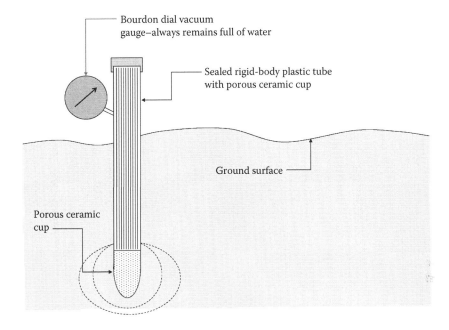

Bourdon dial vacuum
gauge–always remains full of water

Sealed rigid-body plastic tube
with porous ceramic cup

Ground surface

Porous ceramic
cup

FIGURE 3.14
Illustration of soil–water tensiometer for field use.

Since water and solutes can pass through the porous ceramic cup which separates the tensiometer from the surrounding soil, measurements obtained with the tensiometer include the effects of the dissolved solutes. At equilibrium, the condition for the soil water (porewater) will be given as

$$\mu_w(\text{pore water}) = \mu_w(\text{tensiometer}) = \mu_w^o + \frac{1}{\rho_w^o}(P_t - P^o) \qquad (3.20)$$

where μ_w is the chemical potential of the water, the superscript o refers to the standard state for the respective parameters, and P_t is the pressure in the tensiometer. If τ_w represents the porewater tension that is measured by the tensiometer at gauge pressure in pascals or atmospheres, the following is obtained (Sposito, 1981):

$$\tau_w = P^o - P_t = -\rho_w^o \left[\psi_p(P,\theta) + \psi_m(P^o,\theta) \right] \qquad (3.21)$$

3.4.2.2 Tensiometer-Type Device for Laboratory Use

The Buchner-type apparatus shown in the right-hand sketch of Figure 3.15 is the equivalent of the tensiometer used for in situ measurements. This

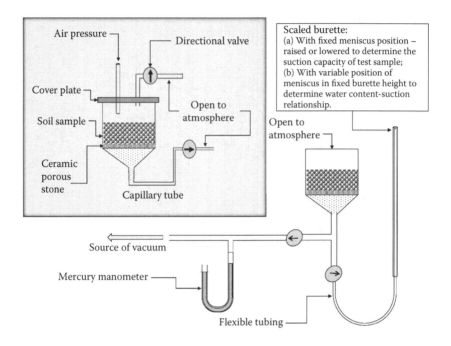

FIGURE 3.15
Buchner-type apparatus for determination of soil water-holding capacity using the pressure method (top left inset) and in the form of a tensiometer with two options for applying the negative pressure (right-hand sketch).

tensiometer method of measurement is a modified form of the system used by Haines (1930), who is credited with first reporting on the hysteretic effect between soil water content and *capillary pressure*. There are two methods of applying negative pressure to the specimen:

a. In the technique using the scaled burette at the right side of the Buchner funnel (with the vacuum valve on the left side closed off), there are two ways of determining the soil–water potential or its equivalent soil suction. These are:

- Using the scaled burette, which can be either raised and lowered, with a fixed meniscus position in the burette, to determine the maximum amount of water held in tension by the test specimen (soil suction) at the specified water content.

- Maintaining the scaled burette in a fixed position, allowing water uptake by the specimen and noting the change in position of the meniscus in the burette. This technique determines the suction in relation to the water content of the test sample.

b. Using the vacuum system on the left-side of the Buchner funnel to apply the negative pressure. It is not possible to apply a negative pressure of greater than –1 atmosphere gauge pressure, or zero absolute pressure—thus determining the limit of applicability of this method.

Since the test sample, which rests on a membrane/ceramic porous stone, is exposed to the atmosphere whilst the pressure in the water below the membrane is kept below atmospheric pressure, nucleation of air bubbles of dissolved air is possible when one approaches the limiting tension. It is important for water to be continuous from the sample through the underlying membrane or ceramic porous stone and into the scaled burette.

3.4.2.3 Pressure–Membrane Techniques

The pressure membrane techniques presently in use are limited to laboratory determination of soil–water potential. They use air pressure to drive the porewater from the sample contained in the pressure cell. A single cell (left inset in Figure 3.15) and cells accommodating multiple samples (Figure 3.16) have been used. The upper air pressure limit for each of these cells is determined by the structural strength of the cell chamber and the air-entry value of the ceramic porous stone at the base of the cell. Ceramic bases are usually used up to several hundred kPa, and cellulose membranes are generally used for higher pressures because of their higher air-entry values. For the Buchner-type apparatus shown in the left inset in Figure 3.15, since the apparatus is made of glass, the maximum air pressure that can be applied is severely limited. More robust single cells have been used as single-cell pressure systems. The Buchner-type system has been used by researchers interested in comparing the differences in soil–water potential obtained by using the air pressure system (pressure test), or by connecting the bottom tube to a vacuum system to apply suction to the sample (tension test). This will be discussed in the last section of this chapter.

The multiple-sample cell shown in Figure 3.16 reflects the type of *pressure membrane apparatus* developed by Richards (1949) and others in the 1940s. It is also called the *pressure plate apparatus*. Although soil samples can be placed in the pressure chamber without lateral support, a generally accepted procedure is to contain individual soil samples in lucite collars as shown in the figure. Soil samples are usually 1 to 2 cm high and 5 to 7 cm in diameter. Under an applied pressure, medium- to coarse-grained soils usually reach equilibrium in one or two days, whereas clays may take a week or more to reach equilibrium. Vapour losses can occur with long periods required for equilibration. Under such circumstances, the controlled vapour pressure technique (discussed in the next subsection) for determination of the soil–water potential might be more appropriate.

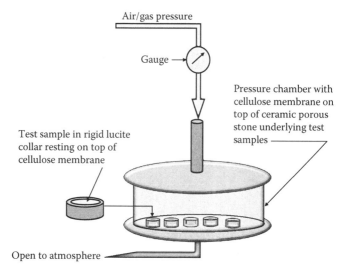

Air/gas pressure

Gauge →

Pressure chamber with
cellulose membrane on
top of ceramic porous
stone underlying test
samples

Test sample in rigid lucite
collar resting on top of
cellulose membrane

Open to atmosphere

FIGURE 3.16
Pressure membrane apparatus used for determination of soil–water potential for silts and clays.

For each applied pressure to the pressure membrane, when equilibrium is reached, the water remaining in the sample is considered to be held within the soil by internal forces (i.e., forces originating within the soil) that are greater than those applied by the air pressure introduced into the chamber. Figure 3.12 shows some typical soil–water potential relationships between three kinds of soils, generally obtained with pressure membrane techniques.

A thermodynamic analysis of the processes associated with the pressure membrane procedure shows that if a sample in the chamber is fully saturated at the onset of a test where the initial pressure P_i in the pressure membrane apparatus is zero, that is, $P_i = 0$, then the applied pressure P_w is a direct measure of the matric and solute potentials, ψ_m and ψ_s, respectively; that is, P_w provides a measurement of $\psi = \psi_m + \psi_s$. Similar to the case of measurements with the tensiometer, the effect of dissolved solutes is included in the measurements obtained with the pressure membrane technique. Accordingly, it is necessary to keep this in mind and to be sure to distinguish between measurements that do or do not include the effect of dissolved solutes. There are at least two different schools of thought concerning the matric potential ψ_m. These revolve around whether the matric potential does or does not include the effects of solutes. For nonswelling soils, this issue is not particularly significant. However, for swelling soils, if one maintains that the solute potential is the osmotic potential (i.e., $\psi_s = \psi_\pi$), it becomes necessary to account for this potential. The following relationship can be used: $\psi_s = -RTC_s$, where R is the universal gas constant, T is the absolute temperature, and C_s is the concentration of solutes. So long as one is careful in differentiating between the various effects, either concept is acceptable.

3.4.2.4 Vapour Pressure Technique

Extended periods required for equilibration of samples under pressure may result in vapour losses in the samples. A way in which this can be avoided is to use the controlled vapour pressure method for determination of soil–water potential. In this method, a controlled vapour pressure of water is introduced to the chamber. The soil samples in the chamber will either gain or lose water until the potential of the soil water is the same as that of the surrounding air. At this juncture, the sample water content can be determined (Figure 3.17). A simple method consists of placing soil samples in a desiccator containing sulphuric acid. Since sulphuric acid maintains a definite vapour pressure which can be regulated by regulating the concentration, this permits one to use the hypsometric relationship which relates the vapour pressure to soil–water potential to determine the soil–water potential of the samples in the desiccator. Strictly speaking, it should be noted that the sulphuric acid is sorbed in the soil sample.

3.4.2.5 Thermocouple Psychrometer Measurements

The thermocouple psychrometer, which can be used in laboratory studies or as a field instrument, measures vapour pressure. This is a well-defined

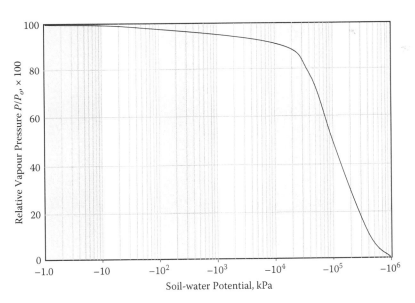

FIGURE 3.17
Soil–water potential at different relative vapour pressures with which soil is in equilibrium.

thermodynamic quantity which depends on both the matric ψ_m and solute ψ_s (osmotic) components of the soil–water potential ψ_w. The psychrometer probe essentially consists of a small ceramic bulb within which the thermocouple end or juncture is embedded (Figure 3.18). Cooling of the juncture is obtained by passing an electrical current through it (Peltier effect). Cooling of the juncture below the dew point will result in condensation at the juncture. The condensed water will evaporate when the electrical current is removed or discontinued. There is an inverse relationship between the rate of evaporation of the condensed water and the vapour pressure in the psychrometer bulb. Evaporation of the condensed water at the juncture will result in a drop in the temperature, the magnitude of which will depend on the relative humidity and temperature of the immediate volume surrounding the psychrometer. The drier the surroundings, the faster the evaporation rate, and hence the greater the temperature drop. The drop in temperature is measured as the voltage output of the thermocouple. The relationship between ψ_w and the relative humidity is given as

$$\psi_w = \frac{RT}{V_m} \ln \frac{p}{p^o} \tag{3.22}$$

where R is the universal gas constant, T represents the absolute temperature, V_m is the molal volume of water, p is the vapour pressure of the air in the soil

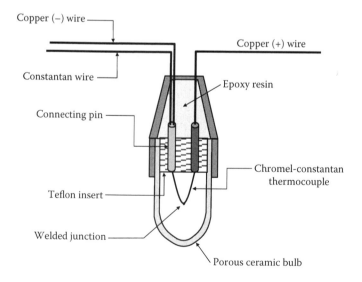

FIGURE 3.18

Schematic illustration of thermocouple psychrometer cross-section showing principles of Peltier cooling technique for determination of soil–water potential.

voids, p^o is the vapour pressure of saturated air at the same temperature, and the ratio of p/p^o is the relative humidity.

3.4.3 Hysteresis of Soil–water Potential Relationships

As illustrated in the left-hand sketches in Figure 3.2, soil pores include different masses of water in the drying process and in the wetting process at the same capillary potential. This is because the demonstrated capillary force is a function of only the structure or geometry of the pores, as witness the left-hand illustration of the pore radii involved in the wetting and drying processes. It follows that the matric potential ψ_m relationship for volumetric water content θ, which is defined as the *water characteristic curve* or *water retention curve*, varies between the drying process and the wetting process. This phenomenon, which is defined as *hysteresis* of the water characteristic curve, is more predominant in sands as opposed to clays.

When a soil begins the drying process in the fully saturated water content state, and continues onward until the water content reaches the fully dried state, the resultant water characteristic curve is defined as the *primary drying curve* (Figure 3.19). Conversely, when a fully dried soil begins its wetting process and continues until it reaches the apparent saturated state, the resultant water characteristic curve is defined as the *primary wetting curve*. The final

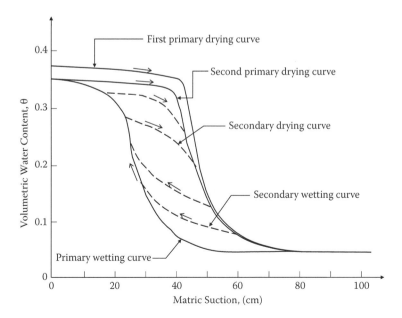

FIGURE 3.19
Hysteretic phenomena observed for water characteristic curves for sand. Arrows indicate wetting and drying processes.

saturated water content obtained in this wetting process is less than that of the fully saturated water content since the soil will entrap air bubbles in the wetting process. When drying of the apparently saturated soil (which includes the entrapped air bubbles) occurs, the measured water characteristic curve will differ from the primary boundary curves. When the wetting or drying occurs in the middle of the drying or wetting process, the developed characteristic curves, which are defined as the *secondary wetting* or *secondary drying* curves, will be different from the primary curves. The water characteristic curves are important pieces of information that reflect the changing water content processes, and are useful in analyses of water uptake, drainage, and water flow in engineering projects.

The hysteretic phenomena of water characteristic curves can be analyzed and simulated quantitatively using two kinds of pore models since it is a function of the characteristics of the soil pore structure:

- Blown tube model: The pore geometry is simplified by assuming a shape similar to the pore structure model illustrated by the left-hand sketches in Figure 3.2 in Section 3.2.1 (Nakano, 1980).
- Sphere packing model: The pore geometry is assumed to be the space formed in an ideal packing of spheres of similar radii (Zou, 2007).

Hysteretic performance in clays is not easy to explain with a simple model. This is because of the ink-bottle effect (as shown in Figure 3.2) and also because of changes in the nature and sizes of the various microstructural units. Movement of particles in these microstructural units occurs in the wetting and drying processes, resulting in fabric distortion and "plastic readjustment." When volume change occurs, the fabric changes accompanying volume changes will produce corresponding changes in pore size distributions and clay–water interactions arising from physicochemical forces. Interparticle contacts and forces at the points of contacts differ on wetting and drying.

3.5 Water Uptake and Transfer

The discussion in this section directs its attention to the principles governing movement of moisture into partly saturated soils. The discussion on water movement in fully saturated soils will be found in the chapters and sections dealing with the hydraulic properties and performance of soils (Chapter 7). The term *water* is generally considered to mean *liquid water*, whereas the term *moisture* is used as a more inclusive term, that is, liquid water and vapour. Where it is necessary to avoid confusion, the term *liquid water* will be used in place of *water*. The term *partly saturated* soil is used in preference to *unsaturated* soil.

The water content of a soil is a significant property of a soil because of its role in establishing the properties of soils. It is not a static quantity. Additions of water come naturally from rainfall, snow-melt, subsurface flow, and condensation, and natural depletion of water content occurs as water losses due to evaporation, transpiration, and drainage. Additions and depletions of water content also occur due to anthropogenic activities such as irrigation in agricultural practices, alteration of surface hydrology features, land reclamation, and so forth. The terms *water uptake* and *migration* refer specifically to water entering a dry or relatively dry soil (*water uptake*), and water transfer in partly saturated soils after moisture uptake (*water migration*).

3.5.1 Moisture Transfer

In natural circumstances, most of the water movement in soils is due to gradients of matric potential ψ_m or capillary potential ψ_c that arise from differences in water content. Concentration gradients (differences in concentration of solutes in porewater in different parts of the soil) will also provoke water transfer. The osmometer experiment shown as a schematic in Figure 3.20 illustrates this phenomenon. In this example, the soil sample in the left-hand cell is fully saturated with ionic solutes in its porewater. The total potential in the soil is $\psi = \psi_s$. For a partly saturated, the potential will be, $\psi = \psi_m + \psi_s$.

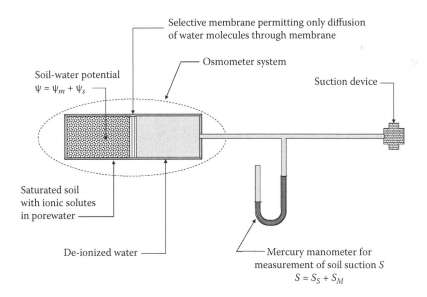

Selective membrane permitting only diffusion of water molecules through membrane

Osmometer system

Soil-water potential
$\psi = \psi_m + \psi_s$

Suction device

Saturated soil with ionic solutes in porewater

De-ionized water

Mercury manometer for measurement of soil suction S
$S = S_S + S_M$

FIGURE 3.20

Osmometer-type cell showing development of suction required to counter flow of water into the left-hand side chambers because of the total potential ψ in the saturated soil. S, S_S, and S_M refer to the total suction, solute suction, and matric suction, respectively.

The right-hand cell in the osmometer system contains deionized water which is separated from the left-hand cell by a fixed-position selective membrane that permits only diffusion of water molecules through the membrane. When the concentration of a solution differs from that at another point, there is a tendency for the more dilute liquid to diffuse into the region of higher concentration. This is the case for the left-hand and right-hand cells shown in Figure 3.20. The potentials in the soil in the left-hand cell produce gradients that will induce the deionized water in the right-hand cell to diffuse into the soil to attain a more uniform ionic concentration. To restrain diffusion of water in the right-hand cell through the selective membrane to the left-hand cell, one needs to apply suction to the water in the right-hand cell. The suction required to restrain diffusion, S, can be considered to consist of S_S and S_M, where S_S and S_M are the solute suction and matric suction, respectively.

3.5.1.1 Water Transfer and Wetting Front

Water movement in soil above the water table occurs when both water and air are present in the voids. This is the partly saturated soil zone of interest in many soil and geoenvironmental engineering projects. Water transfer in partly saturated soils is of considerable interest also to the agro industry. Commonly accepted terminology that describes water transfer in partly saturated soils as *unsaturated flow* will be used in this book. The characteristics of unsaturated flow in soils are demonstrated in the portrayal of *wetting-front advance* results obtained from a horizontal soil column permeation experiment (Figure 3.21). Beginning with a dry soil, a constant head source of water is supplied by the double Mariotte flask. The ability to locate the Mariotte flask air entry position (up or down) allows one to conduct the permeation experiment with various values of negative or positive heads.

The profile depicted in the diagram, which is called a wetting front profile, consists of a wetting zone ahead of a transmission zone and behind the wetting front. The wetting zone and wetting front combine to form characteristic shapes that can inform one on the nature of water diffusion into the soil.

In general, moisture transfer without convective flow is expressed using a Darcy-type equation as follows:

$$v = -k(\theta)\frac{\partial \psi}{\partial x} \tag{3.23}$$

where v is the flux, $k(\theta)$ is the Darcy coefficient, θ is the volumetric water content, x is the spatial distance from the source of water, and ψ is the soil–water potential—that is, the sum of the various potentials associated with the various forces operating in the soil–water system (see Section 3.4.1).

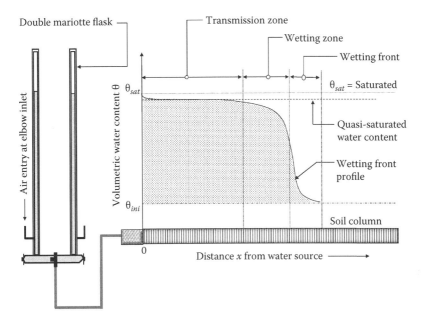

FIGURE 3.21
Characteristics of a wetting front profile. The shaded area represents the volumetric water content in the zone behind the wetting front. The double Mariotte flask provides a source of water with a constant head defined by the position of the elbow inlet. The flask can be raised or lowered to provide different constant heads, from negative to positive.

The total soil–water potential ψ is generally considered to be given as $\psi = \psi_m + \psi_\pi + \psi_g$ in analyses of moisture transfer in soils, where ψ_m is the matric potential, ψ_π is the osmotic potential or solute potential, and ψ_g is the gravitational potential that is ignored in the case of horizontal flow, as shown in Figure 3.21. A detailed discussion of the Darcy coefficient will be given in Chapter 7. When one expresses each potential as the energy per unit weight of soil–water, that is, hydraulic head, the commonly used one-dimensional equation with the gravitational term for analyses of water transfer is described as follows:

$$v = -k(\theta)\frac{\partial \psi_w}{\partial x} - k(\theta) \qquad (3.24)$$

where ψ_w is the water potential, that is, $\psi_w = \psi_m + \psi_\pi$. In the case of flow in granular soils, the osmotic potential can be ignored, and the water potential is expressed only in terms of the matric potential, that is, $\psi_w = \psi_m$.

There are at least three types of unsaturated fluid flow in soils. These relate directly to what happens to the soil during or as a result of the unsaturated flow, and can be classified by tracking the status of the fabric of the soil subject to the unsaturated flow. They are

- No change in soil volume and no change in soil fabric during and as a result of unsaturated flow, analogous to a rigid porous block.
- No change in soil volume but change in soil fabric (i.e., change in pore geometry) during and as a result of unsaturated flow. This is the most likely case for nonswelling soils. It can also be the case for swelling soils under confinement.
- Change in soil volume and change in soil fabric, that is, change in pore geometry and porosity during and as a result of unsaturated flow. This case will be considered in detail in Chapter 4.

For the case of no change in soil volume and little-to-no change in pore geometry and porosity, unsaturated fluid flow in such soils can be generally determined in terms of changes in the volumetric water content θ at a point by the mass conservation law. The equation of continuity, which states that the flow of water into and out of a unit volume of soil is equal to the rate of change of the volumetric water content, is given as:

$$\frac{\partial \theta}{\partial t} = -\frac{\partial v}{\partial x} \tag{3.25}$$

where t = time. Combining Equations (3.24) and (3.25), one obtains the equation with the gravitational term for water movement in water uptake (including the film flow) in the absence of external forces such as overburden forces, as follows:

$$\frac{\partial \theta}{\partial t} = \frac{\partial}{\partial x}\left(k(\theta)\frac{\partial \psi_w}{\partial x}\right) + \frac{\partial k(\theta)}{\partial x} \tag{3.26}$$

or

$$C(\psi_w)\frac{\partial \psi_w}{\partial t} = \frac{\partial}{\partial x}\left(k(\psi_w)\frac{\partial \psi_w}{\partial x}\right) + \frac{\partial k(\psi_w)}{\partial x} \tag{3.27}$$

where $C(\psi_w)$ refers to water capacity, that is, $C(\psi_w) = d\theta/d\psi_w$.

In the case of fully saturated flow, since the potential ψ responsible for flow consists primarily of the pressure potential ψ_p and gravitational potential ψ_g, that is, $\psi = \psi_p + \psi_g$, one obtains:

$$\frac{\partial n}{\partial t} = \frac{\partial}{\partial x}\left(k(n)\frac{\partial \psi_p}{\partial x}\right) + \frac{\partial k(n)}{\partial x} \tag{3.28}$$

where n refers to porosity. When porosity n and permeability coefficient k (n) are constant, Equation (3.28) takes the form of a Laplace equation with constant permeability coefficient k:

$$\frac{\partial}{\partial x}\left(k\frac{\partial \psi_p}{\partial x}\right) = 0$$

Introducing the water diffusivity coefficient $D(\theta) = k(\theta)(d\psi_w /d\theta)$ into Equation (3.26), one obtains the commonly cited equation for partly saturated flow as follows:

$$\frac{\partial \theta}{\partial t} = \frac{\partial}{\partial x}\left(D(\theta)\frac{\partial \theta}{\partial x}\right) + \frac{\partial k(\theta)}{\partial x} \tag{3.29}$$

Equation (3.29) is used because water content gradients are sometime easier to measure and also because water flow is more easily solved with diffusivity rather than conductivity.

The gravitational term is ignored for horizontal flow (Figure 3.21), and also in the case of flow in clays because the osmotic and matric potentials are dominant in comparison to the gravitational potential. In this case, the following relationship is commonly used for analysis of water transport:

$$\frac{\partial \theta}{\partial t} = \frac{\partial}{\partial x}\left(D(\theta)\frac{\partial \theta}{\partial x}\right) \tag{3.30}$$

The theoretical moisture profiles predicted from an evaluation of the relationship given in Equation (3.30) can be seen in Figure 3.22, with initial and boundary conditions: $\theta = \theta_i$ (constant) at $x > 0$, $t = 0$ and $\theta = \theta_{sat}$. at $x = 0$, $t > 0$, where θ_i is the initial water content and θ_{sat} is the saturated water content. In reality, there will always be some air voids trapped in the void spaces in "saturated" soils at soil surface and hence, the quasi-saturated soil condition is generally assumed to be the fully saturated condition for most practical applications.

Time-wetting studies show that a linear relationship exists between the wetting front distance x from the source and the square root of time ($\sqrt{t}$) required for the wet front to reach x, as shown, for example, by tests on a kaolinite (kaolin clay) at three different densities under the no-volume change condition (Figure 3.23). Determination of water entry and the characteristics of water transfer in soils is facilitated by using a system that allows one to observe and record the rate of advance of the wetting front into the soil such as the system shown in Figure 3.21. As we have indicated previously, the advantage of using the movable double Mariotte water source system shown

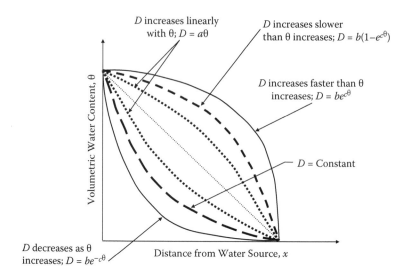

FIGURE 3.22
Theoretical moisture profiles predicted from evaluation of the diffusion equation (Equation 3.30), showing the nature of the D relationships relative to the profiles obtained.

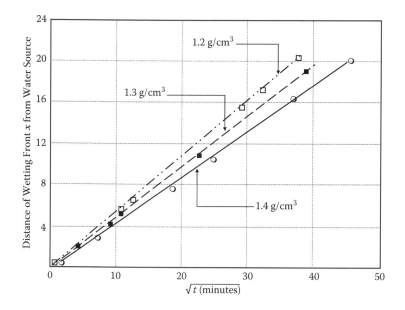

FIGURE 3.23
Wetting front advance, x, in relation to square root of time t for a kaolinite clay at three different densities, 1.2, 1.3, and 1.4 g/cm³.

in the left-hand side of Figure 3.21 lies in its ability to impose both positive and negative hydraulic heads to the sample. The tops of the tubes are sealed. This means that exposure to atmosphere is only at the elbow junction at the bottom end of the tubes. By this means, a constant head can be maintained at the level of the elbow atmospheric outlets. The zero hydraulic head situation shown in the illustration depicted in Figure 3.21 is best used to test the water uptake capability of the test sample, that is, the capability of the soil–water potential ψ_w (in soil engineering terminology, this would be *soil suction S*) to draw water into the soil.

For the conditions of no-volume change and little-to-no change in pore geometry (rigid porous block), the linear relationship between x and $\sqrt{t}$, square root of time, shown in Figure 3.23, confirms the use of a similarity solution technique for Equation (3.30).

Using the Boltzmann transform $\lambda = \lambda(\theta) = \dfrac{x}{\sqrt{t}}$, Equation (3.30) can be reduced to an ordinary differential equation:

$$\frac{\lambda}{2}\frac{d\theta}{d\lambda} = \frac{d}{d\lambda}\left(D\frac{d\theta}{d\lambda}\right) \tag{3.31}$$

Subject to the usual boundary conditions:

$$t = 0, 0 < x < \infty \text{ (thus, } \lambda = \infty), \theta = \theta_i$$

$$t > 0. \; x = 0 \text{ (thus, } \lambda = 0), \theta = \theta_o,$$

the amount of water entering per unit area, $(q_t)_o$, at the plane of $x = 0$ can be obtained using the Boltzmann transform as follows:

$$(q_t)_o = t^{\frac{1}{2}} \int_{\theta_i}^{\theta_{sar}} \lambda(\theta)d\theta = \int_{\theta_i}^{\theta_{sat}} x\,d\theta \tag{3.32}$$

3.5.1.2 Vapour Transfer

In partly saturated soils, the mechanism for moisture transfer will depend on whether the soil is dry, relatively dry, or relatively wet. For dry (anhydrous) soils, vapour transfer is greater than liquid water transfer, especially if large temperature gradients exist. Vapour movement can also occur under isothermal conditions. This happens when there is a moisture content gradient. Movement in the vapour phase is by convective (bulk) flow of the soil air, or by diffusion of water molecules, in the direction of decreasing vapour pressure. A vapour-pressure gradient can be developed by such factors as temperature, salt concentration, or differential suction within the soil. Of these factors, the temperature gradient is by far the most important. Differences in salt concentration result in a reduction in the vapour pressure of the porewater in proportion to the salt concentration. This causes a vapour pressure gradient, which in turn will result in vapour transfer.

Diffusion of water vapour is generally modelled as a Fickian diffusion process, and under isothermal conditions, this is given as $q_v = -D_v \partial c_{\theta v}/\partial x$, where q_v is the vapour flux, D_v is the vapour diffusion coefficient, x is the spatial coordinate, and $c_{\theta v}$ is the concentration of vapour in the gaseous phase. Considering the state of vapour to obey the equation of state of an ideal gas, the concentration of vapour will be given as a function of pressure and temperature. For nonisothermal condition, a temperature factor needs to be added to the Fickian diffusion equation (Nakano and Miyazaki, 1979).

Under isothermal conditions, in the absence of a gravitational term and convective flow currents and assuming that linear superposition could be applied to water transfer, we can combine this Fickian process with the relationship given in Equation (3.30) to obtain the combined water transfer (water movement and vapour transfer) relationship for partly saturated clays as follows:

$$\frac{\partial \theta}{\partial t} = \frac{\partial}{\partial x}\left((D_{\theta v} + D_{\theta l}) \frac{\partial \theta}{\partial x} \right) \tag{3.33}$$

where $D_{\theta v} = D_v \partial c_{\theta v}/\partial \theta$, and $D_{\theta l}$ is equal to $D(\theta)$ in Equation (3.30) and the total water diffusivity coefficient D_θ is defined by $D_\theta = D_{\theta v} + D_{\theta l}$. Since the water content at the surface will change with time, the initial and boundary conditions are given as follows:

$\theta = \theta_i = \text{const. at } x > 0, t = 0, \text{ and } \theta = f(t) \text{ at } x = 0, t > 0$

Figure 3.24 shows the total water diffusivity coefficient D_θ calculated for a sample of Kunigel bentonite–sand mixture (bentonite mixed with 30% sand), based on Equation (3.33). The calculated values are for changes of volumetric water content with time in unsteady infiltration experiments conducted at 25°C, reported by Chijimatsu et al. (2000). The total water diffusivity coefficient ($D_\theta = D_{\theta v} + D_{\theta l}$), which is larger at low water contents close to air-dried condition and at high water contents near the fully saturated condition, is due to (a) the larger $D_{\theta v}$ value, which reflects the predominant vapour flow in the unsaturated region, and (b) the larger $D_{\theta l}$ value, reflective of the predominant liquid water flow in the near-saturated region of the soil. The results shown in the figure indicate that vapour flow occurs up to a volumetric water content θ of about 0.25 (i.e., about 25%), which is attributable to the presence of macropores developed because of the sand component in the bentonite–sand mixture. The influence of temperature on the production of vapour flow will be discussed in Chapter 4.

The general relationship for the isothermal vapour diffusivity $D_{\theta v}$ has been given as

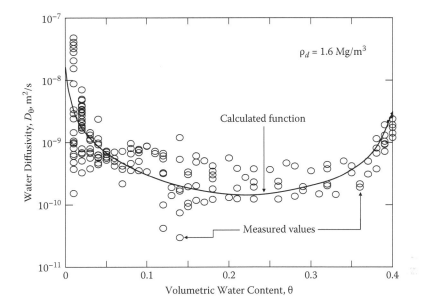

FIGURE 3.24
The total water diffusivity coefficient D_θ calculated for a sample of Kunigel bentonite–sand mixture (bentonite mixed with 30% sand) at a dry density of 1.6 Mg/m³. The calculated values use the results of changes of volumetric water content with time in unsteady infiltration experiments at 25°C reported by Chijimatsu et al. (2000).

$$D_{\theta v} = {}_{a\omega}D_{atm}\,\gamma g \rho_v/\rho_{LRT}\,\partial\psi_w/\partial\theta,$$

where ψ_w is the soil–water potential, α denotes tortuosity, ω is the volumetric air content, D_{atm} is the molecular diffusion coefficient, γ is the mass flow factor, g is gravitational acceleration, ρ_v represents the density of vapour, ρ_L is the density of liquid, and R is the universal gas constant (Yong et al., 2010). The dependence of the soil–water potential ψ on temperature T has been previously given by Philip and deVries (1957) as $\partial\psi_w/\partial T = \psi_{w/\sigma}\,\partial_\sigma/\partial T$. In general, the molecular diffusion coefficient at temperature T is given as follows: $D_{atm} = D_0 p_0/p\,(T/T_0)^n$, where D_0 is the molecular diffusion coefficient at a reference condition, p is the air pressure, p_0 is the reference pressure, T_0 is the reference temperature, and n is a constant that is 2.3 for vapour (Rollins et al., 1954).

3.6 Chemical Reactions in Porewater

The surface functional groups associated with the various soil fractions, discussed in Chapter 2, and the ions and other dissolved solutes such as naturally occurring salts in the porewater of a soil will react chemically when brought together as a wet soil mass (soil–water system). The chemistry of the porewater is linked to the chemistry of the reactive surfaces of the soil solids. Interactions between the solutes in porewater and soil particles involve many different sets of chemical reactions, including biologically mediated chemical reactions. The pH of the soil–water system and the various other dissolved solutes in the pore water influence the various interaction mechanisms such as acid-base reactions, speciation, complexation, precipitation, and fixation.

3.6.1 Acids, Bases, and pH

The porewater in a wet soil, without dissolved solutes, is a solvent that can be either a *protophillic* or a *protogenic* solvent. This means to say that it can function as an acid or as a base. Through self-ionization, it can produce a conjugate base OH^- and a conjugate acid H_3O^+.

$$2H_2O \text{ (solvent)} \Leftrightarrow H_3O^+ \text{ (acid)} + OH^- \text{(base)}$$

In the standard Arrhenius definition of an *acid*, this is defined as an aqueous substance which dissociates to produce H^+ ions, and a *base* as an aqueous substance which dissociates to produce OH^- ions. In other words, acids are substances that produce hydrogen ions, H^+, in solution, and bases are substances that produce hydroxide ions, OH^-, in solution. The pH scale, which was developed by Sörenson in 1909 in his studies of the *Arrhenius theory of electrolytic dissociation* as a means to identify the degree of acidity, determines the pH of a solution to be the negative logarithm to the base ten of the molar hydrogen ion concentration. This concept of acids and bases, first proposed by Arrhenius in 1881 and refined in 1887, has been shown to be useful in application to aqueous solutions, but faces difficulties in application to soil–water systems, for example, application of the Arrhenius concept to solvents and to the specific requirement for the presence of OH^- ions in a base.

For soil–water systems, because of the presence of reactive surfaces of soil particles, it is more convenient to consider the definitions of *acids* and *bases* in terms of proton donation or acceptance. By releasing one from considering acids and bases as aqueous substances, this allows one to categorize nonaqueous substances in terms of acids and bases. The Brønsted–Lowry definition of acids and bases expands the concepts of Arrhenius (Brønsted, 1923). In the Brønsted–Lowry definition, an *acid* is a substance that can donate a proton (H^+), and a *base* is a substance that can accept a proton. This (Brønsted–Lowry) acid-base concept considers an *acid* as a *proton donor (protogenic* substance) and a *base* as a *proton acceptor (protophillic* substance). Since water has the capability to both donate and accept protons (i.e., both protogenic and protophillic), it is called an *amphiprotic* substance.

In the Lewis (1923) concept, an *acid* is a substance that is capable of accepting an electron pair from a base, and a *base* is a substance that is capable of donating an electron pair. This means that substances with unshared pairs of electrons are bases, and substances lacking an octet are acids. With this set of definitions, metal ions M^{n+} are Lewis acids, and theoretically, all cations are Lewis acids. Water is both a Lewis base and a Brønsted base. Whilst Lewis bases are also Brønsted bases, it does not follow that Lewis acids are Brønsted acids. This is because Lewis acids include substances that are not proton donors. The Lewis concept of acids and bases allows one to treat metal–ligand bonding as an acid-base reaction.

Pearson (1963) classified a whole range of atoms, ions, and molecules into three groups of acids and bases as hard, borderline, and soft Lewis acids and Lewis bases. This classification is known by the acronym HSAB (hard, soft, acid base) classification. With this classification, Pearson noted that hard acids prefer to bind with hard bases, and soft acids prefer to bind with soft bases. This classification scheme is useful when one considers examines and analyzes contaminant–soil interactions and transport in soils. The following summary list from Yong et al. (2010) shows some of the more common contaminants found in soils as a result of facilities and activities associated with humans:

- Lewis hard acids are generally small in size, with high positive charge, high electronegativity, low polarizability, and do not have unshared pairs of electrons in their valence shells. The acids include aluminium chloride, arsenic (III) ion, chloride ion, iron (III) ion, magnesium ion, manganese (II) ion to uranium (IV) ion, and zirconium ion.

- Lewis borderline acids range from antimony (III) ion, copper (II) ion, iron (II) ion, to sulphur dioxide and zinc ion.

- Lewis soft acids are generally large in size with low positive charge and low electronegativity, in contrast to Lewis hard acids. They have unshared pairs of electrons in their valence shells. The acids range from borane, cadmium ion, hydroxyl cation, and iodine, to thallium (III) ion, and 1,3,5-trinitrobenzene.

- Lewis hard bases usually have high electronegativity, low polarizability, and are difficult to oxidize. The bases range from acetate ion, ammonia, carbonate ion, chloride ion, to hydroxide ion, nitrate ion, and water.

- Lewis borderline bases include aniline, bromide ion, nitrogen, and sulphide ion.

- Lewis soft bases usually have low electronegativity, high polarizability, and are easy to oxidize, in contrast to the hard bases. The bases range from benzene, ethylene, hydride ion, and iodide ion to trimethylphosphite.

3.6.2 Acid-Base Reactions, Hydrolysis

The chemical reactions in porewater are acid-base reactions. Acid-base reactions are defined as the proton transfer reactions between a proton donor (acid) and a proton acceptor (base). Proton loss is called *protolysis*, and the proton transfer reaction is a *protolytic reaction*. Hydrolysis is a neutralization process, and technically speaking classifies as an acid-base reaction inasmuch as it (hydrolysis) refers to the reaction of H^+ and OH^- ions of water with the dissolved solutes and other constituents present in the porewater. For example, the hydrolysis of iron in porewater is shown as

$$Fe^0 \Leftrightarrow Fe^{2+} + 2e^-$$

$$2H_2O + 2e^- \Leftrightarrow 2OH^- + 2H^+$$

Then, $$Fe^0 + 2H_2O \Leftrightarrow Fe^{2+} + 2OH^- + 2H^+$$

or $$Fe^{2+} + 2H_2O \Leftrightarrow Fe(OH)_2 + 2H^+$$

The presence of ionized cations and anions associated with the soil particles in a soil–water system results in pH levels in the soil–water system that vary from below neutral to above neutral pH values dependent on the strength of ionization of the ions. Hydrolysis reactions in a soil–water system will continue so long as the reaction products are removed from the system through processes associated with precipitation, complexation, and sorption by the soil particles. Hydrolysis reactions of metal ions in the porewater are influenced by (a) pH of the active system, (b) type, concentration, and oxidation state of the metal cations, (c) redox environment, and (d) temperature. Favourable circumstances for hydrolysis reactions include high temperatures, low organic contents, low pH environment, and low redox potentials.

3.6.3 Oxidation-Reduction (Redox) Reactions

The presence of solutes and microorganisms in the porewater of soils mean that abiotic and biotic oxidation-reduction (redox) reactions will occur in the porewater. Biotic redox reactions are of greater significance than abiotic redox reactions. Oxidation-reduction reactions involve the transfer of electrons between the reactants. The activity of the electron e^- in the chemical system typified by the reactive soil particles and solutes in the porewater is of particular importance. Generally speaking, the transfer of electrons in a redox reaction is accompanied by proton transfer. In the case of inorganic solutes in the porewater, redox reactions result in the decrease or increase of the oxidation state of an atom—a matter of some significance for those ions that have multiple oxidation states. A measure of the electron activity is the *redox potential Eh.* It allows one to determine the potential for oxidation-reduction reactions in a clay–water system and is given as $Eh = pE(2.3RT/F)$, where E is defined as the electrode potential, R = universal gas constant, T = absolute temperature, and F is the Faraday constant. The mathematical term pE is the negative logarithm of the electron activity e^-. The relationship between Eh and pE at a standard temperature of 25°C is

$$Eh = 0.0591pE = E^0 + \left(\frac{RT}{nF}\right)\ln\frac{a_{i,ox}}{a_{i,red}} \tag{3.34}$$

where E^0 refers to the standard reference potential, n is the number of electrons, a is the activity, and the subscripts for the activity a refer to the activity of the ith species in the oxidized (*ox*) or reduced (*red*) states.

The redox capacity is a measure of the amount of electrons that can be added or removed from the soil–water system without a measurable change in the Eh or pE, and is comparable to the *buffering capacity*, which measures the amount of acid or base that can be added to a soil–water system without any measurable change in the system pH. The maximum amount of acid or base that can be added to a soil–water system without any measurable change in the Eh or PE of the system will establish the redox capacity of the soil. In a sense, this concept is similar to the buffering capacity concept of clays. Many chemical reactions occurring in a soil–water system are dependent on temperature, concentration of solutes in the porewater, ligands in the porewater, and on Eh and also *pH*, with the latter two being important because proton transfer is neutralized by electrons. The Nernst equation, which is similar to Equation (3.34), demonstrates this influence:

$$pE = 16.92Eh = E^0 + \left(\frac{RT}{nF}\right)\ln\frac{[A]^a[H_2O]^w}{[B]^b[H^+]^h} \tag{3.35}$$

where the superscripts a, b, w, and h refer to the number of moles of reactant, product, water and hydrogen ions, respectively.

3.7 Physical Reactions and Hydration

The discussion in this section follows the discussion given in Yong et al. (2010). The character of the hydration layer surrounding the surfaces of soil particles is different from that of bulk water. The charged surfaces of soil particles attract water molecules, resulting in the formation of hydrated soil particles. The ions in porewater, which are also hydrated (with water molecules), are bound to hydrated clay minerals by such forces as Coulomb, van der Waals, short-range repulsion forces, and covalent bonds, as described in Chapter 2. In general, the hydration state of ions or the association state of ions with clay mineral surfaces can be demonstrated by observing the status of oxygen around the ions in porewater, using neutron diffraction (ND), the x-ray diffraction (XRD) method, and the extended x-ray absorption fine structure spectroscopy (EXAFS) method (Nakano et al., 2004).

3.7.1 Hydrated Cations and Clay Minerals—From EXAFS Analyses

The results from EXAFS analyses are summarized in Table 3.1 (Nakano et al., 2004; Nakano and Kawamura, 2006). The results show the coordination number of water molecules around Cs^+, Ba^{2+}, and Sr^{2+} in porewater and their bond distances, determined by extended *x-ray adsorption fine structure* (EXAFS) analyses using the two-shell fitting technique on oxygen around metal ions in porewater. In addition, the number of the clay mineral oxygens associated with the metals and the bond distances between the metals and the clay mineral oxygen are shown for air-dried samples in comparison with that for the paste samples. The bond distance between the metals and the clay mineral oxygen expresses the distance between metal and clay mineral surface. In other words, Table 3.1 shows the hydration state of Cs^+, Ba^{2+}, and Sr^{2+} in porewater and the mode that the ion is attracted to the hydrated clay mineral surface. The hydration numbers of Cs^+, Ba^{2+}, and Sr^{2+} of 6, 8, and 8 in bulk solution, respectively, have been measured using *neutron diffraction* (ND) or *x-ray diffraction* (XRD) methods (Ohtomo and Arakawa, 1979; Albrigh, 1972). In comparison with these results, the hydration number of metals in porewater is smaller for air-dried clays, whilst the hydration number for the paste samples is nearly equal

TABLE 3.1

Hydration of ions and their association with clay minerals

Sample	pH	Cs air-dried paste		Ba air-dried paste		Sr air-dried paste	
(a) Hydration of metal ion in	4.5	4.5	5.6	5.1	6.9	5.6	6.8
porewater, number of water molecules coordinated	10	4.5	6.1	5.7	6.8	5.4	7.2
Distance between metal and	4.5	3.17	3.19	2.88	2.87	2.58	2.59
water molecule (Å)	10	3.18	3.18	2.85	2.87	2.58	2.61
(b) Association of metal ion	4.5	5.9	3.9	3.7	3.7	2.7	2.4
with clay mineral, number of mineral oxygens associated with metal	10	6.3	4.2	4.3	4.1	2.5	2.5
Distance between metal and	4.5	3.55	3.62	3.09	3.11	2.77	2.80
oxygen on mineral (Å)	10	3.56	3.61	3.07	3.11	2.76	2.83

Note: The data have been obtained from the two-shell fitting technique of EXAFS spectra.
(a) *denotes the first shell of oxygen around the metal ion,* (b) *denotes the second shell of oxygen around the metal ion.*

to that in bulk solution. The number of clay mineral oxygens associated with the metals is about 3 to 4 except for that of the *Cs* ion in the air-dried state. *Cs* ions specifically associate with the mineral oxygen of 6, indicating their strong bonding to mineral surfaces in the air-dried state. The evidence suggests that metals in the porewater hover over clay mineral surfaces due to the hydration of clay minerals. This conclusion arises from the observation that the distances between metal and mineral oxygen is considered to be slightly larger than the bond distances of metal-oxygen in the hydrated metals in porewater. Accordingly, the metals are assumed to be easy to partition in porewater when water will migrate to the air-dried clays from the surrounding region.

3.8 Concluding Remarks

There are some important observations that need to be highlighted:

- The energy status of a soil–water system is an important characteristic of soils. Measurement of the energy state of water in a soil (soil water) requires techniques and procedures that produce results that are often operationally defined. Take, for example, the pressure membrane test system, which is a common laboratory technique used to determine such a characteristic. For pressure membrane tests, it is difficult to obtain representative undisturbed samples in small sizes to fit the lucite collars. Because of vapour losses and difficulties in obtaining representative samples, measurements from pressure membrane are less satisfactory for clays than for loams or sands. Air bubbles almost always accumulate in the water below the ceramic porous stone or membrane, due possibly to leaks or movement of air through the water phase due to a pressure gradient. The air will interfere with the measurement of water being discharged, and will also carry away water vapour from the sample in the chamber.

- The results of tests using the Haines-type (suction) system and a pressure system using the Buchner-type apparatus described in Figure 3.15 are shown in Figure 3.25. Note the importance of the initial state of the sample under test. The differences in results are due to two important factors: (a) pore geometry effects between sorption and desorption (wetting and drying) as illustrated in Figure 3.2 and (b) entrapped air.

- Hysteretic phenomena observed in water characteristic curves predominate in sands rather than in clays. The water characteristic curves are important pieces of information that reflect the changing water content processes, and are useful in analyses of water uptake, drainage, and water flow in engineering projects.

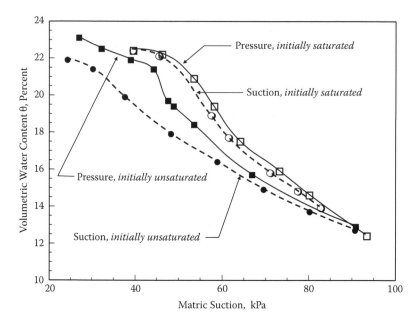

FIGURE 3.25
Soil–water characteristic curves for a fine silt—showing differences in characteristic values depending on whether the samples are initially saturated or initially unsaturated, and also dependent on whether the Haines-type (suction) method or the pressure method is used.

- Water movement in partly saturated soils occurs along film boundaries in soil pore spaces that are not completely filled with water, and as pore channel flow for those pore spaces that are completely filled with water. Initially, water uptake is through hydration processes. Further water transport into the soil is from film boundary transport. As further water is drawn in, the air within the soil must escape to the surface. At full saturation, if trapped air in some of the voids cannot escape, it will remain in the microstructural unit.

- Pore channel flow has been modelled as saturated flow, which is obviously inappropriate. Obviously, we cannot evaluate or analyze unsaturated flow using two separate and different analytical models; that is, it is not practical to perform film boundary flow analysis in conjunction or combination with saturated flow analysis since we have no means to determine the proportion of film boundary or saturated flow contributing to the total flow. Instead, we have to rely on a deterministic analysis of the combined unsaturated flow phenomenon. Because bacteria in the soil utilize oxidation-reduction reactions as a means to extract the energy required for growth, they essentially function as catalysts for reactions involving molecular oxygen and soil organic matter and organic chemicals. Electron transfer in a redox reaction is generally accompanied by proton transfer. Microorganisms play a significant role in catalyzing redox reactions. Their utilization of redox reactions as a means of extraction of the energy required for growth serves as a catalyst for reactions involving molecular oxygen and organic matter (and organic chemicals) in the clays.

References

Alammawi, A.M., 1988, Some aspects of hydration and interaction energies of mont-morillnite clay, Ph.D. thesis, McGill University, Montreal, Canada.

Albrigh, J.N., 1972, X-ray diffraction studies of alkaline earth chloride solutions, *J. Chem. Phys.*, 56, 3783–3786.

Bolt, G.H. 1956, Physico-chemical analysis of compressibility of pure clays, *Geotechnique*, 6:86-93.

Brønsted, J., 1923, Some remarks on the concept of acids and bases, *Recueil des Travaux Chimiques des Pays-Bas*, 42:718–728.

Buckingham, E., 1907, Studies on the movement of soil moisture, *U.S. Dept. Agric. Bur. Soils, Bulletin* 38, 61 pp.

Chijimatsu, M., Fujita, T., Kobayashi, A., and Nakano, M., 2000, Experiment and validation of numerical simulation of coupled thermal, hydraulic and mechanical behaviour in engineered buffer materials, *Int. J. Numer. Anal. Methods in Geomechanics*, 24(4):403–424.

Ferris, A.P., and Jepson, W.B., 1975, The exchange capacities of kaolinite and the preparation of homoionic clays, *J. Colloid Interface Sci.*, 51:245–251.

Flegmann, A.W., Goodwin, J.W., and Ottewill, R.H., 1969, Rheological studies on kaolinite suspensions, *Prof. Brit. Ceramic Soc.*, pp. 31–44.

Grahame, D.C., 1947, The electrical double layer and the theory of electocapillarity, *Chem. Rev.*, 41:441–501.

Haines, W.B., 1930, Studies in the physical properties of soils, V: The hysteresis effect in capillary properties and the modes of moisture distribution associated therewith, *J. Agric. Sci.*, 20:97–116.

Hogg, R., Healy, T.W., and Fuerstenay, D.W., 1966, Mutual coagulation of colloidal dispersions, *Trans. Faraday Soc.*, 62:1638–1651.

Iwata, S., Tabuchi, T., and Warkentin, B.P., 1995, *Soil-Water Interaction, Mechanisms Add Applications*, 2nd edition, Marcel Dekker, New York, 440 pp.

Kruyt, H.R., 1952, *Colloid Science*, Vol. I, Elsevier, Amsterdam, 389 pp.

Langmuir, I., 1938, Repulsive forces between charged surfaces in water and the cause of the Jones-Ray effect, *Science*, 88:430–432.

Lewis, G.N., 1923, *Valence and the Structure of Atoms and Molecules*, The Chemical Catalogue, New York.

Nakano, M., and Miyazaki, T., 1979, The diffusion and non-equilibrium, thermodynamic equations of water vapour in soils under temperature gradients, *Soil Sci.*, 128(3):184–188.

Nakano, M., 1980, Pore volume distribution and curve of water content versus suction of porous body, *Soil Sci.*, 130: 7–10.

Nakano, M., Kawamura, K., and Emura, S., 2004, Local structural information from EXAFS analyses and adsorption mode of strontium on smectite, *Clay Science*, 12:311–319.

Nakano, M., and Kawamura, K., 2006, Adsorption sites of Cs on smectite by EXAFS analyses and molecular dynamics simulations, *Clay Sci.*, 12(Suppl. 2):76–81.

Ohtomo, N., and Arakawa, K., 1979, Neutron diffraction study of aqueous ionic solutions. 1. Aqueous solutions of lithium chloride and cesium chloride, *Bull. Chem. Soc. Jpn.*, 52:2744–2759.

Pearson, R.G., 1963, Hard and soft acids and bases, *J. Amer. Chem. Soc.*, 85:3533–3539.

Philip, J.R., and de Vries, D.A., 1957, Moisture movement in porous materials under temperature gradients, *Trans. Am. Geophys. Union*, 38:222–232.

Richards, L.A., 1928, *The usefulness of capillary potential to soil moisture and plant investigators. J. Agric. Res.* (Cambridge), 37:719–742.

Richards, L.A., and Gardner, W., 1936, Tensiometers for measuring the capillary tension of soil water, *J. Amer. Soc. Agron.*, 28:352–358.

Richards, L.A., 1949, Methods of measuring soil moisture tension, *Soil Sci.*, 68:95–112.

Ritchie, G.S.P., and Sposito, G., 2002, Speciation in soils, in A.M. Ure and C.M. Davidson (Eds.), *Chemical Speciation in the Environment*, Blackwell Science Ltd., U.K.

Rollins, F.L., Spangler, M.G., and Kirkham, D., 1954, Movement of soil moisture under a thermal gradient, *Proc. Highway Res. Board.*, 33:492–508.

Sörenson, S.P.L., 1909, Enzyme studies II: The measurement and meaning of hydrogen ion concentration in enzymatic processes, *Biochemische Zeitschrift*, 21:131–200.

Sposito, G., 1981, *The Thermodynamics of Soil Solutions*, Oxford University Press, New York, 223 pp.

Sposito, G., 1984, *The Surface Chemistry of Soils*, Oxford University Press, New York, 234 pp.

Stern, O., 1924, Zur theorie der Electrolytichen Doppelschicht, Z. *Electrochem.*, 30:508–514

van Olphen, H., 1963, *An introduction to clay colloid chemistry*, Interscience. New York, N.Y, 301p.

van Olphen, H., 1977, *An introduction to clay colloid chemistry*, 2nd edition, Wiley, New York.

Yong, R.N., and Mohamed, A.M.O., 1992. A study of particle interaction energies in wetting of unsaturated expansive clays, *Can. Geotech.J.* 29:1060–1020.

Yong, R.N., and Warkentin, B.P., 1959, A physico-chemical analysis of high swelling clays subject to loading, *Proc. 1st. Pan Amer. Conf. on Soil Mech. Found. Engr.*, 2:867–888.

Yong, R.N., and Warkentin, B.P., 1975, *Soil Properties and Behaviour*, Elsevier Scientific Publishing Co., Amsterdam, 449 pp.

Yong, R.N., Sethi, A.J., Ludwig, H.P., and Jorgensen, M.A., 1979, Interparticle action and rheology of dispersive clays, *J. Geotech. Engineering Division*, GT10, pp. 1193–1209.

Yong, R.N., 2001, *Geoenvironmental Engineering: Contaminated Soils, Pollutant Fate and Mitigation*, CRC Press, Boca Raton, FL, 307 pp.

Yong, R.N., Pusch, R., and Nakano, M., 2010, *Containment of High-Level Radioactive and Hazardous Solid Wastes with Clay Barriers*, Spon Press, Taylor & Francis, London, 468 pp.

Zou, Y., 2007, A hysteresis model for the soil-water characteristic curves of simple granular soils, *Soils and Foundation*, 47:337–348.

4

Swelling Clays

4.1 Introduction

Clays generally undergo changes in volume with corresponding changes in their water contents. When they are dried, shrinkage and cracking occur, and if they are rewetted after drying, swelling occurs. The amount of swelling after rewetting in the presence of available free water depends primarily on the type of clay minerals in the clay. Clays containing montmorillonite show an almost reversible swelling and shrinking on rewetting and redrying, whereas clays containing kaolinite or illite show an initial large volume decrease on drying with only limited swelling on rewetting. Table 4.1 shows the average particle size and swelling capability of these minerals. The former (high volume change) clays are often referred to as *high swelling* clays and the latter (limited swelling on rewetting) are generally identified as *low swelling* clays.

When swelling clays are rigidly confined, they will develop swelling pressures against their rigid confinement apparatus upon water uptake. This pressure is called the swelling *pressure*, the magnitude of which depends on the type of clay minerals, and the species and concentration of cations in the interlayer water. Volume changes, swelling pressure, shrinkage, and cracking of a clay will affect its properties and behaviour, such as hydraulic conductivity, strength, and deformation. The underlying factor required to obtain swelling upon water uptake is the presence of forces of repulsion, which act to separate and move particles apart from each other (see Section 3.3 in Chapter 3). In this chapter, we will be discussing (a) the phenomena and mechanisms of swelling in swelling clays and (b) the knowledge required to understand swelling clays, with particular attention to the methods and results of experiments and analyses. The important role of swelling clays in soil engineering and geoenvironmental engineering projects, coupled with some very unique compositional and behavioural features of these clays, make it useful to devote a separate chapter to a discussion of the fundamental aspects of these clays.

TABLE 4.1

Size and Swelling of Some Clay Minerals

Mineral	Approximate Particle Thickness, (nm)	Average Specific Surface Area, (m²/g)	Observed Volume Change
Montmorillonite	2.0	700–800	High
Illite	20	80–120	Medium
Kaolinite	100	10–15	Low

4.2 Swelling Phenomena

4.2.1 Pertinent Soil Characteristics

The amount of soil swelling upon water uptake depends on several factors related directly to the clay minerals in the soil and the soil porewater chemistry. In respect to clay minerals, these include (a) types of clay minerals, (b) proportions of the various clay minerals in the total soil mass, and (c) arrangement of clay particles (particle orientation and microstructure). Cementation between particles by iron hydroxides, carbonates, and various organic molecules is a major factor in limiting volume increase of clays on swelling (Figure 4.1). The factors of importance related to porewater chemistry and clay minerals are (a) physicochemical properties, (b) valences and concentration of the various exchangeable ions, and (c) types and effectiveness of bonds between particles.

Everything else being equal, soil swelling increases with increasing surface area of the soil particles. The total surface area of a clay depends on the number, shape, and size of the particles constituting the clay, the arrangement of particles, and the types of microstructural units. In general, soils possess a variety of particles that include primary and secondary minerals and also clay minerals (see Figure 2.1 in Chapter 2). By and large, the soil particles that contribute most to soil swelling phenomena are the clay-sized particles, and more specifically the clay minerals. From Section 2.2 in Chapter 2, we learnt about different specific surface areas, cation exchange capacities, and surface functional groups of different kinds of clay minerals. Clay minerals ranging from kaolinite to montmorillonite have vastly different particle sizes and surface areas, as seen in Table 4.1. The mechanisms of clay swelling will be discussed in detail in the next section in conjunction with the discussion on the why and how montmorillonite endows a clay with high-swelling properties.

Monovalent exchangeable cations such as sodium cause greater swelling than divalent calcium ions as can be deduced from the discussions in Chapter 2 concerning the diffuse double-layer (DDL) phenomena, and as seen in Figure 4.2. Highly acidic clays have polyvalent aluminium as exchangeable cations, with consequent low swelling. An increase in porewater salt concentration decreases swelling of high-swelling clays, particularly if monovalent

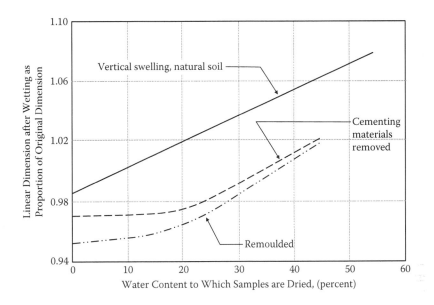

FIGURE 4.1
Measured swelling of a clay from Central Canada as influenced by remoulding and removal of cementing materials. (Adapted from Yong, R.N., and Warkentin, B.P., 1975, *Soil Properties and Behaviour*, Elsevier Scientific Publishers, Amsterdam, 449 pp.)

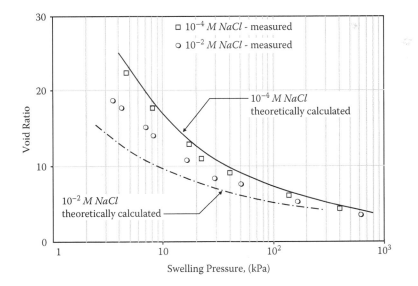

FIGURE 4.2
Experimentally measured swelling pressure of *Na*-montmorillonite, with two different concentrations of *NaCl* (10^{-4} and 10^{-2} *M NaCl*) and theoretically calculated pressures.

ions are present. Swelling and dispersion of soils with exchangeable sodium are reduced at higher salt concentrations.

In regard to particle arrangement and swelling, it is seen that swelling against restraint is greatest for parallel particle orientation for high swelling clays such as sodium montmorillonite. For most clays that do not fall into the class of high-swelling clays, the swelling volume and swelling pressure is greatest for random or flocculated particle arrangement. Higher swelling pressures (against restraint) are generally obtained with closer packing of particles so long as particles are not bonded to each other by cementing materials. The dependence of swelling pressure on volume change and the physical (experimental) method of determination makes a precise measurement of swelling pressure difficult. This is because of the controlling role played by particle orientation and packing of particles (soil fabric).

There are essentially two direct means of laboratory measurement of swelling pressure of a clay sample: (a) applying direct pressure on the sample and determining the equilibrium resistance pressure at various some water contents, and (b) allowing water uptake into a sample and measuring the swelling pressure by restraining volume change during water uptake. The direct pressure method for determination of swelling pressures is essentially similar in respect to the technique used in consolidation experiments. The "swelling pressure" results obtained with this method can be construed to be the consolidation pressures and may not be the same as those obtained with the water uptake method. The reasons for the differences are explained in a later section dealing with measurement of swelling pressures.

4.2.2 Mechanisms of Swelling

Soil swelling upon water uptake, beyond the phenomenon of volume expansion due to hydration of clay particles, is the consequence of effective repulsion between particles developed as a result of the interaction between contiguous clay particles (see Figure 3.9 in Chapter 3). From thermodynamic considerations, water will move into the soil so long as the free energy of the water in the soil is less than that of free water. The greater the proportion of clay particles—especially the clay mineral montmorillonite—the greater is the swelling of the soil upon water uptake. Montmorillonite is the dominant mineral in clays identified as bentonites and smectites.

A notable feature in swelling clays made up of the 2:1 unit layer structure of montmorillonites saturated with different exchangeable cations is the water uptake or hydration characteristics of the clay minerals. In water, hydration proceeds until the maximum number of interlayer hydrates are formed, provided that there is no geometrical restraint. In humid air, the number of hydrates obtained depends on the relative humidity The basal spacing $d(001)$ of 0.95 nm to 1.0 nm for the anhydrous montmorillonite will expand from 1.25 nm to 1.92 nm depending on the amount of water intake to satisfy the hydration process. The hydration of the clay particles depends on the charge and coordination of adsorbed cations and water molecules. Whilst the crystal lattice constitution determines the charge and coordination of adsorbed interlayer cations and water molecules, the size and charge of the cations and the charge distribution in the crystal lattice are important factors in the development of hydration and clay swelling. Except for *Li* and *Na* as exchangeable cations in the interlayer or interlamellar space, expansion of the basal spacing for montmorillonites that contain other exchangeable cations appears to reach a maximum value of about 1.92 nm at $p/p_o = 1$. This corresponds to the thickness of four layers of water. (The terms *interlayer* and *interlamellar* have been used in the literature to mean the same thing.) By and large, the interlayer hydrates (that is, hydration layers in the interlayer spaces) are responsible for the swelling of montmorillonites. Table 4.2 gives the various expansions of the basal spacings of Wyoming montmorillonites with different exchangeable cations in the interlayer spaces, and in respect to different relative humidities.

Many researchers (e.g., van Olphen, 1977) maintain that the dominant mechanism responsible for interlayer spacing of up to about 1 nm upon water uptake (hydration) is due to the actions resulting from the adsorption energy of water at the mineral particle surfaces. One needs to distinguish between crystalline swelling, that is, swelling due to sorption of the first two to three water layers between the unit layers of the 2:1 dioctahedral series of alumino-silicate clays, and interparticle sorption or water uptake (Figure 4.3). In short, interlayer expansion or clay swelling upon water uptake is determined by the layer charge, interlayer cations, properties of adsorbed liquid, and particle size. Expansion of the interlayer distance beyond hydration is due to mechanisms represented by the diffuse double-layer models discussed in Chapter 3.

The basal spacings at the 100% humidity level for the Wyoming montmorillonite shown in Table 4.2 corresponds to values also measured by Suquet et al. (1975) and by Quirk (1968), who showed *Li* and *Na* swelling as > 4 nm instead of "M" (macroscopic swelling). Mooney et al. (1952) indicated that the 1.24–1.25 nm spacing corresponded to one monolayer of water in interlayer region, and that basal spacings of 1.5, 1.9, and 2.2 nm corresponded to 2, 3, and 4 layers of water between each alumino-silicate layer. The swelling in the interlayer between 1.0 nm and 2.2 nm is generally identified as *crystalline swelling* if a definite hydration structure of water exists. Swelling beyond 2.2 nm, as is the case for the *Li* and *Na* montmorillonites shown in Table 4.2 is most often associated with double-layer swelling.

The studies of Morodome and Kawamura (2009) on the effects of high temperature on the basal spacing of both Na- and Ca-montmorillonites show that whereas the basal spacing for Na-montmorillonite is independent of temperature in range of 70°C to 110°C, that of Ca-montmorillonite decreases with increasing temperature, in range of relative humidity less than 40%, as they keep about two layers of hydration at about 40% of relative humidity and 50°C, as shown in Figure 4.4.

TABLE 4.2

Basal Spacings in nm for Wyoming Montmorillonite–Water Complexes Saturated with Various Cations at 25°C and Equilibrated at Different Relative Humidities (p/p_o)

Exchangeable Cations	p/p_o			
	0	0.5	0.7	1
Li	0.95	1.24	d	M
Na	0.95	1.24	1.51	M
NH$_4$	1	d	d	1.5
K	1	1.24	d	1.5
Cs	1.2	1.28	1.28	1.38
Mg	0.95	1.43	d	1.92
Ca	0.95	1.5	1.5	1.89
Ba	0.98	1.26	1.62	1.89

Source: Information from Norrish, K., 1954, *Faraday Soc.*, 18:120–134.
Note: d = diffuse reflections, M = macroscopic swelling, that is, >4 nm.

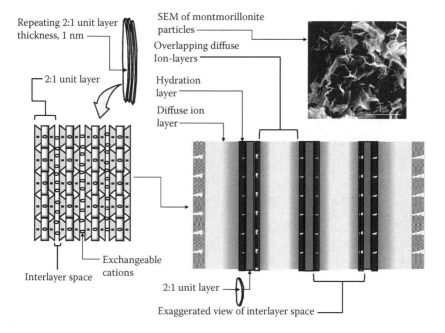

FIGURE 4.3

Basic repeating 2:1 unit layers for montmorillonite showing expanded view of repeating layers and exaggerated view of interlayer space with hydration layers and overlapping diffuse ion layers. Note that many researchers consider swelling due to the role of hydration layers as "crystalline swelling," and swelling due to actions from overlapping diffuse ion-layers is called *diffuse double-layer* (DDL) *swelling*.

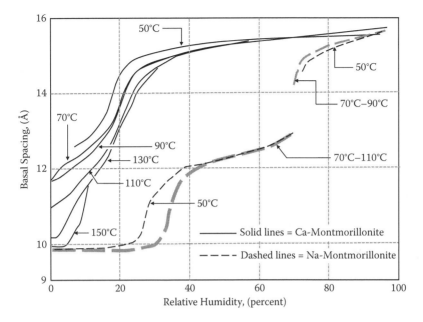

FIGURE 4.4

Effects of temperature on the basal spacing of montmorillonite $[Na_{0.42}Ca_{0.068}K_{0.08}(Al_{1.56}Mg_{0.31}$ $Fe(III)_{0.09}Fe(II)_{0.01})(OH)_2(Si_{3.91}Al_{0.09})O_{10}]$ in Kunimine bentonite in relation to relative humidity. Whereas one-layer hydration appears clearly for *Na*-montmorillonite, it is obscure for *Ca*-montmorillonite, and whilst the basal spacing for *Na*-montmorillonite is independent of temperature, it decreases with increasing temperature for *Ca*-montmorillonite until two-layer hydration appears. (Adapted from information in Morodome, K., and Kawamura, K., 2009, *Clay Clay Miner*, 57(2):150–160.)

4.2.2.1 Scoping Calculations for Swelling Pressures

The total amount of swelling in a soil exposed to available water is difficult to calculate because of the need to obtain accurate information on the forces of attraction and repulsion. Interparticle repulsion decreases as swelling volume increases until a balance is reached between forces of repulsion and attraction. Calculating swelling pressures at less than a completely swollen state of the soil is easier to handle because the developed swelling pressures depend on the net forces of repulsion. The swelling pressure is calculated as the osmotic pressure due to the difference in concentration of ions between clay particles and in the outside porewater. Osmotic swelling can be calculated with certain simplifying assumptions, discussed in Section 3.3.2. These assumptions severely restrict the application of interparticle repulsion in quantitative predictions of swelling of clays, but are considered satisfactory as *scoping* calculations.

The model used for calculations of osmotic pressure due to difference in concentration of ions between clay particles and outside porewater gives only the forces of repulsion. The forces of attraction are generally neglected. The model assumes that the charged clay particles are tabular in shape and are in parallel arrangement with diffuse layers of exchangeable ions which overlap. As shown in Figure 4.3 and Figure 3.9 in Chapter 3, the overlapping of the diffuse layers of exchangeable ions results in ion concentrations between particles that are higher than that at distances far removed from the particles. The concentration of ions midway between the interacting parallel particles has been given as Equation (3.17):

$$c_c = \frac{\pi^2 c_o}{K^2 x^2},$$

(repeat, 3.17)

where $K = \sqrt{\varepsilon \kappa T / 8 \pi e^2 z_i^2 \, n_i}$, and z_i are the concentration and valence of the ith species of ion in the outside porewater (bulk solution); e, κ, and T represent the electronic charge, Boltzmann constant, and the absolute temperature, respectively; c_c and c_o are the ion concentrations midway between particles and in the outside porewater; and x is the distance. The swelling pressure P can be calculated from the Van't Hoff relationship, if the swelling pressure is defined with osmotic pressure, as

$$P = RT(c_c - c_o)$$

(4.1)

where R is the gas constant and T is the absolute temperature. If one uses thermodynamic concepts, the swelling pressure P is expressed as $P = (\mu_o - \mu_{in})/v_w$, according to the relationship $\pi = CRT = -\Delta\mu/v_w$, where μ_o and μ_{in} are the chemical potentials of the outside porewater and water between particles, respectively v_w is the specific volume of water π is the osmotic pressure, C is the molar ion concentration, and $\Delta\mu$ is the chemical potential difference.

Figure 4.2 shows that calculations using the Van't Hoff relationship for high-swelling sodium montmorillonite at low salt concentrations match measured values. However, at higher salt concentrations, the measured swelling pressures exceed the calculated values. Whilst some minor flocculation might be expected, the major culprit contributing to the difference between theoretically calculated and measured values probably results from errors in using concentrations rather than activities of the exchangeable ions, and from neglecting the tactoid structure of the clay. Since the activity coefficients are not obtainable for situations where the concentration of cations exceeds that of the anions, as compared to the case for the region between contiguous particles, the errors in using the Van't Hoff relationship cannot be strictly quantified. Bolt (1956), however, states that the magnitude of error obtained by using the Van't Hoff relationship is about 10%. One could therefore expect that calculated swelling pressures would undercalculate pressures in the high-pressure region, and overcalculate pressures in the low-pressure region.

Application of the theory of the diffuse double layer (DDL) to calculate swelling pressures requires acceptance of the ideal behaviour or the ions in the double layer (ion layer). The Boltzmann relationship discussed in Section 3.3.1 in Chapter 3 is based on the principle that the work needed to bring an ion from the bulk solution into the area of the charged clay particle can be attributed solely to the potential energy of the ion in the field of the charged particle. However, the following issues and factors are not considered: (a) the polarization energy of the ion in the electric field, (b) the energy of interaction between the ion and the surrounding ions or water molecules, (c) the energy of interaction between the ion and the particle surface, and (d) the influence of dielectric saturation in the diffuse ion layer.

The energies of interaction between ion and surrounding ions or water molecules, and the energy of interaction between the ion and particle surface have been addressed in Section 3.3.2, insofar as inclusion into an energy interaction model for calculations of swelling pressure. In regard to dielectric saturation, Grahame (1947) indicates that this will not materially affect calculations based on a constant dielectric constant—a conclusion based on the agreements obtained between his capacity measurements of mercury and calculations using the Gouy-Chapman model.

4.2.3 Swelling—Effect of Salt Concentration and Porewater Composition

4.2.3.1 Higher Valences and Mixtures of Cations

The use of the Gouy-Chapman theory and the Van't Hoff relationship for calculations of osmotic pressure would predict that an increase in porewater salt concentration would decrease the swelling of high-swelling clays. This prediction is in accord with calculations that take into account the reduction in the diffuse ion-layer thickness. However, increasing salt concentration does not have the quantitatively predicted effect in decreasing swelling pressure. This is thought to be the result of low swelling in water combined with errors in using concentrations rather than activities. Furthermore, because divalent ions and higher valence ions are present mainly in the Stern layer, it follows that calculations of swelling pressure using the DDL (diffuse double-layer) model will not accord with measured values. For montmorillonite, there is an uncertainty about the relevant surface area to use in the calculations because the particles are arranged or grouped together into tactoids, resulting thereby in swelling only between the tactoids and not between individual particles. The studies of Norrish (1954) and Blackmore and Miller (1961) show that divalent cations tend to restrict the swelling of montmorillonite to a maximum of about 0.9 nm because of stabilization of particles into tactoids or domains. Attempts made to modify calculations by taking into account tactoid surface areas have produced some level of agreement between theoretically calculated and measured values. Shainberg et al. (1971) have suggested that DDL calculations can account for such behaviour if one makes allowance for the interparticle water in the tactoids, and if the analysis is based on the external ion distributions.

Because the swelling potential of swelling clays containing mixtures of cations is closely related to the content of Na ions in the free water relative to the divalent species of cations in the same free water, quantification of swelling can be undertaken in terms of the exchangeable sodium percentage (ESP) on the clay particle surfaces, or the sodium adsorption ratio (SAR) in the porewater. The ESP and SAR are given as follows:

ESP = [(Exchangeable Na^+)/CEC]·100, where the units for Na^+ and CEC are Meq/100 g.

SAR = $Na^+/[(Ca^{2+} + Mg^{2+})/2]^{1/2}$, where all concentrations are in Meq/L.

Soils with low salt concentrations and high SAR and ESP values will show high swelling characteristics. Swelling is generally not significant unless the ESP is at least 25 or more (McNeal et al., 1968; Quirk, 1968; Aylmore and Quirk, 1959).

4.2.3.2 Influence of Exchangeable Cations

The influence of exchangeable ions on water uptake in the interlayer and swelling of montmorillonites shows decreasing uptake in the order of Li > Na > K > Rb > Cs. The decrease in interlayer swelling with increasing atomic number of alkali ions has been explained by Michaels (1959) as being due to the increasing polarisability of the cations. This means to say that swelling increases with increasing hydration capability of the cation. The individual lattice-layers of swelling clays in the anhydrous state are assumed to be bonded to one another by secondary valence-force interactions between adjacent lattice-layer surfaces, and also by bridging action due to the exchangeable cations. Exposure of the anhydrous clay to water will result in separation of the various lattice-layers, with the degree of separation being dependent on the affinity of the cations for water relative to their affinity for the lattice-layer surfaces. If the hydration tendency of the cation is high, there will be a cluster of molecules around the cation, resulting in minimizing or reducing the original bridging role of the cations. The outcome will be high water sorption and swelling of the clay. It appears that as the hydration energies of the cations increase, the degree of swelling of the corresponding montmorillonite also increases.

The reciprocal thickness K of the diffuse double-layer is a function of the ionic valence z_i and concentration n_i. Figure 4.5 shows the influence of changes in valence of the ions present the diffuse ion layer in relation to the development or determination of the total potential ψ as a function of distance x from a negatively charged clay particle surface (left panel). The right panel in the figure shows the resultant potential obtained between two interacting parallel clay particles. The curve A illustrates the situation where

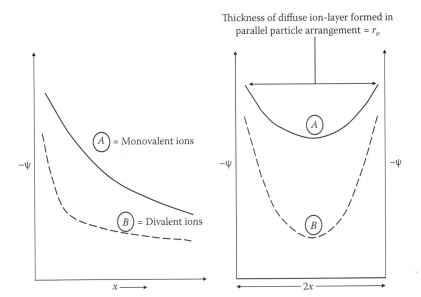

FIGURE 4.5
Influence of valency of ionic species in the diffuse ion-layers in development of total potential ψ. Left panel shows the distribution of potential ψ in relation to distance x from the surface of a net negatively charged surface of a clay particle. Right panel shows the distribution of potential ψ between two net negatively charged clay particle surfaces in parallel alignment and separated by a distance of $2x$.

the clay particle is in equilibrium with a dilute salt solution consisting of monovalent ions. Substituting the monovalent ions with divalent ions, or increasing the salt concentration in the solution, will result in a potential distribution ψ represented by curve B.

4.2.4 Compositional Effects

4.2.4.1 Surface Charge and Attractive Forces

The types and proportion of clay and nonclay minerals in a soil have significant effect on the surface properties of the soil such as specific surface area (SSA), cation exchange capacity (CEC), and the nature and location of charge imbalance due to isomorphous substitution or broken bond charges. Although surface charge density is directly proportional to CEC and inversely proportional to SSA, it is relatively constant for the various clay minerals. Swelling pressure is affected to a small degree by the magnitude of surface charge density (Low, 1968), and since short range attractive forces are theoretically proportional to surface charge density (MacEwan, 1954), one would expect that at low interparticle spacings (less than 4 nm) the effect of surface charge density would be significant for swelling of clays.

Brindley and MacEwan (1952) have shown that, as the surface charge density decreases, the swelling ability of a swelling clay increases, and clays with octahedral substitution swell more than clays with tetrahedral substitution. This is based on the fact that the attractive short range force of interaction between the negatively charges of the clay particle and the interlayer cation is greater than the repulsive force resulting from short range hydration. When the surface charge density is low, and substitutions are mainly in the octahedral layer, the short range attractive forces are small in comparison to long range repulsive forces.

The attractive force between negatively charged face-to-face surfaces and the exchangeable cations contained therein has been studied by Norrish (1954). Assuming that the relationship of the cation to the surface is similar to that of a point charge to a thick conductor, the attractive energy E_a is given as $E_a = \sigma z e / 2x\varepsilon$, where σ, z, e, x, and ε are surface charge density, ionic valence, electronic charge, distance between ions and surface, and dielectric constant, respectively. If, on the other hand, the interparticle cations and charged surfaces are considered as plates of a parallel plate condenser, the attractive energy E_a would be given as $E_a = 2\pi\sigma^2 x/\varepsilon$. Low (1968) has suggested that when the cations are far apart, the E_a obtained with the point-charge assumption is appropriate, and when the cations are close, the E_a obtained using the parallel plate condenser analogy is more appropriate.

The MacEwan (1954) analysis assumes parallel particle arrangement and that at particle surface separations less than 1 nm, the interparticle cations are arranged in a single layer. This results in the cations attracting the opposing surfaces electrically. At greater distances, with greater space availability, segregation of the cations occurs, resulting in the formation of two layers of cations—one layer being attached to each clay particle surface. For maximum separation, the cations in these layers are assumed to be arranged in hexagonal positions. Adjustment of the relative positions of the two lattices to make the mutual potential energy a minimum, the resultant attraction and attractive energy can be expressed approximately as

$$E_a = 2\sqrt{\frac{\sigma^2 ez}{\varepsilon}} \exp\left(-2\pi r_o \sqrt{\frac{\sigma}{ez}}\right), \qquad (4.2)$$

where r_o is the separation distance between cationic layers.

The E_a calculated with the above equation can be considered as the attractive interaction between the Stern layers associated with each interacting particle. Taking the distance between the interacting charged clay particles as $2x$, the Stern layer thickness δ (see Figure 3.5 in Chapter 3) is given as $\delta = 4zc_c/\kappa^2\Gamma$, where c_c and Γ are total salt concentration and surface charge density of the clay particle. Accordingly, the thickness of diffuse ion-layer formed in parallel particle arrangement of clay minerals r_o will be $r_o = 2x - 8zc_c/\kappa^2\Gamma$ (Figure 4.5).

4.2.4.2 Mineralogical Effect

The major influence of mineralogical composition in a soil is through the specific surface area and particle size. The DDL model predicts higher swelling pressures for a clay with higher specific surface area for a given composition of porewater and identical porosity. This is the outcome of the higher volume of water associated with a clay particle with higher surface area for a given interparticle spacing, resulting in a higher porosity. The relationships postulated by Dzyaloshinskii et al. (1961) suggest that the magnitude of the long-range attractive forces is not significantly influenced by mineralogy. However, mineralogical composition affects the value of the clay particle's static dielectric constant. This value varies between 4 and 8 for various clay minerals (Grim, 1953). There is a 20% decrease in the magnitude of attractive forces between the upper and lower limit values of the dielectric constant.

Mineralogy controls the nature of the resultant charge imbalance in the clay particle and whether the imbalance is from isomorphous substitution or broken bond charges. The imbalance impacts directly on the particle's proximity to cations external to the particle. There will be greater electrostatic attraction between external cations and the negatively charged mineral particle surface if there is tetrahedral substitution in which the charge imbalance is located closer to the mineral particle surface. The cations are less able to disassociate from the mineral particle surface, resulting in a lesser thickness of the diffuse ion-layer.

Broken bonds result in unsatisfied charges on the edge of the clay particle, the nature of which (negative or positive) will depend on the pH of the porewater. This type of charge imbalance is particularly important in kaolinite systems where under acid conditions, a positively charged edge may exist. The same is most likely true for illitic clays and to a much lesser extent montmorillonites (Low, 1968). It is thus possible under acidic conditions (pH < 5) for electrostatic attraction to exist between positive edges and negative planar surfaces of clay particles. It has been shown that the effect of anion adsorption on these edge sites prevents edge-edge flocculation, thereby increasing the swelling and dispersion potential of a soil (Yong et al., 1978). The adsorption of montmorillonite flakes on kaolinite particle edges may also block the positive edge charge sites, resulting in creating a soil that is more susceptible to dispersion due to disruption of edge-face association in kaolinite.

In situations where edge charge development is significant, as in kaolinite systems or where salt species are divalent (and more) and salt concentrations are high, preferred particle interactions may be nonparallel, in accordance with potential energy concepts. These conditions result in particle crosslinking, tactoid formation, and greater physical interaction between particles, resulting in "dead volume" pore spaces and water contents higher than theoretical water contents for a given value of swelling pressure (Norrish, 1954;, Shainberg et al., 1971). El-Swaify and Henderson (1967) have shown that very fine fractions (<0.2 μm) of illite, kaolinite, and vermiculite may swell in reasonable accord with theoretical predictions, provided tactoid formation is taken into account.

Free swelling of a swelling clay depends on isomorphic substitution in dioctahedral structures, with decreased swell being associated with increasing substitution (Foster, 1953, 1955). Enhanced octahedral substitution causes less cation dissociation and hence less swell. Isomorphous substitutions are related to the b-dimension of the clay unit cell (Radoslovich, 1962). Davidtz and Low (1970) have reported that maximal swelling is linearly related to the b-dimension, and that this b-dimension increases as the water content of the montmorillonite increases. The evidence shows that there is a direct clay–water interaction, which is related to the configuration of the crystal lattice of the clay.

4.3 Water Uptake and Swelling

The sequence of water uptake in a swelling clay depends on the nature of the exchangeable cations in the interlayer spaces and the initial water content of the partly saturated clay. Upon first exposure to water (or water vapour), hydration processes dominate, and water sorption is an interlayer phenomenon. For swelling clays containing Li or Na as the exchangeable cation, continued uptake beyond hydration water status occurs in response to double-layer forces. Farmer (1978) indicates that most Li and Na-smectites swell in dilute solution or in water into gel-like state with average interlayer separation in proportion to $1/\sqrt{c}$, where c is the electrolyte concentration in the liquid phase.

The solvation shell surrounding small monovalent cations consist of about 6 water molecules if the solution is dilute (Sposito, 1984). This reduces to about 3 for concentrated solutions. Primary and secondary solvation shells obtained in the case of divalent cations move together as a solvation complex. The primary shell consists of about 6 to 8 molecules whilst the secondary shell contains about 15 water molecules. Sposito (1984) reports the relationship for a sodium montmorillonite suspended in $NaCl$ as follows:

$$V_{ex} = 0.5524 + 0.3046 c^{-\frac{1}{2}},$$

where the exclusion volume V_{ex} is a measure of the volume change per unit mass of soil due to the osmotic activity of the ions of concentration c in the solution. This apparent interlayer separation relationship with c and the fact that the interlayer space contains no discernable hydration states, allows for a separation to be made between crystalline swelling and double-layer swelling phenomena. This means to say that expansion beyond four layers of water is most likely due to osmotic forces causing dilution of the ionic concentration in the interlayers.

The different physical nature of interlayer water in montmorillonites containing different species of cations is believed to be because the stacks of layers in *Ca*-montmorillonite, for example, are much stronger and less easily disrupted than in *Na*-montmorillonite. In general, interlayer water is believed to be more viscous than ordinary water and to be largely immobile under normal hydraulic gradients. As pointed out earlier, the crystal lattice constitution determines the charge and coordination of adsorbed cations and water molecules, and the amounts of hydration sites. We can distinguish between interlayer and extralayer water by undertaking spin-echo proton NMR measurements. Results from such measurements suggest that the spin–spin relaxation time T_2 is around 20–40 μs for protons in interlayer water in montmorillonite as compared to 2.3 s in free water (Pusch, 1993). This difference indicates strong structuring and the very limited mobility of interlayer water, and the small difference in proton mobility of 1 and 3 hydration layers indicates that the interlayer water possesses approximately the same physical properties, irrespective of the number of hydration layers.

4.3.1 Nature of Interlayer Water—From Molecular Dynamics Simulation

The nature of interlayer water has been the subject of study by soil scientists over the ages. A recent technique uses molecular dynamics (MD) simulations of a clay–water system to gather information on molecular positions and velocities at the microscopic level. For clay–water systems, the information gathered includes the clay mineral with its multiple interlayer water molecules. The following discussion on MD studies of the nature of interlayer water takes its cue from the presentation given in Yong et al. (2010).

MD simulation consists of solution of the Newtonian motion equation for all the atoms in the system to obtain their time-dependent behaviour. Introduction of the proper relationships between macroscopic properties and the position and velocity of all the atoms in the system allows one to deduce the various macroscopic properties of a clay–water system such as free energy, heat capacity, and adsorption sites, as well as density, viscosity, and so forth. Solution of the Newtonian equation requires one to have knowledge of the forces acting for all the atoms. Since these forces are assumed to be derived from the potential energies of interaction, it is very important to choose the proper form of interaction potential energies for all atoms. The interactions associated with certain water molecules in a clay–water system are assumed to come (a) from all atoms around the water molecule, (b) from atoms associated with the clay minerals, and (c) from the surrounding water molecules.

The interaction energy acting between two atoms is called a *pair interaction potential energy*. The pair interaction potential energy in water system by Kawamura (1992) and Kumagai et al. (1994) has been applied to a clay–water system by Ichikawa et al. (1999) and Nakano and Kawamura (2006). The total interaction energy for a certain atom in a clay–water system is given as the sum of the pair interaction potential energy that consists of the five components: (a) Coulomb, (b) van der Waals, (c) the short-range repulsion energy, (d) the radial covalent bond energy between O-H, Si-O and Al-O, and (e) the angular covalent bond for H-O-H bond of water molecule and the Si-O-Si bond of clay minerals.

The interaction potential energy for a pair of atoms is given as follows:

$$u_{ij}(r_{ij}) = \frac{z_i z_j e^2}{4\pi\varepsilon_o r_{ij}} - \frac{c_i c_j}{r_{ij}^6} + f_o(b_i + b_j)\exp\left(\frac{a_i + a_j - r_{ij}}{b_i + b_j}\right)$$

$$+ D_{1ij}\exp(-\beta_{1ij}r_{ij}) + D_{2ij}\exp(-\beta_{2ij}r_{ij}) + D_{3ij}\exp[-\beta_{3ij}(r_{ij} - r_{3ij})^2] \quad (4.3)$$

where r_{ij} is the distance between two atoms; ε_o is the dielectric constant of vacuum; z is ionic valence; e is the elementary electric charge; f_0 is a constant for unit adaptations between each term (6.9511×10^{-11} N); a, b, and c are parameters defined for each atom; D_1, D_2, and D_3 are parameters establishing the magnitude of each potential; r_{3ij} is the threshold distance activating the dissociation of the OH bond; and β_1, β_2, and β_3 are parameters. The first term represents the Coulomb potential, and the second is van der Waals potential. The third is the short-range repulsion potential. The fourth, fifth, and sixth terms represent each component of the radial covalent bond potential. The fourth and fifth are repulsive and attractive terms, respectively. The sixth term is introduced only to stabilize the covalent bond of OH.

The angular covalent bond potential energy for the H-O-H bond of water molecule and the Si-O-Si bond of clay minerals is given as follows:

$$u_{jik}(\theta_{jij}, r_{ij}, r_{ik}) = -f_k\left\{\cos[2(\theta_{jij} - \theta_o)] - 1\right\}\sqrt{k_1 k_2} \quad (4.4)$$

$$k_l = \frac{1}{\exp[g_r(r_{ijl} - r_m)] + 1}, \quad (4.5)$$

where θ_{jij} represents the angle of H_w-O_w-H_w and Si-O_c-Si; θ_0 is the adjustment parameter to determine the angles of H_w-O_w-H_w or Si-O_c-Si; k_1 and k_2 are introduced to provide the effective range of the three-body potential; subscripts 1 and 2 imply each of two H_w or two Si atoms, r_{ijl} represents the inter-atomic distance between O-H or O-Si; r_m is an adjustment parameter for determination of the bond length of O-H or O-Si; g_r is an adjustment parameter for determination of the magnitude of the effective range of the three-body potential; and f_k is a parameter.

Application of potential energy to MD simulations requires determination of the proper values of the parameters. These parameters, which are considered to differ slightly with the components of atoms in minerals, have to be determined for each mineral, using, for example, trial-and-error methods that consider the physical and chemical characteristics of the minerals. An example of the values for these parameters can be seen in the study by Nakano and Kawamura (2006) for caesium beidelite.

4.3.1.1 Interlayer Water

Interlayer water possesses physical properties that are different from those of bulk water, especially at locations near the surface of clay minerals because of the strong interactions between clay minerals and water molecules. The interaction energies involved include (a) Coulomb, (b) van der Waals, (c) short-range repulsion, and (d) covalent bond energy acting between atoms, that is, O, H, Si, Al, and metals of clay minerals, as well as O and H of interlayer water. The results shown in Figure 4.6 obtained with the help of

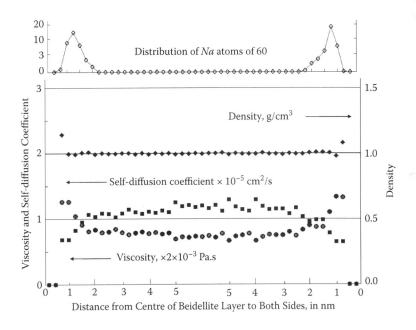

FIGURE 4.6
Density, viscosity, self-diffusion coefficient, and distribution of Na in interlayer water of Na adsorbed beidellite ($Na_{0.5}Al_2(OH)_2(Si_{3.5}Al_{0.5}O_{10})$) that have been calculated using molecular dynamics (MD) simulations. Simulations have been performed for a clay–water system that includes a half layer of beidellite at both sides of the interlayer water of about 10.7 nm in thickness in a box of $51.9 \times 53.8 \times 116.8$ Å. The total number of atoms was 32,460; the number of H_2O molecules was 10,000 and number of adsorbed Na atoms required to maintain electric neutralization of system was 60. The number of O, Si, Al, and H consisting of a clay mineral were 1,440, 420, 300, and 240, respectively. (Results obtained and offered by Prof. Katsuyuki Kawamura, Tokyo Institute of Technology.)

molecular dynamics (MD) simulations, describe the density, viscosity, and self-diffusion coefficient of interlayer water at a distance from surface of clay minerals. The results, obtained by Kawamura (2010), were calculated for the clay mineral beidellite for the case where the thickness of the interlayer is 10.7 nm. At the surface of a clay mineral, the density and the viscosity are larger. The self-diffusion coefficient is smaller than those located about 2.0 nm away, and at distances beyond 2.0 nm, the coefficients have the same values as those of bulk water. No adsorbed *Na* atoms are found beyond the 2.0 nm distance. The results indicate that the thickness of the electric diffuse double layer formed by the adsorbed metals will be about 2.0 nm, and that the flow of bulk water will occur in about 70% of the interlayer space, with the rest being taken up by the diffuse double layers at both sides of the inter-layer space. It is important to realize that since the chemical components in minerals differ with each type of clay, the thickness of diffuse double layer will depend on the clay minerals involved.

4.4 Water Movement

4.4.1 Wetting Front Advance

The linear relationship for *rate of wetting front advance* shown in Figure 3.23 (Chapter 3) for water uptake and advance in soils does not apply in the case of swelling clays. Linearity in the rate of wetting front advance is applicable for situations where no volume change occurs in the soil upon water uptake and advance. The wetting front advance into a natural clay loam containing a small proportion of montmorillonite in respect to the square root of time is shown in Figure 4.7. For a no-volume change sample, the relationship for the rate of wetting front advance vis-a-vis the square root of time should be linear, as shown previously in Figure 3.23 and also in Figure 4.6 (rigid porous material). Absence of linearity, which allows one to use the Boltzmann trans-form $\lambda = x/\sqrt{t} = \lambda(\theta)$ in the similarity solution technique to reduce Equation (3.27) in Chapter 3 into an ordinary differential equation, makes it improper to use this transformation to seek solutions to the unsaturated flow relation-ship stated as Equation (3.27).

When free swelling of the soil occurs during water uptake and advance, departure from linearity between wet-front advance and the square root of time occurs. As one would expect, the higher the swelling capability (poten-tial) of the sample, the greater is this departure, as shown by the results for the use of mixtures of the clay with sterile glass beads in Figure 4.6. Restrictions placed on swelling volume change have direct impact on the development of the wetting front profile, as shown in Figure 4.8. Comparison of the wetting front profiles between free-swell and no-swell (completely

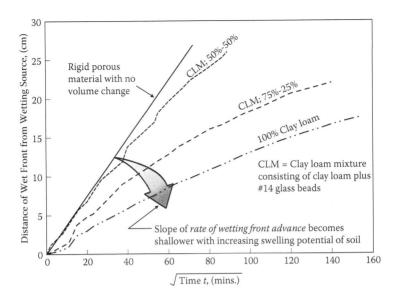

FIGURE 4.7

Influence of swelling clay content on *rate of wetting front advance* of a clay loam containing a small proportion of montmorillonite. Results show clay loam wetting front advance with different proportions of glass beads (50% clay and 50% glass beads; 75% clay and 25% glass beads) compared with 100% clay loam itself.

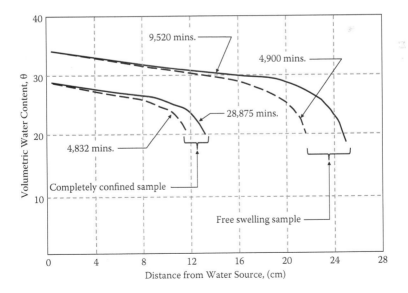

FIGURE 4.8

Comparison of free-swelling and confined sample wetting profiles for a clay loam. Both samples are similar except for a slightly higher initial density for the completely confined sample.

confined) show that the rate of advance of the wet front is reduced significantly for a completely confined sample. The higher the initial density of the clay, the greater will be the reduction in rate of advance of the wet front, with the same type of swelling clay and with the condition of *no volume change*.

4.4.1.1 Analysis Using Material Coordinate System

The relationship given in Equation (3.27) in Chapter 3 does not apply when measurable volume change of the clay occurs during wetting. The material coordinate system used by Philip and Smiles (1969), which allows one to consider volume changes during water uptake and advance, uses a relationship between the spatial coordinate x and a material coordinate m. This material coordinate m is defined as $dm/dx=1/(1+e)$ where e is the void ratio. The Darcy model used in this treatment considers the velocity of water relative to the moving clay particles v_{ws} as $v_{ws} = -k\partial\psi/\partial x$. The external pressure that acts on the soil particles is the total potential ψ. This total potential includes the overburden potential Ω due to any surface load. In the system of three phases consisting of water, pore (gaseous), and solid, the continuity condition is described as

$$\left[\frac{\partial}{\partial t}\left(\left(1+e\right)\theta\right)\right]_m = -\frac{\partial v_{ws}}{\partial m} \tag{4.6}$$

It follows therefore, using the chain rule of derivatives (Philip and Smiles, 1969):

$$\frac{\partial}{\partial t}\left(\left(1+e\right)\theta\right) = \frac{\partial}{\partial m}\left(\frac{k}{1+e}\frac{\partial\psi}{\partial m}\right)$$
$$= \frac{\partial}{\partial m}\left(D_m\frac{\partial e}{\partial m}\right) \tag{4.7}$$

where $D_m = (k/(1+e))\,(d\psi/de)$.

For water movement in a two-component system, that is, a fully saturated state, the relation $e=\theta/(1-\theta)$ is determined for a fully saturated condition. Assuming that, $D(\psi) = k(\psi)(\partial\psi/\partial\theta)$ D_m reduces to the following form,

$$D_m = \frac{D(\psi)}{\left(1+e\right)^3} = (1-\theta)^3 D(\psi) \tag{4.8}$$

Using the water ratio $\vartheta = (1+e)\theta$ to refer to the volume of water per unit volume of solid, Equation (4.7) reduces to:

$$\frac{\partial \vartheta}{\partial t} = \frac{\partial}{\partial m}\left(D_\vartheta \frac{\partial \vartheta}{\partial m}\right) \tag{4.9}$$

where $D_\vartheta = (k/(1+e))\,(d\psi/d\vartheta)$, and ϑdm is the volume of water per unit cross section of dm.

In a clay–water system where the initial water content of the clay is uniform, the water content at the surface in contact with incoming water will change as time progresses due to swelling constraints at the far end. The initial and boundary conditions are given as follows:

$$\vartheta = \vartheta_i = \text{const. at } (dm/dx) > 0 \; t = 0$$
$$\vartheta = f(t) \text{ at } (dm/dx)_{x=0} = 0 \; t > 0$$

where ϑ_i is the initial water ratio.

4.4.1.2 Soil Particle Diffusivity Analysis

Analysis or prediction of clay swelling upon water uptake and advance can be undertaken by considering the swelling in terms of volumetric strains. Considering the status of an elemental clay volume of dimensions dx, dy, and dz, and one-dimensional unsaturated flow in the x-direction, clay particles flowing into and out of the elemental volume must satisfy conservation principles. We use the designation of v_{cx} as the velocity of clay particles in the x direction, that is, clay particle flux in the x direction. Accordingly, we can write

$$\frac{\partial v_{cx}}{\partial x} = \frac{1}{1-n}\frac{\partial n}{\partial t} = \frac{\partial \varsigma}{\partial t} \tag{4.10}$$

where n is the porosity and ς represents the volumetric strain. Designating Φ as the internal pressure responsible for movement of the clay particles because of the developing fluid flow, one obtains:

$$v_{cx} = k_c \frac{\partial \Phi}{\partial x} = +k_c \frac{\partial \Phi}{\partial \varsigma}\frac{\partial \varsigma}{\partial x}$$
$$= D_\sigma \frac{\partial \varsigma}{\partial x} \tag{4.11}$$

where $D_\sigma = k_c(\partial \Phi/\partial \varsigma)$ is the clay particle diffusivity coefficient, and k_c is the clay particle conductivity coefficient. Substituting Equation (4.11) into Equation (4.10) gives us when D_σ is independent of the spatial coordinate x:

$$\frac{\partial \varsigma}{\partial t} = D_\sigma \frac{\partial^2 \varsigma}{\partial x^2} \tag{4.12}$$

Subject to the initial and boundary condition:

$$\varsigma = \varsigma_o \quad x = 0 \quad t > 0$$
$$\varsigma = 0 \quad x > 0 \quad t = 0$$

The analytical solution will be obtained as:

$$\varsigma(x,t) = \varsigma_o \left[1 - erf\left(\frac{x}{2\sqrt{D_\sigma t}} \right) \right] \tag{4.13}$$

The same clay loam used to obtain the results shown in Figures 4.7 and 4.8 has been used to obtain the results shown in Figure 4.9. In this case, the experimentally obtained volumetric strains are for the 50–50 clay loam mixture. Theoretical predictions of the volumetric strain at time $t = 2{,}625$ minutes after water entry are shown in the figure, seeking to match the measured values at that same time interval. The calculations are made for a constant D_σ and also a variable D_σ.

4.4.2 Water and Solid Content Profiles in Confined Swelling Clays

The water content profiles for water uptake into air-dried confined swelling clays are significantly different from those of nonswelling soils. The reasons are that (1) water in the liquid state and vapour move simultaneously in the

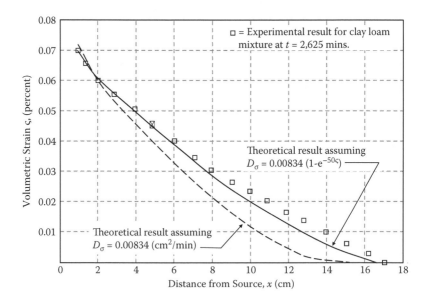

FIGURE 4.9
Comparison of measured and theoretically calculated volumetric strains in a clay loam mixture (50% clay and 50% glass beads) during unsaturated flow.

clay mass, (2) the rate of liquid water movement is relatively low in comparison with the rate of vapour movement, and (3) vapour transfer which precedes liquid water movement reaches farther positions in the clay mass, condenses onto the surfaces of clay particles, and rapidly increases the water content of the farther location.

The profiles of water content that are formed with the combined vapour-and-water movement—with vapour transfer ahead of water movement—will show ill-defined (obscure) wetting fronts. At the same time, significant changes in the dry density will occur in the clay mass because of swelling resulting from transfer of both liquid and vapour. The dry density at the location adjacent to the water supply will decrease markedly, whereas the part distant from the water supply increases gradually when the clay mass is completely confined at the boundaries of the mass as shown in Figure 4.10. The measurable dry density changes can be analyzed using a concept of the flow of clay particles since the changes are large enough to fall outside the bounds of mechanical deformation theory. The changes in dry density with time can be simulated using the clay particle diffusivity coefficient D_σ that

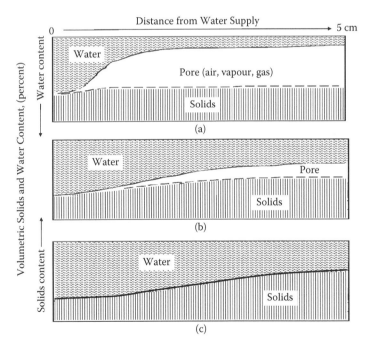

FIGURE 4.10
Distribution of water content and solids content observed in a confined bentonite (Kunigel-V1) with initial dry density of 0.846 g/cm³ and initial water content of 0.078 g/cm³ (air-dried condition). The illustration shows the relative proportions of solid, water, and pore air or vapour or gas after a defined period of water uptake. Beginning from the top, the diagram shows the distribution of solid, water, and pore air–vapour–gas (a) after one day of water uptake, (b) after 26 days of water uptake, and (c) the profile estimated at infinite time ($t = \infty$).

has been calculated from the changes of solid profiles with time obtained in unsteady state experiments (Nakano et al., 1986).

On the theory that clay particles will flow due to the gradients of the total soil–water potential ψ, it is assumed that: (a) $v_{cp} = -k_{cp}(\partial\psi_{cp}/\partial x)$, where subscript cp expresses the amount of clay particles, (b) $(\partial\psi_{cp}/\partial x) = -\delta(\partial\psi_w/\partial x)$, and (c) δ is a coefficient. In combination with the continuity equation, one obtains:

$$\frac{\partial\sigma}{\partial t} = \frac{\partial}{\partial x}\left(D_\sigma(\sigma,\theta,D_\theta)\frac{\partial\sigma}{\partial x}\right) \tag{4.14}$$

where σ is the volumetric clay particle content, D_σ is the clay particle diffusivity coefficient, and D_θ is the total water diffusivity coefficient.

The initial and boundary conditions are

$$\sigma = \sigma_i = \text{const. at } x > 0, t = 0$$

$$\sigma = f(t) \qquad \text{at } x = 0, t > 0$$

where σ_i is the initial volumetric clay particle content.

4.5 Concluding Remarks

4.5.1 Swelling and Water Movement

Swelling of clays can be visualized as the result of repulsion between contiguous clay particles. Adsorbed water layers in interlayer spaces and around particles are responsible for the swelling of most swelling clays, and extended swelling beyond the adsorbed water layers can be ascribed to osmotic swelling due to DDL forces. When this occurs, the potential associated with this double-layer expansion is the osmotic potential ψ_π. Interlayer expansion due to adsorbed water is determined by the layer charge, interlayer cations, properties of adsorbed liquid, and particle size. The mechanism for sorption from very low water contents or from the anhydrous state is due to (a) the presence of charge sites and exchangeable cations in the interlayers, and (b) the attractions between water molecules and the polar surface groups.

When particles are in random orientation, repulsion occurs at the points of closest approach, but considerable water is also held by capillary forces in pores between microstructural units. For this reason, swelling volume is greater than for parallel arrangement of particles if osmotic swelling is small. The amount of osmotic swelling and the swelling pressure are however dependent on the amount of surfaces available for interaction. For high-swelling clays, the parallel particle orientation results in the greatest swelling.

The movement of water in swelling clays is a complicated phenomenon to describe because of the interactions between particles, unit layers, and water. Water can move in either the liquid or vapour phase, with the latter being more important for relatively dry clays. The most common flow condition is unsaturated flow, where the water is at negative potentials or when air is present in the larger voids. The existence of the microstructural units provides for sinks/sources, which will affect total movement of water through the clay. To account for these, and to take into consideration the various sets of driving forces associated with the various thermodynamic potentials at the micro level, it is necessary to adopt a micro-mechanistic approach to the modelling of water movement in the system.

Water uptake and advance in swelling clays can occur when the hydraulic heads are negative. This means to say that one does not need to have positive water pressures acting on a swelling clay for water to move into the clay. Water uptake by swelling clays and particularly by swelling clays is due to suction forces associated with the matric ψ_m and osmotic ψ_π potentials. The literature has used the terms *matric potential* and *matric suction* to mean the same thing. Hillel (1998) cites the definition of the *matric suction of soil* in accord with the ISSS (International Soil Science Society) Terminology Committee, "as the negative gauge pressure, relative to the external gas pressure on soil water, to which a solution identical in composition with the soil solution must be subjected in order to be in equilibrium through a porous membrane wall with the water in the soil." By this definition, one assumes that the physical measurements of suction obtained are matric suctions, and by extension of this reasoning, it has been generally argued that the matric potential ψ_m or the matric suction can be considered to be the total soil–water potential ψ (total suction).

For high swelling clays, because a major contributor to clay swelling is due to osmotic phenomena, it is not clear that one could equate the matric potential ψ_m with the total soil–water potential ψ. The simple osmometre experiment shown in Figure 3.20 in Chapter 3 demonstrates that the solute potential (solute suction) is an active component of the total suction. As we have indicated in Chapter 3, the terms solute potential ψ_s and the osmotic potential ψ_π are used interchangeable. For low-swelling clays, it is possible to ignore the contribution of the osmotic potential ψ_π to the total soil–water potential, without significant loss of accuracy. However, for high swelling clays, it is reasonably argued that ψ_π cannot be ignored, especially when swelling of the clay is a major issue and more so when swelling is restrained.

4.5.2 Laboratory Measurement of Swelling

The dependence of swelling pressure on volume change makes a precise measurement of swelling pressure difficult. It is not uncommon for miniscule imperceptible volume changes to occur in laboratory swelling pressure devices during swelling pressure tests, resulting in measured pressures that will underestimate the real (actual) swelling pressures. A pressure-measuring device actuated by very small volume changes is required. There is no standard universally accepted device for measuring swelling pressures. Amongst the strict requirements for a laboratory swelling pressure apparatus are (a) a rigid robust confining cell and (b) access to water or solution through a micro-porous ceramic stone or water-permeable membrane at one end and at the other end a solid or fluid piston which allows for sensing of pressures. A typical example of such a device is shown in Figure 4.11. The procedure for determination of swelling pressure will depend on whether the test sample is initially wet or initially dry. The two test procedures can be identified as the *consolidation technique* or the *swelling technique*. These are shown in Figure 4.12. Whilst the curves in the right-hand portion of Figure 4.12 might resemble consolidation-rebound curves, they are by no means the same.

Beginning with low water contents or in the dry state, exposure to water for clay swelling does not produce instantaneous full swelling pressures. Swelling pressure increases very slowly because of the mechanisms of water uptake and movement in the interlayer spaces and between particles. If an initially air-dry sample is being tested, the complete hydration of the unit layer and particle surfaces will require some considerable time. Full

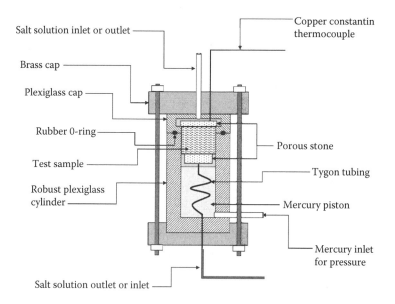

FIGURE 4.11

Cross-section of swelling pressure device. Note that the mercury inlet for pressure application has a capillary tube attachment that allows one to monitor potential miniscule movements of mercury piston.

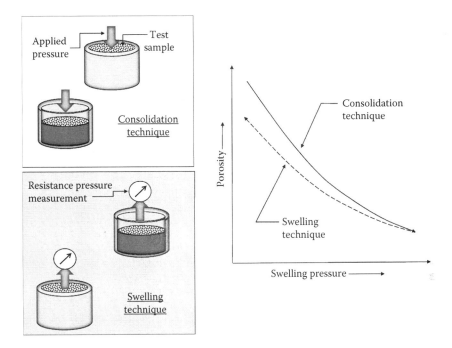

FIGURE 4.12
Consolidation and swelling techniques for determination of swelling pressure of soil sample. Note that the typical graphical results shown are not unlike those obtained in compression-rebound tests. The different results between the two techniques can be attributed in large measure to soil structural influence.

swelling pressure will not be developed until hydration is complete. The swelling pressure in such type of devices is not measured instantaneously. Most swelling pressure devices depend upon some volume change for actuation. This expansion of the test sample requires that water moves into the sample, and since the permeability of swelling clays is low, the necessary distribution of water under small pressure gradients will require some considerable time.

In the schematic illustration shown in Figure 4.12, the upper sketch shows the sample under an equilibrating consolidation pressure for measurement of the corresponding water content. The corresponding equilibrium consolidation pressures are considered as the swelling pressures for that particular water content or porosity. The lower sketch shows the swelling technique where a compact sample is exposed to available water but is restrained from swelling. Allowing controlled swelling to various equilibrium volumes or water contents provides one with measurements of corresponding restraining or resistance pressures, which are considered as swelling pressures. The graphical sketch shows the corresponding types of results obtained. It is more accurate to describe the swelling pressure measurement of a soil as an operationally defined measurement, meaning that the results obtained are dependent on how the measurements are made.

References

Aylmore, L.A.G., and Quirk, J.P., 1959, Swelling of clay systems, *Nature*, 183:1752–1753.

Blackmore, A.V., and Miller, R.D., 1961, Tactoid size and osmotic swelling calcium montmorillonite, *Soil Science Soc. Am. Proc.*, 25:169–173.

Bolt, G.H., 1956, Physico-chemical analysis of the compressibility of pure clay, *Geotechnique* 6:86–93.

Brindley, G.W., and MacEwan, D., 1952, Structural aspects of the mineralogy of clays and related silicates in ceramics, *Proc. of Symposium, British Ceramic Society*, pp. 15–19.

Davidtz, D.L., and Low, P.F., 1970, Relation between crystal lattice configuration and swelling on montmorillonites, *Clay and Clay Minerals*, 18:325–332.

Dzyaloshinskii, I.E., Lifshitz, E.M., and Pitaevskii, L.P., 1961, General theory of van der Waals forces, *Soviet Physics, Vspeaki*, 73:381–422.

El-Swaify, S.A., and Henderson, D.W., 1967, Water retention and osmotic swelling of certain colloidal clays with varying ionic composition, *J. Soil Sci.*, 18:223–232.

Farmer, V.C., 1978, Water on particle surfaces, in Greenland D.J., and Hayes M.H.B. (Eds.), *The Chemistry of Soil Constituents*, chap. 6, Wiley, New York, pp. 405–448.

Grahame, D.C., 1947, The electrical double layer and the theory of electrocapillarity *Chem. Rev.*, 41:441–501.

Grim, R.E., 1953, *Clay Mineralogy*, McGraw Hill, New York, 384 pp.

Hillel, D., 1998, *Environmental Soil Physics*, Academic Press, San Diego, 771 pp.

Ichikawa, Y., Kawamura, K., Nakano, M., Kitayama, K., and Kawamura, H., 1999, Unified molecular dynamics and homogenization analysis for bentonite behaviour: current results and future possibilities, *Eng. Geol.*, 54:21–31.

Kawamura, K., 1992, Interaction potential models for molecular dynamics simulations of multi-component oxides, F. Yonezawa, (Ed.) *Molecular Dynamics Simulations*, Springer Series in Solid State Sciences, 103:88–97.

Kawamura, K., 2010, Private communication to M. Nakano on Na-adsorbed beidellite results from MD simulation study (see Yong et al. 2010).

Kumagai, N., Kawamura, K., and Yokokawa, T., 1994, An interatomic potential model for H_2O systems and the molecular dynamics applications to water and ice polymorphs, *Mol. Simulations*, 12(3–6):177–186.

Low, P.F., 1968, Mineralogical data requirements in soil physical investigations, *Soil Sci. Am.* Special Publ. 3, pp. 1–34.

MacEwan, D.M.L., 1954, Short range electrical forces between charged colloid particles, *Nature*, 174:39–40.

McNeal, B.L., Layfield, D.A., Norwell, W.A., and Rhoades, J.D., 1968, Factors influencing hydraulic conductivity of soils in the presence of mixed salt solutions, *Soil Science Soc. Amer. Prof.*, 32:187–193.

Michaels, A.S., 1959, Physico- chemical properties of soils: Soil water systems, *J. Soil Mech. Found. Div., Proc. Amer. Soc. Civil Eng.*, 85:91–102

Mooney, R.W., Keenan, A.C., and Wood, L.A., 1952, Adsorption of water vapour by montmorillonite, II: Effect of exchangeable ions and lattice swelling as measured by x-ray diffraction, *J. Amer. Chem. Soc.*, 74:1371–1374.

Morodome, K., and Kawamura, K., 2009, Swelling behaviour of Na- and Ca-montmorillonite up to 150°C by in situ X-ray diffraction experiments, *Clay Clay Miner*, 57(2):150–160.

Nakano, M., Amemiya, Y., and Fujii, K., 1986, Saturated and unsaturated hydraulic conductivity of swelling clays, *Soil Sci.*, 141(1):1–6.

Nakano, M., and Kawamura, K., 2006, Adsorption sites of Cs on smectite by EXAFS analyses and molecular dynamics simulations, *Clay Sci.,* 12(Suppl. 2):76–81.

Norrish, K., 1954, The swelling of montmorillonite, Discussion, *Faraday Soc.*, 18:120–134.

Philip, J.R., and Smiles, D.E., 1969, Kinetics of sorption and volume change in three-component systems, *Aust. J. Soil Research*, 7:1–19.

Pusch, R., 1993, Evolution of models for conversion of smectite to non-expandable minerals, SKB Technical Report, TR 93-33, SKB, Stockholm.

Quirk, J.P., 1968, Particle interaction and soil swelling, *Isr. J. Chem.*, 6:213–234.

Radoslovich, E.W., 1962, The cell dimensions and symmetry of layer lattice silicates, *Am. Mineralogist*, 47:617–636.

Shainberg, I., Bresler, E., and Klausner, Y., 1971, Studies on Na/Ca montmorillonite systems, I, The swelling pressure, *Soil Sci.*, 3:214–219.

Sposito, G., 1981, *The Thermodynamics of Soil Solutions*, Oxford Press, New York.

Sposito, G., 1984, *The Surface Chemistry of Soils*, Oxford Press, New York, 234 pp.

Suquet, H., de la Calle, C., and Pezerat, H., 1975, Swelling and structural organization of saponite, *Clay Clay Miner*, 23:1–9.

Yong, R.N., and Warkentin, B.P., 1975, *Soil Properties and Behaviour*, Elsevier Scientific Publishers, Amsterdam, 449 pp.

Yong, R.N., Sethi, A.J., Ludwig, H.P., and Jorgensen, M.A., 1978, *Physical Chemistry of Dispersive Clay Particle Interaction*, ASCE, Preprint 3379.

Yong, R.N., Pusch, R., and Nakano, M., 2010, *Containment of High-Level Radioactive and Hazardous Solid Wastes with Clay Barriers*, Taylor & Francis, Spon Press, London, 468 pp.

5

Stressors, Impacts, and Soil Functionality

5.1 Introduction

The stressors discussed in this chapter are *soil environment stressors*. The two basic questions of interest are "What constitutes soil environment stressors, or where do stressors come from?" and "What are their impacts on short- and long-term soil properties and behaviour?" These questions provide the basis for the discussions provided in the various sections of this chapter. The central issue in seeking answers to the two basic questions revolves around the problem of determining a soil parameter that can describe the quality of a soil. The parameter chosen for such a task—to provide a qualitative and/or quantitative index of soil quality—is the *soil functionality index (SFI)*.

The two immediate questions of interest are: (a) what constitutes a soil functionality index (*SFI*)? and (b) will soil functionality be unchanged, degraded, or enhanced by soil environment stressors acting on a soil mass of interest? From Section 1.5.1 of Chapter 1, *soil functionality* has been defined as the capacity of a specific soil to function under designed circumstances to meet its planned intentions or requirement without any loss of its original functional capability. Geoenvironmental engineering projects involving in situ and prepared soils or transported soils (buffers and clay barriers, embankments, etc.), and other projects involving interactions with soils, are generally designed or planned with the assumption that the soil properties and characteristics of the candidate soils remain relatively unchanged over the working or service life of the soil for the particular project (i.e., soil functionality remains unchanged). Should changes in soil functionality occur, it is important to be able to track the changes and to make the necessary adjustments or provide the required tools to accommodate the changes.

Soils are living dynamic systems with characteristics and properties that are the products of their: (a) origin (i.e., source material), (b) type and level of maturation (i.e., evolution), (c) location, (d) climate and environmental envelope, and (e) bio-ecosystem and human activities. Starting from their initial formation, they encounter various processes and undergo various kinds of reactions such as mechanical, geochemical, and biogeochemical, the nature and extent of which are functions of (a) their geologic and hydrogeologic

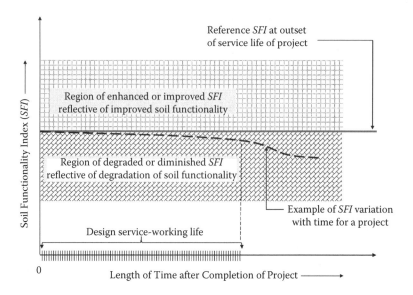

FIGURE 5.1
Example of time-*SFI* variation for an imaginary project. Over the course of the "design service-working life" shown on the abscissa, the *SFI* shows some level of degradation. The question that needs to be asked is whether the degraded *SFI* is at a level that can be tolerated.

settings, (b) their climatic and environmental envelopes, and (c) biological and anthropogenic activities. The sum total of all of these processes and factors is responsible for (1) the nature of a soil at any one instance of time, and (2) the changing nature of soils with time, that is, the evolution or maturation of soils.

Figure 5.1 shows a simple schematic that illustrates the problem at hand, using the SFI as a measure of the consequences of impacts from the soil environment stressors. The SFI, which utilizes soil functionality indicators, will be discussed in detail in a later section. The time-SFI variation shown in the schematic diagram for an imaginary project points out that some degradation in SFI occurs during the design-service life of the project. The question that needs to be asked is whether the various soil functionality indicators have degraded to such an extent that the integrity of the project is threatened.

5.2 Stressor Sources and Stressors

5.2.1 Soil Environment

To fully appreciate the nature and impact of soil environment stressors that can act on a soil mass, it is important to recall that the integrity and status

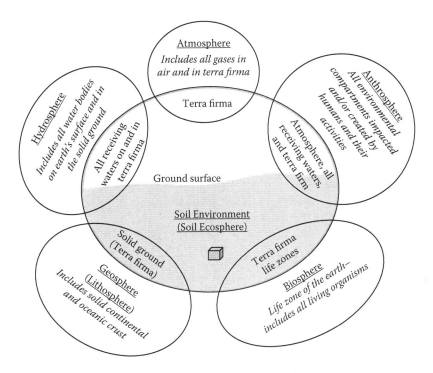

FIGURE 5.2
Venn diagram showing the various elements from the atmosphere, hydrosphere, biosphere, anthrosphere, and geosphere that form the soil environment.

of a soil mass is defined by its environment. The Venn diagram in Figure 5.2 shows the various elements from atmosphere, the geosphere, hydrosphere, biosphere, and anthrosphere that make up the soil environment. The soil environment includes the various soil fractions discussed previously in Chapter 2. Soil water, the aqueous elements (porewater, aquifers, and groundwater), solutes, and microorganisms interact with the various soil fractions and all the other soil environment elements that define the status of a soil mass, its integrity, and its properties. Any change in any of the components or elements because of stresses caused by soil environment stressors will obviously affect the integrity and properties of a soil mass in the soil environment. In turn, soil behaviour will be affected and the design service or working life will be imperiled if soil functionality indices are reduced.

5.2.2 Types of Stressor Sources and Stressors

Simply speaking, stresses on soils are due to various agents and sources that include physical, thermal, hydraulic, and chemical loads and forces; pressures; heat; water; foreign matters; and others. These stresses, which act on a particular piece of soil mass, originate from: (a) natural environmental events such as volcanic eruption, forest fire, landslides, hurricanes, and so forth, and (b) human-related activities such as soil contamination from waste leachates and loading from constructed facilities. The actions that result in imposition of stresses in the soil environment are called *soil environment stressors*—or simply, stressors. In the context of soils in the ground, we consider: (a) stressor sources originating from a variety of different agents, and (b) soil stressors generating different types of stresses. Since environmental stressors and stresses are terms used in many different disciplines and also to describe particular situations, we should not confuse their concept as they are applied to a soil environment. Examples of stressor sources and stressors in the soil environment are shown in Table 5.1. Each stressor shown in Table 5.1 imposes stresses on a soil mass.

The types of soil environment stressors acting on a particular soil mass can be mechanical, hydraulic, thermal, chemical, physicochemical, electrical, gaseous, radioactive, and so forth, with sources that are either natural or human-related. The Venn diagram in Figure 5.2 shows that the natural stressor sources delivering stressors can be elements from the atmosphere, anthrosphere, biosphere, and hydrosphere. In turn, it should be noted that stressors generated in the individual spheres by actions associated with human activities will affect the soil ecosphere directly. Since the soil environment is an integral part of the soil ecosphere, it follows that some form of stressor in the individual sphere will be transferred to the soil environment.

Given the manner in which soils are formed and located in the landscape, a soil mass is subject to soil environment stresses by stressors all the time. The simplest example of this is the case of overburden pressure acting on a particular soil mass located at some depth below ground surface. In this instance, this overburden pressure would be considered as a natural setting and classified as part of the group of base stressor values in the reference base—assuming no human intervention.

5.2.3 Natural Stressor Sources and Stressors

The two categories of soil environment stressor sources are *natural* and *anthropogenic* as illustrated in Table 5.1. Except for well-defined events, it is not often easy to distinguish between natural soil environment stressor sources and those associated with or related to human activities and products. The well-defined natural events that readily classify as natural soil environment stressor sources generally fall under the category of *natural disasters*. These include such events as hurricanes, earthquakes, tornadoes, volcanic eruptions, rainstorms, and blizzards, which generate various types of stressors in a soil mass. For example, earthquakes generate dynamic forces, for example, seismic force, to a soil mass through two types of body waves—primary or *P* waves and secondary or *S* waves.

TABLE 5.1

Examples of Soil Environment Stressors, Stresses, Soil Properties, and Behaviour

Stressor Sources	Stressors for Soils	Kinetics Induced in Soils	Affected Soil Properties and Behaviour
Natural sources			
Volcanic eruption	Ejecta, heat	Alteration of soil	Soil elements, water retention
		Heat transfer	Thermal and hydraulic properties
Forest fire, deforestation	Heat	Heat and water transfer	Soil structure, water retention
		Alteration of soil elements	Thermal and hydraulic properties
Landslides, floods, and avalanches	Soil elements, water	Mass runoff	Soil elements, soil structure
Cyclical temperature and drying	Heat, water	Heat and water transfer	Soil structure, thermal and hydraulic properties
Anthropogenic sources			
Industrial effluents	Leachates	Adsorption, chemical reaction	
Mining and metal processing	Contaminants	Water and solute transfer	Chemical composition, soil structure, and hydraulic properties
Constructed facilities			
Buried waste barrier	Water, solutes, heat	Water, solute, and heat transfer	Hydraulic, chemical, and thermal properties
Structural loads	Pressure	Shear stress, consolidation	Mechanical properties and behaviour
Agricultural land use			
Fertilizing	Chemical compounds	Adsorption, chemical reaction	Chemical composition, soil structure,
Irrigation and drainage	Water	Water and solute transfer	Salt accumulation, water retention
River embankments	Overburden pressure	Mass runoff, shear stress, consolidation	Hydraulic properties, mechanical behaviour
Underground water use	Water, solutes	Adsorption and desorption	Chemical composition, salt accumulation
Abnormal climate			
Gas component in air	CO_2, CH_4, N_2O, etc.	Adsorption, gas transfer	Chemical composition, soil structure
Air temperature	Heat	Heat transfer	Water retention
Acid rain	Acid precipitation	Chemical reaction, solute transfer	Hydraulic properties

Volcanic eruption and forest fires are also classified as typical natural disasters that will generate various types of stressors. They cover land surface with ejecta and ashes, apply extreme high temperature or heat, and chemical components constituting ejecta and ashes to soil surface that will migrate into soil and underground water through percolating rainwater. Since such heat and chemical compounds are defined as stressors, volcanic eruption and forest fires can be classed as stressor sources.

Landslides and floods, which are often classified as natural disasters, can be classified as soil environment stressor sources. They are disasters that result from situational vulnerability to provocative events such as earthquakes and rainstorms. Landslides and floods can also occur in instances where human activities have created vulnerable circumstances. Deforestation of slopes could make them more susceptible to landslides, and harvesting of groundwater in low-lying coastal regions can lower ground surfaces to levels that invite flooding. Landslides and floods transport various matters, such as soil components and relevant chemicals, to downstream areas. It is more appropriate to classify landslides and floods as stressor sources responsible for generating such stressors as chemical elements, in the soil environment.

Cyclical events such as regular temperature (cold–hot) cycles and nonregular wetting–drying cycles that fall under the category as natural events do not normally classify as natural disasters. Regular cyclical events are events such as winter–summer cycles, and nonregular events are wetting–drying events where repeat cycles do not adhere to any time-calendar schedule. It would be a mistake to exclude any of these cyclical events from consideration as soil environment stressor sources. Take, for example, the freezing temperatures that accompany winter seasons. In many jurisdictions, under the right conditions, ground uplift due to frost heaving causes considerable damage to overlying structures. The stressor source is subzero freezing temperatures (winter), and the soil environment stressors associated with this stressor source are thermal in nature. In the case of wetting–drying events, one can consider the normal rainfall-drying periods as one type of cyclical event, or the more extreme cases of deluge and drought, where the deluge results in severe flooding and the drought event will result in parched-earth conditions. The types of stressors generated in the affected soil mass include thermal, hydraulic, chemical, and mechanical.

5.2.4 Anthropogenic Stressor Sources and Stressors

Anthropogenic soil environment stressors are very diverse and perhaps too numerous to individually classify or identify. The sources of stressors that are classified as anthropogenic include various industries such as mining and resource exploitation, and human activities such as constructed facilities, disposal and land management of waste, as well as agricultural land use, underground water use, and abnormal climate, as shown in Table 5.1. For this discussion, we will consider two distinctly different types of anthropogenic sources of soil environment stressors: (a) an upstream–downstream set of industries and agricultural land use where the stressors act directly on the soil environment, and (b) a constructed facility that utilizes a specific type of soil mass and where the facility itself is the source of the stressors acting on the soil mass.

5.2.4.1 Industry Sources of Stressors

The various kinds of industries that produce goods for human consumption include both those that produce raw goods and those that provide finished products. The activities and industries can be broadly grouped into two groups: upstream and downstream. The following definitions apply:

- Upstream industries are those industries that produce the raw goods and source materials for downstream industries. Production of food, that is, agricultural production of food (wheat, corn, barley, livestock, etc.), and mining and processing of metal ores are good examples of upstream activities and industries. The raw products are used as resource material for downstream industries.

- Downstream industries include (a) those industries that use the raw goods produced by the upstream industries as resource material for other downstream industries, and (b) production and assembly plants and industries that produce consumer goods and products that are directly utilized by the individual and collective consumers such as buildings, bridges, automobiles, leather goods, newsprint, electronic products, processed food, and so forth.

There are various stages in the operations of both upstream and downstream industries that serve as soil environment stressor sources. Some of the more prominent stressor sources are shown in the illustration in Figure 5.3 for an upstream–downstream set of activities involving metal mining, processing, fabrication, and preparation operation. The main stressor sources for the upstream metal-mining industries include (a) operations associated with quarrying and mining shown in the left side of the figure, (b) the various process waste streams associated with recovery, milling, digestion, clarification, separation, thickening, beneficiation, and so forth that are discharged in liquid sludge and tailings form into holding ponds, and (c) rejects, debris, oversize, and the like that are piled into *tips* that provide inviting sources of acid leachates, giving rise to acid mine drainage problems.

Figure 5.4 shows the processes involved in the development of acid mine drainage—for reactions with pyrite (FeS_2). For the downstream metal fabrication and preparation industries, the main stressor sources include (a) wastewater from metal fabrication processes, cutting, and forming, and (b) sludges and wastewater from processes and operations associated with anodizing, electroplating, chemical conversion, plating, painting, and so forth, discharged into holding ponds.

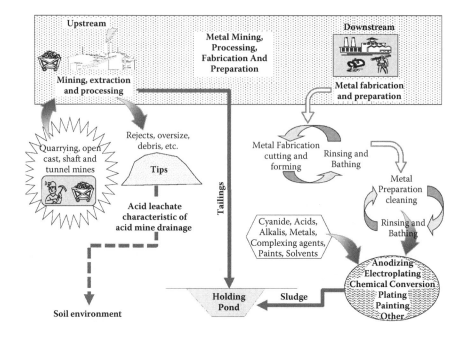

FIGURE 5.3
Example of some of the more prominent types of stressors originating from metal mining, processing, fabrication, and preparation. Not shown in the illustration are the various wastewater discharges at the various stages of extraction-processing and rinsing-bathing sequences.

The types or form of stresses developed in the soil environment take their cue from the types of stressors generated by the various stages of processing and production in an industrial activity. The examples shown in Figures 5.3 and 5.4 show (a) that upstream metal mining operation creates stressors that will generate chemical reactions in the soil through the production of acidic iron and sulphate-rich fluid, and (b) that sludges derived from downstream industry operations will include many of the agents such as cyanide, acids, alkalis, metals, paints, solvents, and complexing agents used in the final preparation of the metals. These are inorganic and organic chemicals, and heavy metals—meaning that the stressors in the soil environment will be chemical and physicochemical in nature.

5.2.4.2 Constructed Facility—Geologic Containment of HLW

A good example of constructed facilities as sources of stressors is the facility contemplated for containment and isolation of high-level radioactive waste (HLW). One of the proposed schemes for disposal of spent fuel from nuclear reactors, considered as high-level radioactive waste, involves construction of deep repositories in competent geologic rock formations. The deep geologic burial of high-level radioactive wastes requires buffer separation between

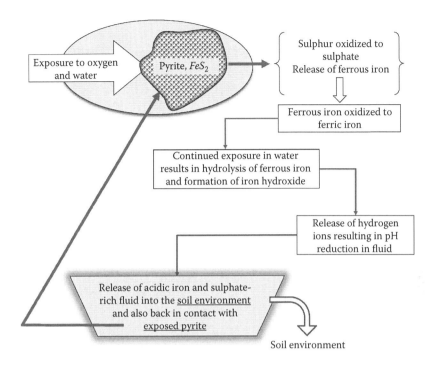

FIGURE 5.4
Effect of exposure of pyrite to oxygen and water. Continued exposure to water will result in generation of iron hydroxide (yellowboy) and acidic solution that will be released to the soil environment. Similar reactions shown in the diagram will also occur for sulphides of copper, lead, arsenic, cadmium, and zinc.

the spent fuel canisters and the host rock that serves as a repository. The schematic illustration in Figure 5.5 shows the deep burial location of tunnels in a host rock. Spent high-level radioactive waste canisters are placed within regular-spaced boreholes in the tunnels—as shown in the bottom right-hand portion of the figure. The top right-hand sketch shows how the smectite clay in the borehole is used as a buffer separation between the canister and the host rock.

For deep geologic burial of high-level radioactive wastes, there are three significant sources of stressors that act on the smectite clay buffer: (a) long-lived radioactive nuclides in the spent fuel canisters that would become stressors on the buffer if the canister should somehow lose some integrity—giving rise to fugitive radionuclides, (b) long-lived heat emission from the spent fuel, (c) groundwater input from the surrounding host rock, and (d) pressure imposed from the surrounding host rocks such as overburden pressures and seismic forces. Since both radioactivity and heat emissions from the spent fuel last for more than 100,000 years, design specifications require considerations of corrosion products from the canisters, loss of integrity of the canisters, saturation of the initially unsaturated

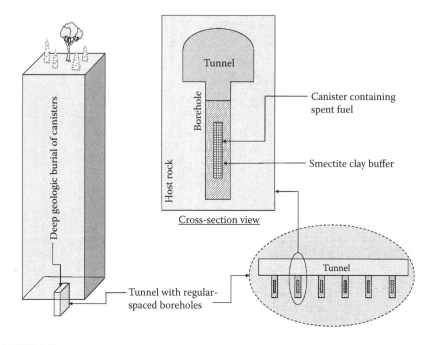

FIGURE 5.5
Illustration of deep geologic burial of spent fuel canisters. The bottom right-hand corner shows
a tunnel with regularly spaced boreholes with contained canisters, and the middle drawing
shows the cross-section view of the containment system. The smectite clay buffer separates the
canister from the host rock, and the tunnel is generally in-filled with a clay that could be mixed
with other granular materials.

smectite clay buffer, swelling pressures exerted on the canister, transformation of the clay minerals, and so forth. The schematic diagram shown
in Figure 5.6 illustrates some of the major stressor sources. The types of
stressors generated in the smectite clay buffer over the short and long term
include (a) thermal, (b) chemical (e.g., chemicals in groundwater, corrosion
products, and fugitive radionuclides), (c) mechanical (e.g., overburden pressures, seismic forces, and swelling pressures generated from smectite clay
buffer itself due to migration of underground water), and (d) radioactive—
from fugitive radionuclides.

5.3 Stressor Impacts

5.3.1 Impacts on Soil Environment

Stresses in the soil environment due to the action of various stressors act on
all the constituents in the soil environment. In effect, the soil environment

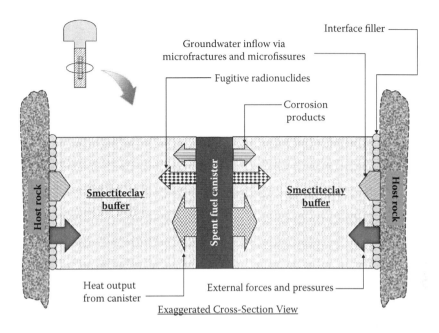

FIGURE 5.6
Illustration of stressor types acting on the smectite clay buffer. Whilst the host medium is a competent rock, it is expected that over the long term, the presence of microfissures, and so forth will permit inflow of groundwater into the initially unsaturated clay buffer.

needs to be viewed in terms of the soil ecosphere (pedosphere). One needs to examine stress effects or impacts on not only the soil solids but also on the fluid phase and the microorganisms in the soil–water system—since the imposed stresses can affect their habitat and nutrient sources. Different stressors produce different responses in the soil environment—for example, reactions, transfer processes, and mechanical dynamics—and generate alteration and changes of the soil environment—for example, changes of soil properties and behaviour as illustrated in Table 5.1. The alteration and changes of soil environment are called *impact* on the soil environment, leading to serious natural disasters, damage of constructed facilities, changes of land use, and degradation of food production and human health.

Figure 5.7 provides a pictorial view of some of the major short- and long-term reactions, processes, and impacts in the clay barrier and sub-base environment of an engineered multibarrier landfill liner system, generated by the waste pile as the principal stressor source. The stressor impact scenario assumes that the clay barrier will be stressed because of failure of the overlying multibarrier system, that is, the barriers (filter, double-membrane, and leachate collection system) above the clay barrier, to completely fulfil their required functions. Such being the case, leachates emanating from the waste pile will be the primary stressors, and the major impacts on the clay barrier or the sub-base soil will be hydraulic, mechanical, chemical, physicochemical,

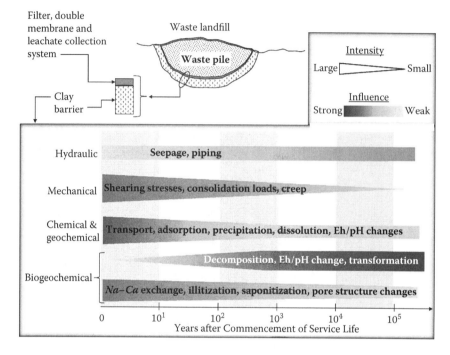

FIGURE 5.7

Example of some of the major stressors, stresses, and impacts in a barrier-liner system. Diagram at top of figure shows a waste landfill with an engineered liner-barrier system consisting of a filter, double membrane, and leachate collection barrier system overlying an engineered clay barrier.

and biogeochemical in nature—in relation to degradation and decrease in barrier functionality.

5.3.2 Impacts on Soil Mass

Because of the countless numbers and types of stressors and the variety of applications involving a soil mass, it is not feasible or practical to consider the nature of individual impacts generated or developed by each type of stressor in terms of alteration and changes in soil properties and behaviour—as illustrated in Table 5.1. Instead, it is more practical to consider the actions or phenomena as impacts triggered by stressors—generically and generally—on a soil mass. By and large, actions, events, or phenomena will generate reactions and processes in the soil by stressor groups that can be categorized as (a) hydraulic, (b) mechanical, (c) thermal, (d) chemical, (e) geochemical, and (f) biologically mediated. Because the range of impacts and the specific type of impact on a soil mass will depend on situational circumstances and specific *soil use* application, the following discussion can only list some of the main impacts generated by the types of stressor groups.

5.3.2.1 Hydraulic

The stressors that are classified under the *hydraulic* group are directly related to water and its actions in soils. The types of actions generated are primarily pressures that act directly on the soil solids and affect water movements. The outcome of the various types of actions could:

- Initiate piping that could undermine the stability of overlying structures and facilities
- Trigger erosion, landslides, and *quick* soil conditions from excessive porewater pressures, self-detachment of soil particles
- Restructure affected soils because of the pressures resulting in changes in soil properties and behaviour
- Dilute or decrease contaminant concentration in contaminated soils
- Initiate or increase advective transport of contaminants in the soil
- Detach sorbed contaminants and contribute to the environmental mobility of the contaminants
- Affect biological processes through changes in natural habitat and energy sources
- Influence natural soil-weathering processes leading to alteration or transformation of susceptible minerals

Other actions classed under the hydraulic stressor grouping could include those associated with floods, droughts (lack of water as a hydraulic stressor), excessive rainfall, and water availability detrimental to agricultural productivity, and others.

5.3.2.2 Mechanical

In respect to soils, the main types of stressors included in the *mechanical* group are those whose actions give rise to pressures and stresses in a soil mass. Natural stressor sources are earthquakes and avalanches, and anthropogenic sources include constructed facilities. The actions and phenomena generated by these stressors in the mechanical group include

- Direct loading from a solid mass such as an overlying structure—for example, a bridge abutment and foundation footings for a structure, or facility. Resultant effects in the affected soil include:
  - Collapse of overlying structure due to failure of the soil to support the applied load
  - Settlement of the overlying structure due to consolidation, secondary compression, and/or creep of the supporting soil

- Pressures on a soil mass as a result of actions related to water movement—for example, swelling pressure in confined swelling soils and pressures developed in the soil in unstable slopes:
  - For swelling soils under overlying structures, swelling pressure could undermine stable support of the structures, thereby causing collapse of the structure.
  - For unstable slopes, instability of the slope will result in slope movement or slope failure.
- Soil freezing and frost heaving pressures resulting from formation of ice lenses. The consequent effects are similar in principle to those experienced with swelling soils. In the case of ice lens formation, the consequent effects from subsequent thawing of the ice lens can be severe if ice lenses formed in the freezing stage have created significant frost heaving in the soil.

5.3.2.3 Thermal

Stressors classifying under the *thermal* group generate heat or cooling in soils. The obvious natural stressor source is the summer–winter cycle. A significant anthropogenic source is canisters containing high-level radioactive wastes embedded in underground repositories generating heat over long periods of time. Two particular sets of issues deserve close consideration: (a) high temperatures, and (b) freezing temperatures. Actions or phenomena from these include

- Water and vapour transfer in soils resulting in changes in soil properties and behaviour
- Types and rates of chemical reactions and biological processes resulting in changes in the nature and energy status of the affected soil—consistent with soil weathering processes
- Freeze-thaw phenomena and development of ice lenses in frost susceptible soils with water availability and thaw subsidence from disappearing ice lenses

5.3.2.4 Chemical

Organic acids obtained from decomposing surficial organic matter, in combination with the natural chemical constituents in soils, constitute natural stressors. Waste landfills, discharge of contaminated wastewater, acidic leachates from mine heaps, acid rain, and others are stressors attributable to human activities. The impact of actions and/or events in the chemical environment in soils will be felt in terms of changes in the nature and properties of the soil because of changes in (a) interactions between soil particles and between particles and water, and soil-water energy characteristics (Sections

3.3 and 3.4 in Chapter 3), (b) biological processes in the soil, leading to changes in the nature and character of the soil (Section 2.8 in Chapter 2), and (c) chemical reactions and processes (Section 3.6 in Chapter 3).

5.3.2.4.1 Geochemical

The stressors in the geochemical category are leachates or contaminants from mine heaps, constructed facilities and excess fertilization in agriculture. The impact of actions or phenomena in a soil mass are (a) decomposition of soil minerals and chemical constituents in soils, and (b) changes of soil particle surface functional groups (Section 2.4.2, Chapter 2). These decompositions and changes result in alteration of soil structure and chemical constituents in soil mass, thus affecting water, solute, and heat transfer phenomena in soils.

5.3.2.4.2 Biologically Mediated

In this category, the stressors are microorganisms associated with chemical constituents and their interactions in soils. The impact of their interactions is decomposition of organic matter and alteration or decomposition of clay minerals and chemical constituents of soil. In turn, changes in soil structure would occur, resulting in corresponding changes in not only hydraulic, mechanical, and thermal phenomena in soils but also the relevant soil properties.

5.4 Soil Functionality Index (SFI)

Soils in nature and of constructed soil facilities will (a) continue to change their properties and mode of behaviour under stresses imposed by stressors, or (b) undergo various kinds of effects from the stressors. The original qualities and functionalities of a soil, determined at a certain time, will not be maintained as time progresses. This is the essence of soil maturation (i.e., the evolution of soil in the context of its environmental stressors). In the real world, soils must continue to perform their specific design or planned function. This means that utilization of soils for specific purposes or projects must take into account soil functionality, and especially long-term soil functionality, recognizing that maturation processes will result in evolution of the soil, and hence, changes in soil functionality.

In Chapter 1, *soil functionality* has been defined as the capacity of a specific soil to function under designed circumstances to meet its planned intentions or requirement without any loss of its original functional capability. The concept of using soil functional capability, that is, the ability of a soil or a site to function according to design or service requirements, is, in a sense, a novel concept, in that it addresses the performance aspects of a specific piece of soil or a soil mass. The soil attributes used in assessment of soil functionality are (a) properties and characteristics of the particular soil under consideration,

TABLE 5.2

Examples of Soil Functionality Indicators (*SFIs*)

	Soil Functionality Indicator
Soil properties and characteristics	
Hydraulic	Particle size, aggregates, density and porosity, water content, colour
Mechanical	Particle size, aggregates, density and porosity, water content
Thermal	Soil constitutions, clay minerals particle size, aggregates, density and porosity, soil constituents
Chemical and geochemical	Chemical components, clay minerals, organic matter, colour
Biological	Surface functional groups of soil constituents, aggregates, density and porosity, organic matter species
Performance requirement for soils	
Hydraulic	Permeability, unsaturated conductivity, water diffusion coefficient
Mechanical	Deformation coefficients, elasticity and plasticity coefficients, settling and subsidence speed
Thermal	Thermal or heat conductivity, specific heat capacity
Chemical and geochemical	Diffusion/dispersion coefficient pH, Eh, electric conductivity
Biological	Cation/anion exchange capacity, biomass (phospholipid), species (DNA), soil respiration, biodiversity

and (b) performance requirements of the soil to meet design or service specifications over the short- and long-term periods. To determine soil functional capability—that is, soil functionality—one needs to introduce the use of the *soil functionality Indicators* (SFI) as an assessment tool. To appreciate how the SFI can be usefully utilized as a tool in assessment of soil functional capability (a) over the long term, or (b) in relation to stresses generated by soil environment stressors, one needs to specify soil functionality indicators such as the examples given in Table 5.2. These indicators constitute the parameters for computing soil functionality, using, for example, single-parameter analysis, dimensional analysis, multivariant analytical techniques, risk-based analysis, fuzzy logic, or lumped-parameter analysis.

5.4.1 Soil Functionality Indicators

Indicators are used in everyday events and situations. Perhaps the most common and familiar type of indicators is the group of sensory indicators such as aural, visual, scent, and sensation. For those who operate vehicles in a traffic situation, green lights at cross-roads indicate that one has the right of way, and red lights indicate that one should stop at the intersection. However, it must be stressed that indicators inform one of the status or nature of the situation at hand. They can also be used to provide guidance or insight into the performance of a system or even a particular piece of equipment. Returning to the operation of a vehicle, the loud clanking sound that one hears emanating from the engine of the vehicle is an indication that something may be wrong with the engine.

In soil science, agriculture, and the earth environment, indicators are often used in soil quality and soil functionality assessments, with particular interest to soil health in relation to agricultural productivity, human health, and preservation/improvement of habitats and biodiversity. Soil quality indicators can range from complex and involved techniques using microorganisms as indicators of soil health, to more simple ones such as the use of soil colour to determine the soil constituents or components, the water content, and water logging status of soil. Bending et al. (2004), for example, used microbial and biochemical soil quality indicators to differentiate between areas under contrasting agricultural management, whilst Blavet et al. (2002) studied the use of soil colour variables as indicators of duration of water logging.

In soil and geotechnical engineering, indicators are derived from soil tests structured to determine various properties and characteristics of soils, and soil behaviour. These include tests for determination of the physical, mechanical, hydraulic, and frost susceptibility—to name a few. However, whilst these properties have been used, as in the design and construction of various kinds of facilities, it has not been a common practice in soil and geotechnical engineering to consider them as *indicators* of soil functionality, except for evaluation of settlement and ground subsidence. The primary reason perhaps is because all the properties determined from laboratory and field testing have been obtained for the purpose of design and construction of a particular facility, and follow-up monitoring. In geoenvironmental engineering, where issues of soil contamination by pollutants and toxicants and land settlement and subsidence by overburden pressure from constructed facilities and excess withdrawal of ground water are of significant concern, indicators are used to monitor the status of soils threatened by pollutants and toxicants, and by ground subsidence. In this case, they are used as signals to inform the stakeholder that the system is functioning well, or to alert the stakeholder to potential problems.

Figure 5.8 illustrates some major soil attributes adopted as indicators in multi-barrier systems used for containment of waste pile in a landfill. The illustration shows that the major indicators are grouped into biological, chemical, and physicomechanical categories, together with clay barrier performance as major indicators.

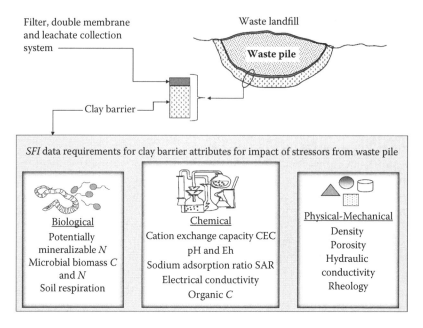

FIGURE 5.8
Major soil attributes required as *SFI* data for clay barrier component of multibarrier system used for containment of waste pile in a landfill. Data from these attributes serve as input for determination of *SFI* of clay barrier.

5.4.1.1 Selection of Indicators

At the present time, in geoenvironmental, soil, and geotechnical engineering, there is no hard and fast rule governing the selection of soil functionality indicators to be used in a project—the primary reasons being the highly diverse types of projects involved, and the lack of a formal structure for development of indicators as a whole. By definition, a *soil functionality indicator* provides one with the basis for assessing the status of the specific function as determined by the attribute characterized by the indicator. There are several levels of specificity (i.e., levels of detail) in selection of soil functionality indicators from the various soil attributes—that is, physical, mechanical, hydraulic, chemical, geochemical, biogeochemical, and other attributes processes and characteristics. The selection process and the number of indicators chosen will be governed by the type of *soil use* project—that is, the role of the soil in the project under consideration—and the specific function or functions that the soil is designed to fulfil. Whilst other nonsoil considerations such as economics, time constraints, and expediency can sometimes enter into the selection process, these should not be allowed to interfere with the soil use considerations, since in most instances, soil use considerations are central to safety assessment requirements.

In many instances, it may not be necessary for one to adopt the entire list of available indicators for assessments relative to design performance or service requirements. This is because of the likelihood that the status of some performance indicators can be traced to the relevant soil properties responsible for the observed soil performance. The hydraulic, mechanical, and thermal indicators required for performance assessment are basically functions of indicators such as particle size, aggregates, dry density, porosity, water content, soil constituents—especially, clay minerals, and organic matter. These basic soil properties and characteristics are the building blocks for the soil performance required for utilization of soil as an agricultural-resource material or for engineering purposes.

5.4.2 Indicators and Soil Functionality Index

Indicators serve as soil functionality parameters in assessment of soil performance. In general, one should use multiple indicators for assessment of soil performance because of the various phenomena that occur simultaneously in soils. For example, to evaluate the possibility or threat of collapse of a constructed soil embankment, one must consider hydraulic attributes and their associated indicators in addition to mechanical attributes and their related indicators in developing the relevant functional status of the soil under consideration. As another example, assessment of soil pollution requires the use of hydraulic indicators and chemical or geochemical indicators for development of the appropriate SFI.

SFIs can be defined as a ratio of the value X_n at time t_n to a reference base soil functionality value X_{base} for each indicator, that is, X_n/X_{base}, where X_{base} is set as a value determined at the outset of the project or as determined by the design requirements. For soil attributes that report values of X_n at t_n to be less than X_{base}, as, for example, loss of soil strength or bearing capacity, values of the index (*SFI*) below one would indicate that soil performance has degraded to a point below *satisfactory* performance. There are some soil attributes that could report values of X_n at t_n to be greater than X_{base}, as for example, increase in the permeability of a clay barrier with elapsed time. For such cases, it is obvious that *SFI* values above one would mean performance below *satisfactory*—assuming that design specifications require the clay barrier permeability to be maintained at levels conforming to, or even below, specifications. In most instances, however, soil attributes will report values of X_n at time t_n less than X_{base}. What this says is that one must always be aware of "what constitutes degradation of a particular soil attribute." The limit *SFI* is given by X_{limit}/X_{base}, where X_{limit} is the limiting value for degradation of the particular soil or soil facility.

Assessment of soil performance can take several forms—depending on the particular aspect of performance chosen. For complex interactions of various phenomena such as soil and groundwater pollution by leachates and contaminants from landfills, mining, and various kinds of industrial facilities, the myriad of processes and interactions involved in contaminant transport in soils (see Chapter 9) need hydraulic, chemical, mechanical, and biological indicators to develop the SFI characterizing soil performance. In such cases, it is necessary to seek a unifying indicator for the SFI. For the soil and groundwater pollution problem for example, one could use the soil *assimilative capacity* (AC) as an indicator. There are many ways to structure a unifying indicator such as the AC. One popular method is weighting of individual indicators in the unifying SFI based on their relative importance and relative degree of participation (Figure 5.9). A simple summation of the weighted indicators would yield the unifying SFI. There are other techniques for obtaining unifying indicators for SFIs. These are higher-order techniques such as the ones shown in the bottom portion of Figure 5.9 for assessment of soil performance and functionality. Simulation of relevant phenomena and modelling of reactions or performance could provide one with indicators or parameters with some degree of accuracy.

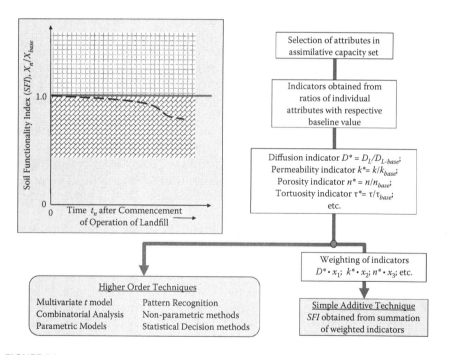

FIGURE 5.9

Examples of procedure for determination of *SFI* variation for functional capability of clay barrier component of a multibarrier system for a landfill depicted in Figure 5.8. The left-hand box shows the simple one-index method for tracking the SFI based on contaminant concentration indicators. The SFI selection protocol shown in the right-hand portion of the figure uses various attributes constituting the assimilative capacity set for determination of indicators and the SFI, with the aid of simple or higher order analytical techniques.

5.4.3 Application of SFI—An Example

Groundwater management and potential soil pollution from a waste landfill serve as useful examples to demonstrate the use of the SFI. In the example of a waste landfill shown in Figure 5.8, there are at least two key groups of functional requirements for the clay barrier component of the multibarrier liner system used for the landfill. These are (a) impermeable supporting layer as interface between in situ base and double-membrane liner system, and (b) contaminant attenuation barrier. For the purpose of demonstrating a procedure for selection of soil functionality indicators, the contaminant attenuation barrier function will be used. The drawing shown in Figure 5.10 illustrates the *performance control* aspect governing the design and construction of multibarrier systems in general, and the clay barrier component and the surrounding host soil in particular. The limit boundary surrounding the landfill establishes the limiting spatial extent of contaminant plume advance. This defines the *control region,* meaning that one must obtain information on the chemistry of the groundwater in the control region for various time periods to determine whether the contaminant plume has reached or breached the limit boundary. In most instances, the indicator for assessment of contaminant plume advance will be contaminant concentration c in designated positions in the control region. The values of c_n at the various time intervals tn, and at the various locations serve as indicators. One could consider these as *soil functionality indicators* since the values of c_n obtained at any time depend on how well the soil functions in (a) contaminant attenuation, and (b) hydraulic conductance control. The limiting contaminant concentration c_{limit} that is generally specified as the control limit establishes the point where contaminant concentrations cannot exceed c_{limit} at all times.

The measured contaminant concentration values c_n obtained by the multisampling monitoring station at x_2 distance from the landfill in time t_n after commencement of landfill operations are, technically speaking, *performance indicators* (see Figure 5.10). Since the soils involved in contaminant plume advance into the surrounding region include both the engineered clay barrier component of the multibarrier system and the surrounding host soil, these performance indicators are the soil functionality indicators for both the performances of clay barrier and the surrounding soil. The choice of c_n as a soil functionality indicator informs one of the status of performance of the clay barrier and the host soil. It does not inform one on how well or how poorly the clay barrier is performing its design function. Nevertheless, it serves as a primary soil functionality indicator since this alerts the operator of the landfill facility of the functional capability of the facility, and to determine whether corrective measures are needed to comply with regulatory safety issues. Discussions on corrective measures such as *pump and treat* and *permeable reactive barriers* (PRBs) as remedial techniques can be found in textbooks dealing with contaminant mitigation and remediation techniques (e.g., Yong, 2001; Yong and Mulligan, 2003).

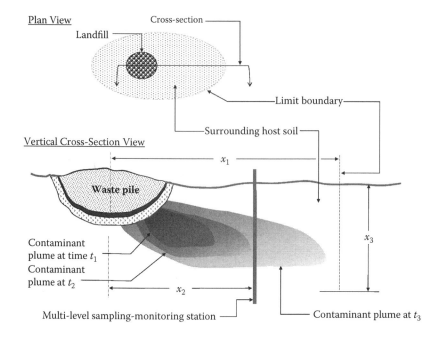

FIGURE 5.10
Example of waste landfill clay barrier component of multibarrier system showing the limit boundary as the performance control boundary. The top sketch is the plan view of the landfill, and the bottom sketch is the vertical cross-section. x_1 is the distance from the centre of the landfill to the limit boundary, x_2 is the distance to the multi-level sampling-monitoring station, and x_3 is the depth to which the limit boundary extends. The limit boundary defines the permissible contaminant plume advance and the limiting target contaminant concentration.

To provide information on the performance of the clay barrier or the host soil, it is necessary to select soil functionality indicators that refer specifically to these soils—in addition to the use of the primary indicator c_n. From c_n, the corresponding soil functionality indicators are determined from back-analysis of a contaminant transport relationship to determine the relevant soil properties that have allowed c_n to reach x_2 in time t_n.

An example of such a relationship is given as follows:

$$\frac{\partial c}{\partial t} = \frac{\partial}{\partial x}\left(D_L \frac{\partial c}{\partial x}\right) - v\frac{\partial c}{\partial x} - \frac{\rho}{n\rho_w}\frac{\partial c^*}{\partial t} \tag{5.1}$$

where: c is the concentration of the contaminant of concern in the plume, t refers to time, D_L is the diffusion–dispersion coefficient (most often called *diffusion coefficient*), v is the advective velocity, x is the spatial coordinate, ρ is the bulk density of the soil, ρ_w is the density of water, n is the porosity of the soil, and c^* is the concentration of contaminants adsorbed by soil fractions (see Chapter 9 for a detailed discussion of contaminant transport). The soils involved include both the clay barrier material and the supporting host soil.

The relevant soil functionality indicators could include one or more of the parameters used in the contaminant transport relationship. Selection of one or more of the parameters as indicators depends on the level of detail sought in determining soil functionality.

The discussions in Section 5.3.2 indicate that there are two simple formats together with sophisticated computation model methods for specification of *soil functionality index (SFI)*:

- *An indicator and its variation* from a base value $c$base.

  In the example shown in Figure 5.10, one could use the value of c_n at time t_n as the indicator. The soil functionality index (SFI) would be c_n/c_{base}. Since performance control for a safe and secure landfill requires that the concentration of contaminants cannot exceed a limit value $c$limit at the limit boundary, which is distant x_1 from the landfill, we can substitute the base value $c$base with $c$limit. This means that the SFI would be determined as $c_n/c_{base} = c_n/c_{limit}$. Values of c_n detected at the limit boundary would indicate advance of the contaminant plume, and as long as c_n/c_{limit} remains below one at this boundary, the landfill can be considered to be operating within specified safe limits. When the SFI reaches one ($c_n/c_{limit} = 1$), this means that the concentration of contaminants has reached the limit concentration values specified as limit-control, and remedial action will be required.
- *A collection of indicators* that form a set associated with a particular soil function.

  In Figure 5.9, for example, the assimilative capacity of the soils (including both clay barrier and host soil) is a major factor in control of the advance of the contaminant plume. The major attributes constituting the set associated with the assimilative capacity (AC) include (a) diffusion coefficient D_L, (b) Darcy coefficient of permeability k, (c) porosity n, (d) wetted surface area of particles S_w, (e) tortuosity of pathways τ, and (f) partition coefficient k_d.

  The indicators derived from the attributes in the *AC* set are obtained as ratios of (actual measured):(base value) for an individual indicator. Weighting of the individual indicators in the *AC* set is based on relative importance and relative degree of participation in development of the assimilative capacity of the soil. The right-hand sketch in Figure 5.9 illustrates the step-wise procedure.
- *The computation models* by which indicators are derived from the attributes in the *AC* set without weighting.

  More sophisticated parametric and statistical decision models are used to obtain the *SFI* for the assimilative capacity of both the clay barrier and the host soil. Figure 5.9 shows an example of these models and methods as higher-order analytical procedures for determination of the *SFI*.

5.5 Time-Related Change of Functionality

Time-related changes refer to changes in the item or items of interest over the course of time following start-up time $t = 0$ due to various factors and influences. As shown in Figure 5.1, the time-related changes of functionality are indicated using the changes of soil attributes with time. We consider that *soil attributes*, which has been stated in Section 5.4.1, refers not only to the performance aspects of soils such as swelling, cation exchange capacity (*CEC*), and water uptake, but also to the state quantities such as dry density, porosity, water content, and others and characteristics such as morphology, texture, colour, smell, and features that describe a soil. Quantification or rating of individual attributes is generally given by constants, coefficients, indices, degrees or ranks, and various parameters (see Table 5.2), which is commonly stated as soil *attribute qualities* and used to indicate soil quality in soil mass or grade of soils for soil utilization in engineering and food production.

The quality of a soil attribute is used (a) to demonstrate design or planned requirements for the functionality of a soil in the engineering projects, and in food production in land use, and (b) to indicate their changes in the planned functional service life of the soil. A higher attribute quality, as for example a higher soil shear strength relative to some initial or reference soil shear strength, would mean a stronger soil (i.e., more resistant to shear) when viewed in terms of its planned functionality (bearing support). Take the case of the soil *hydraulic permeation* as a soil attribute characterized in terms of the Darcy coefficient k. For situations that demand water impermeable clay barriers, very low k values are desirable. The lower the k value, the greater is its quality to fulfil its functional requirements, meaning that one should strive for lower k values as higher attribute qualities in the case of hydraulic permeation of clay barriers. The lesson one learns from this is that soil attribute quality must be evaluated in relation to the attribute's functional requirement.

5.5.1 Time-Related Functionality

In soil evolution, the changing functionality of a soil with time requires one to examine time-related changes of soil attributes. This is perhaps one of the most significant issues in soil engineering and environmental soil behaviour.

To explain what we mean by time-related soil attributes, and to demonstrate why this is an important phenomenon that requires due consideration:

1. Consider a piece of soil under a constant column load P. Consolidation of the soil under the load occurs, the characteristics of which are a function of the physical and mechanical attributes of the soil, which are determined from laboratory tests on soil samples procured from the site. Predictions of time-settlement characteristics are made on the basis of the laboratory test results.

2. Next, consider that the attributes of the soil under the constant load are subject to changes because of the chemical reactions initiated by contaminants brought into the site from groundwater flowing into the site.

3. Because of #2, one of two things will happen: (a) prudent engineering and foresight in site investigation would have discovered that groundwater seepage into the site would bring with it contaminants, and thus the effect of changes in the relevant soil attributes would have been factored into the planning and design of the loaded column; or (b) the ingress of contaminants with the groundwater is either a "new" phenomenon, not originally present, or the ingress has not been considered in the planning of the facility with the constant load, under which circumstances problems (failure of the loaded column, etc.) will arise.

In the preceding illustrative example, time-related soil attributes consider "what might happen" to the attributes of a soil as time progresses. In that sense, item 3(a) demonstrates that prudent engineering has considered time-related soil attributes in its planning process. This is a demonstration of knowledge of *environmental soil behaviour* as will be discussed later in Chapter 10, that is, knowing how soil evolution will affect the time-related soil attributes, and how this (i.e., soil evolution) would affect soil functionality for the specified or planned facility/operation.

5.5.1.1 Time-Related Changes of Soil Attributes

In the case of soils, the time-related changes of importance are the changes of various soil attributes that characterize the soil in question, that is, the soil mass under consideration. Time-related changes in soil attributes refer to changes in the physical, mechanical, hydraulic, chemical, physicochemical, biogeochemical, and thermal attributes of soils. The schematic example in Figure 5.11 has selected permeability, accumulative and buffering capacities, and swelling to represent typical soil attributes and performance as soil attributes. The illustration in the figure has chosen degradative time-related changes in the quality of a soil attribute to

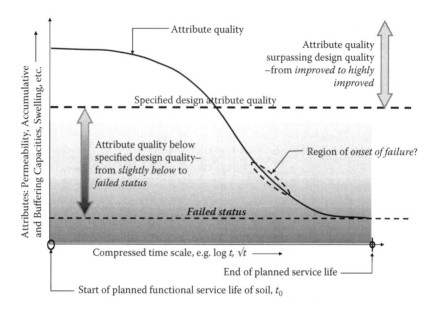

FIGURE 5.11
Illustration of variation of attribute quality in relation to time. It is assumed that the deterioration in attribute quality is due to the negative impacts from time-related changes in soil properties that define the attributes. The time-related changes may be due to stressors imposed by human-made structures or from natural ageing processes.

demonstrate their impact on the planned functional service life of the soil, that is, the capability of the soil to function according to design or planned requirements over the service life of the soil involved in the project. Beginning at time $t = t_0$, the soil attribute qualities should be equal to, or exceed, the specified design attribute qualities required for the project in question. Prudent engineering practice would require the attribute qualities to be in excess of design requirements.

Whilst the schematic example shown in Figure 5.11 has chosen to anticipate deterioration or degradation of qualities of the various soil attributes over the course of the planned functional service life of a particular soil mass—leading to a diminished functional capability of the specific soil mass—it should be emphasized that time-related changes in soil attributes' qualities can also be positive. What must be avoided is a reduction of qualities of the various soil attributes to a level where *onset of failure* occurs, assuming that onset of failure is a progressive event that eventually leads to total failure. The schematic example of a loaded footing in Figure 5.12 shows that as time progresses, improvement in qualities of the soil attributes such as density γ, permeability (coefficient k), shear strength τ, and porosity n are obtained. It is assumed that the improved soil attribute qualities would mean improved soil functionality over the time period under consideration.

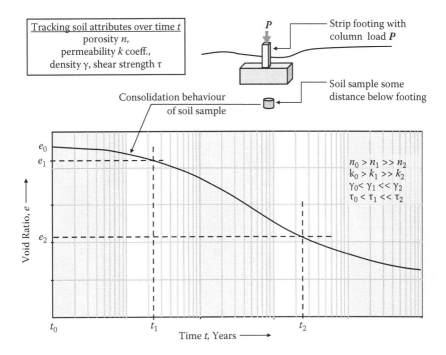

FIGURE 5.12
Schematic illustration of consolidation behaviour of a soil sample some distance below a loaded footing. The soil attributes tracked are soil properties over time t. The effect of consolidation of the soil sample in the ground is felt in terms of the increases in density γ and shear strength τ of the soil, and a decrease in the porosity n and permeability of the soil expressed in terms of the Darcy coefficient k.

5.5.2 Time-Related Changes of Long-Term Functionality

5.5.2.1 Influence of Anthropogenic Stressors

Long-term changes in soil attributes in response to activities generated from stressors in the *anthropogenic group* can be anticipated and accounted for with proper knowledge of soil properties and behaviour, and with appropriate planning and foresight. In the example shown in Figure 5.12, consolidation studies are undertaken on soil samples involved in the project as a requirement for determination of potential settlement of the loaded footing and associated structure.

In most instances, since human-related events require planning of system performance, potential or real time-related changes in soil attributes are considered. Environmental soil behaviour for such cases should be part of the planning and design process. Consider two different types of soil use in projects associated with human-made events: (a) engineered clay barriers used as part of a multibarrier containment system for dangerous wastes, and (b) agricultural-based food production. In the case of engineered clay barriers, *environmental soil performance* refers to the long-term functionality of the clay barrier in the face of chemical, physical, and hydraulic stressors originating from the waste pile. Whilst tracking time-related changes in clay attributes such as permeability, accumulative properties, strength, compressibility, and so forth over the service life of the clay barrier component of the multibarrier containment system can tell one about the functionality of the clay barrier, prudent engineering practice would have studied the problem and would have taken into account potential time-related changes in these clay attributes. In agricultural-based food production, taking fertilizing of chemicals and manure, irrigation, and drainage as the large set of activities in food production, these serve as stressor sources that will act on the soil and occur as time-related changes of chemical, physical, hydraulic and biological soil attributes, resulting in time-related changes of functionality for food production in degradation or improvement.

One can conclude that time-related processes such as compaction, consolidation, piping, erosion, leachate permeation, and others related to actions associated with specific projects, will alter the attributes of a soil during the course of time, that is, soil functionality. Whilst they are generally considered not to be part of the natural ageing processes, they are nevertheless part of the various processes that contribute to the evolution of a soil, that is, the evolution of soil attributes.

5.5.2.2 Influence of Natural Stressors

The effects of regional environment controls on soil formation have been discussed in Chapter 1. We are concerned with *natural ageing* due to internal stressors that arise due to events such as rainfall, flooding, drying, freeze–thaw seasonal cycle, and vegetation changes, that is, those stressors that are associated with the nature of the constituents of a soil. These include the various soil fractions that form the composition of a soil, the porewater and its chemical composition, the gaseous component, and the various microorganisms that inhabit the soil. Geochemical and biogeochemical evolution of natural soils occur gradually as normal time-related ageing processes over very long-term periods, the results of which are demonstrated as changes in the nature of the soil—that is, changes in soil attributes–resulting in changes of soil functionality.

One needs to bear in mind that when human activities are involved, technically speaking, there can be two different soil maturation/ageing time-scales or phenomena. The first is the one that goes with soil formation, as discussed in Chapter 1, and the second is when soils are used as engineering materials, or when soil masses are involved in food production, geoenvironmental projects, and other human-associated activities. In this second time-scale, they can be considered to mature and age from the day of inception of the project that involves the soil material or soil mass. The processes associated with maturation and ageing of the soil are as much dependent on the circumstances related to the use of the soil as they are on its initial formation.

5.6 Concluding Remarks

The items that need attention and scrutiny include (a) the elements that constitute the soil environment, (b) soil environment stressors—types and sources, (c) elements in the soil environment that are impacted by the different stressors, and (d) how all of these impact on soil functionality. To determine the types and sources of soil environment stressors, it is necessary to examine what constitutes a soil environment and how a soil mass relates to its environment.

Connecting the dots between stressor, stress, and impact is not an easy task because (a) there is no straight line between them, and (b) much depends on the amount of detail considered necessary in connecting the dots. The soil environment cannot be considered as being composed solely of soil solids and water. As the Venn diagram in Figure 5.2 tells us, the soil environment is in effect the soil ecosphere or pedosphere. This means to say that it is not proper to ignore or minimize the role of microorganisms in soil, especially in respect to the impact of the various types of stressors on biological processes. These processes, which have led researchers to view alteration and transformation of soil minerals as bioweathering processes, are significant factors in changing the properties and behaviour of soils in the long term. What is not fully appreciated is the fact that in addition to the obvious chemical stressors, hydraulic, thermal, and even mechanical stressors will also impact on biological processes.

It is more practical to consider the actions or phenomena as impacts triggered by stressors—generically and generally—on a soil mass, because of the countless numbers and types of stressors and the variety of applications involving a soil mass. These actions or phenomena will be induced by reactions and processes generated in the soil by stressor groups and they are broken down into the following types: (a) hydraulic, (b) mechanical, (c) thermal, (d) chemical, (e) geochemical, and (f) biologically mediated.

The soil functionality index has been introduced as a tool to facilitate examination of the status of the soil involved in the project under consideration. Data obtained from tests and other kinds of measurements (field and laboratory) of the physical, chemical, and biological attributes of the particular soil under consideration can be used as (a) input to compare with individual attribute indicators, and/or (b) input to statistical and analytical models developed to produce the soil functional index (*SFI*). Application of weighting factors to be applied to indicators is to a large extent based on knowledge of previous behaviour. Development of *SFI* requires consideration of the many different properties and factors that combine to produce the status of the soil under consideration. We must bear in mind that the soil attributes used as soil functional indices must be evaluated in relation to the attributes' functional requirements, and that they are time-related attributes.

References

Bending, G.D., Turner, M.K., Rayns, F., Marx, M-C., and Wood, M., 2004, Microbial and biochemical soil quality indicators and their potential for differentiating areas under contrasting agricultural management regimes, *Soil Biol. Biochem.*, 36:1785–1792.

Blavet, D., Leprun, J.C., Mathe, E., and Pansu, M., 2002, Soil colour variables as simple indicators of the duration of soil waterlogging in a West African catena, *Proc. 17th., World Congress of Soil Science (WCSS) Conference*, Thailand, Paper No. 32.

Yong, R.N., 2001, *Geoenvironmental Engineering: Contaminated Soils, Pollutant Fate, and Mitigation*, CRC Press, Boca Raton, FL, 307 pp.

Yong, R.N., and Mulligan, C.N., 2003, *Natural Attenuation of Contaminants in Soils*, Lewis Publishers, CRC Press, Boca Raton, FL, 319 pp.

6

Mechanical Properties

6.1 Introduction

By and large, the mechanical property of a soil mass is the materialization of the resistance to displacement of soil particles in the soil mass by external stressors. The different kinds of actions resulting from application of the external stressors on a soil mass include such phenomena as compression, shearing, cutting, and twisting (torsion). The three-phase nature of soils (solid, liquid, and gaseous) requires one to consider how the various phases react and interact with each other to provide the response to the external stressors responsible for provoking compression, shearing, cutting, and so forth of the soil mass. Each of the three phases is highly variable, with their compositions and proportions in a soil mass varying continuously because of regional weathering and environmental conditions (see Chapter 10).

The common conventional practice for determining the mechanical property of a soil is through laboratory testing of a representative soil sample, using procedures and techniques that are designed to elicit soil response performance and associated data required for data-reduction models. Relevant mechanical property parameters are obtained from these data-reduction models (e.g., Mohr–Coulomb) for stability and design analyses and/or planning purposes. By virtue of the test techniques and procedures, the laboratory test results obtained are operationally defined, meaning that the type of data and experimental results obtained are dependent on the test constraints, conditions, sample preparation techniques, and various test technique requirements such as drained and undrained conditions, constant rate or incremental loading, and others. Nevertheless, the laboratory test techniques used in conventional practice, together with the procedures adopted in data reduction, have been found to be useful in providing the necessary parameters for the various types of constitutive models used in soil and geotechnical engineering.

A successful constitutive model is one which accurately simulates the response behaviour of a material under a set of stressors in a given scenario. In the case of soils, the challenges are (a) to obtain proper knowledge of how a three-phase system responds mechanically to external loads or forces stressors under various conditions, (b) to construct a constitutive model with parameters that are not only relevant to the anticipated behaviour pattern, but also realizable, that is, they can be obtained from appropriate laboratory or field tests, and (c) to fashion or develop laboratory and other test techniques that will yield the data and the parameters required for the constitutive model. The discussions in this chapter focus on the challenges, (a).

6.2 Mechanical Attributes

The mechanical attributes of soil most often called into play are those pertaining to the strength and integrity of the soil. How these attributes are described or determined is dependent to a large extent on the purpose to which the soil in question is to be used, that is, the role of the soil under consideration. The most commonly sought mechanical attributes include properties such as (a) resistance to compressive forces, (b) shear strength or resistance, (c) stress–strain relationships, and (d) consolidation and creep. The common laboratory procedures for determination of these attributes include two main groups of tests: (a) unconfined uniaxial and axisymmetric (triaxial) compression tests, and (b) consolidation and creep tests. The test apparatus and procedures for the various kinds of tests in each group are well documented in the testing standards set forward by the relevant testing standards agencies, national professional bodies, national licensing bodies, and regulatory agencies, for example, ASTM International, national geotechnical societies, national soil science societies, and environmental protection agencies.

The aim of the various tests in the two main groups is to obtain relevant information concerning the resistance of soil to shear/deformation, either in relation to elapsed time under the external stressor (that is, external loads), or as a direct response to singular or incremental loads. The data obtained on soil sample performance under the external stressors are assessed in relation to the physical properties of the soil. In this manner, one obtains an appreciation of the mechanical properties of a soil for a particular set of physical soil properties. By and large, the information obtained will be in terms of the macroscopic or bulk behaviour of the soil under the prescribed test conditions. Measurements or data obtained in relation to internal or "microscopic" behaviour relate directly to the role of water in the soil. These measurements pertain to the pressures developed in the porewater (porewater pressure), and their impact on inter-granular behaviour. Distinguishing *effective stress*, $\bar{\sigma}$, from *total stress*, σ, is a demonstration of the role of porewater pressure in soils.

6.2.1 Soil Deformation under Load

The basic elements of a constitutive model for soils include some form of the following: (a) stress–strain relationship, (b) yield and failure criteria, (c) plastic potential, and (d) strain-hardening or work-hardening function. The nature of deformation of a material under external loads is a direct function of the type or "make-up" of the material. When a material is stressed in simple tension or compression, its mechanical behaviour is described generally in terms of its stress–strain relationship and its compressive or extensive stress at failure. The shape of the stress–strain "curve" tells one about the nature of the material being tested. Figure 6.1 shows some typical stress–strain curves for some ideal materials. Soils are not ideal materials. They are three-phase materials that have compositional features and attributes that are not there by design but by virtue of their origin and historical development (see Chapters 2 and 10). The various idealized stress–strain curves shown in Figure 6.1 do not readily fit the actual types of stress–strain behaviour of soils. That being said, this fact has not deterred many from trying to "fit" ideal material performance such as those shown in the figure to actual soil behaviour under loading—with some level of success.

6.2.1.1 Yield and Failure in Soils

The term *yield* is used in the field of plasticity to describe the onset of plastic deformation commonly occurring at the upper limit of elastic action, as

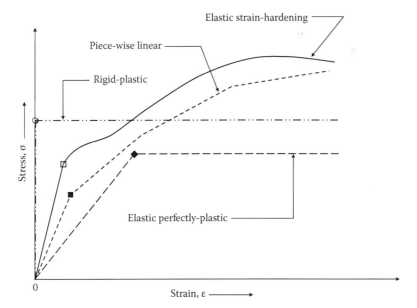

FIGURE 6.1
Stress–strain curves for some ideal materials; σ and ε refer to *stress* and *strain*, respectively.

shown by the various symbols at the junctures of the "curves" shown in Figure 6.1. The upper limit of elastic action is known as the *yield limit*. Except for fracture or rupture, *failure* of a material under stress is not readily defined. This is particularly true for materials that have an element of plastic deformation under loading. Fracture or rupture implies the presence of distinct surfaces of separation in a body, whereas plastic deformation is the result of unrestricted flow. Failure in granular soils can be described by the term *fracture*, especially when particulate behaviour is gravity-controlled.

Yielding in a material can be well defined when the material exhibits initial linearity in stress–strain performance. Initial linearity in the stress–strain curve beginning at the origin indicates elastic behaviour of the material if complete reversibility is obtained upon unloading before the yield point. If complete reversibility is not obtained (i.e., if the rebound slope of the curve is greater than the loading slope), the material cannot be classified as an elastic material. It is more likely to be a type of strain-hardening material. Granular soils (i.e., cohesionless soils) show differing characteristics of strain hardening, depending on the density of the soil at the onset of loading, and on the nature of volume change (of the soil) occurring under load (Figure 6.2). The stress–strain performance of the dense granular soil can be approximated by elastic strain-softening theories, whilst the loose granular soil would

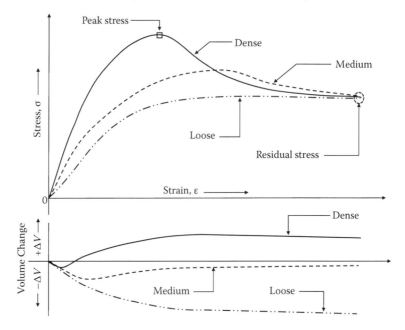

FIGURE 6.2
The top graph shows the stress–strain curves for a granular (cohesionless) soil in relation to initial density. The bottom graph shows the volume changes in the samples in relation to initial density and as a function of strain under the applied stress. The residual stress is the asymptotic stress—commonly known as *large-strain stress*.

more likely be fitted with strain-hardening or work-hardening concepts. The notion of peak and residual stresses can be applied to stress–strain curves that demonstrate noticeable peaks, as shown by the behaviour of the dense granular soil, and to a much lesser extent by the medium density soil.

6.2.2 Plastic Potential

Generally speaking, when a set of generalized stresses σ_1, σ_2, ... and σ_n act on a body, the work dW performed by stresses in relation to infinitesimal strain increments can be expressed as: $dW = \sigma_1 d\varepsilon_1 + \sigma_2 d\varepsilon_1 + ... + \sigma_n d\varepsilon_n$, where $\varepsilon_1, \varepsilon_2, ...$ and ε_n refer to the corresponding set generalized strains. For the elastic perfectly-plastic material whose stress strain curve is shown in Figure 6.1, each strain increment $d\varepsilon_i$ will have an elastic (totally recoverable) and a plastic nonrecoverable strain component, that is, $d\varepsilon_i = d\varepsilon_i^{(e)} + d\varepsilon_i^{(p)}$, where $i = 1, 2, ... n$, and the superscripts (e) and (p) refer to *elastic* and *plastic*, respectively.

Specifying σ_1, ..., σ_n to be the state of stress at the yield limit (i.e., yield point), the stress increments $d\sigma_1$, ..., $d\sigma_n$ imposed will result in the development of a neighbouring state of stress at the yield limit. Both the initial and subsequent states of stress must satisfy the yield criterion (see Sections 6.4.2 and 6.4.4). The relationship between strain rates and stress is given by an associated flow rule where the connection between the flow rule and the yield condition is known as the theory of the *plastic potential* ϕ, where ϕ is a scalar function of stress $f(\sigma_{ij})$, and where plastic strain increments are obtained by partial differentiation of ϕ with respect to the stresses. Note that each yield condition has an *associated flow rule*, that is, a form of flow rule or plastic stress–strain relationship associated with the particular yield condition. If $\sigma_1 ... \sigma_n$ specify a state of stress at the yield limit, and if the imposed stress increments $d\sigma_1 ... d\sigma_n$ will create a neighbouring state of stress at the yield limit, both former and new states of stress must satisfy the yield criterion. This means $\phi = \phi (\sigma_1, ..., \sigma_n) = 0$.

$$dΦ = \frac{\partial Φ}{\partial \sigma_1} d\sigma_1 + ... + \frac{\partial Φ}{\partial \sigma_n} d\sigma_n$$

or (6.1)

$$dΦ = \sum_i \sum_j \frac{\partial Φ}{\partial \sigma_{ij}} d\sigma_{ij}$$

Equation (6.1) describes the orthogonality of vectors $(d\sigma_1 ... d\sigma_n)$ and $[(\partial \phi/\partial \sigma_1) ... (\partial \phi/\partial \sigma_n)]$. In the case of the vector with components $(d\sigma_1 ... d\sigma_n)$, this represents stress increments that join the neighbouring points of the yield locus and is tangential to this locus. The vector with components $\partial \phi/\partial \sigma_1 ... \partial \phi/\partial \sigma_n$ is normal to the yield locus at the stress point under consideration.

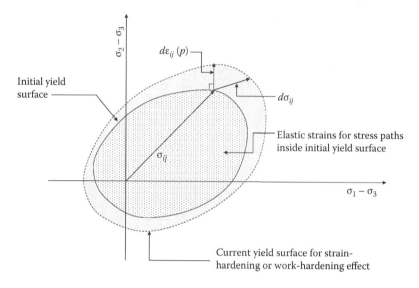

FIGURE 6.3
Yield surface on plane of $(\sigma_1 - \sigma_3)$ and $(\sigma_2 - \sigma_3)$ showing plastic strain-increment vector $d\varepsilon ij(p)$ normal to the yield surface.

The yield surface shown in Figure 6.3 is portrayed on a plane of $(\sigma_1 - \sigma_3)$ and $(\sigma_2 - \sigma_3)$, where σ_1, σ_2, and σ_3, refer to the major, intermediate, and minor principal stresses, respectively. If one plots the total principal strains in principal strain space, and if the principal strain axes are aligned with principal stress axes, the conditions of perfect plastic yielding require the plastic strain-increment vector to be normal to the yield surface. Since the relationship between the strain rates and stress will be given by an associated flow rule, the flow rule that specifies the components of the plastic strain increment for each state of stress at the yield limit may be written as $d\varepsilon_{ij}^{(p)} = d\lambda\, \partial\phi/\partial\sigma_{ij}$, where $d\lambda$ is a nonnegative proportionality constant. The plastic strain increment $d\varepsilon_{ij}^{(p)}$ is normal to the surface defined by ϕ since $\partial\phi/\partial\sigma_{ij}$ is normal to the surface defined by ϕ. When the plastic potential surface ϕ coincides with the yield surface, which is the case for an elastic perfectly-plastic material, the plastic strain-increment vector will assume an outward normal direction to the yield surface. Note that the constant λ is restricted by the condition that plastic flow is always accompanied by dissipation of energy.

6.2.3 Flow Function

Assuming that the deviatoric components of stress and strain are linearly related in the flow region, we can define a stiffness parameter ϕ_s as $\phi_s = \partial\sigma'_{ij}/\partial\varepsilon'_{ij}$, where $\sigma'_{ij} = \sigma_{ij} - (1/3)\,\sigma_{ii}\delta_{ij}$, $\varepsilon'_{ij} = \varepsilon_{ij} - (1/3)\,\varepsilon_{ii}\delta_{ij}$, σ'_1, σ'_2, and σ'_3 are the principal stress components of the stress deviation tensor, respectively, and δ_{ij} is the Kronecker delta (i.e., $\delta_{ij} = 0$, $i \neq j$; $\delta_{ij} = 1$, $i = j$). From the point of view of irreversible thermodynamics, by taking (a) the volume densities of the internal energy, (b) the free energy, (c) the entropy of a volume element surrounding the point of interest in a sample, (d) the components of stress and strain tensor, and (e) the heat received by the volume element in time dt, the internal energy u of system can be expressed as follows:

$$du = dq + \sigma'_{ij}\,d\varepsilon'_{ij} \qquad\qquad Tds = dq + T\xi dt \qquad\qquad (6.2)$$

Combining the two relationships, one obtains:

$$-T\xi dt = du - Tds - \sigma'_{ij}\,d\varepsilon'_{ij} = df - \sigma'_{ij}\,d\varepsilon'_{ij} \qquad\qquad (6.3)$$

where q refers to the heat, T refers to the temperature, ξ denotes the entropy production rate, f refers to Helmholtz free energy of the soil–water system and $df = du - Tds$. Prigogine (1961) has shown, from statistical mechanics, that Equation (6.3) remains valid for reversible or irreversible differential changes from the equilibrium state. Accordingly, one can assume an equation of state of the form $f = f\{s, \xi_2, \xi_3, \ldots, \xi_n\}$, where the entropy $s \equiv \xi_1$, and the other state variables ξ_i are counted from $i = 2, 3 \ldots n$. These variables may contain other parameters to account for physicochemical interactions, electrostatic effects, and so forth between the phases and the generalized forces X_i. Accordingly, Equation (6.3) may be expressed as follows: $-Tds^{(i)} = df - X_i^{(i)}d\xi_i$, where the superscript (i) refers to the irreversible parts of the entropy and generalized forces, respectively.

Denoting $X_i = X_i^{(r)} + X_i^{(i)}$ as the generalized forces, and using superscript (r) to denote the reversible part of the generalized forces, one obtains

$$X_i^{(r)} = \left[\frac{\partial f}{\partial \xi_i}\right]_{s=const.} = a_{ij}\xi_j \quad \text{and} \quad X_i^{(i)} = T\left[\frac{\partial s^{(i)}}{\partial \xi_i}\right]_{V=const.} = b_{ij}\xi_j \qquad (6.4)$$

In accord with Onsager's theorem, the forces are linear functions of the fluxes ξ_j, and the phenomenological coefficients are symmetric tensors ($b_{ij} = b_{ji} \geq 0$). Expanding the free energy density in the form of a power series, and using a two-term approximation, the generalized forces have the form

$$X_i = X_i^{(i)} + a_{ij}\xi_j = a_{ij}\xi_j + b_{ij}\xi_j \qquad\qquad (6.5)$$

The coefficients a_{ij} are symmetric and the condition that $a_{ij} \geq 0$ and $b_{ij} \geq 0$ means that the quadratic form with the numerical factors a_{ij} and b_{ij} are both positive and definite. Solution of Equation (6.5) leads to a strain-time function as follows (Axelrad and Yong, 1966):

$$\varepsilon'_{ij}(t) = a_{ij}\sigma'_{ij}(t) + b_{ij}^{-1}\int_0^t \exp\left[-b_{ij}^{-1}a_{ij}(t-\tau)\right]\sigma'_{ij}(\tau)d\tau, \qquad (6.6)$$

where τ is a time variable. Since a_{ij} and b_{ij} are symmetric tensors, the bracket expression of the integrand can be considered as a thermodynamic impedance function. Under an applied constant stress, and assuming that one can neglect the initial elastic response of a test sample for $t \to 0$, Equation (6.6) can be reduced to the form

$$\varepsilon'_{ij}(t) = \frac{\sigma'_{ij}(t)}{a_{ij}}\left[1 - \exp\left[-b_{ij}^{-1}a_{ij}\cdot t\right]\right] \qquad (6.7)$$

6.3 Concept of Effective Stress

6.3.1 Physical Concept

In a soil–water system where soil particles and porewater coexist to establish the integrity of the soil, application of external pressure to the soil will result in pressures developed in the porewater and soil particles. If the soil–water system is partly water-saturated, pressures will also be developed in the gaseous phase. By definition, the stresses developed between soil particles as a result of stress transfer at the contact points between particles are the *effective stresses*—otherwise known or mechanistically described as intergrain or intergranular or interparticle stresses. To illustrate the concept of effective stress, the schematic illustration shown in Figure 6.4 uses a fully water-saturated discrete granular system to highlight the various elements involved.

In the schematic illustration shown in Figure 6.4, the soil grains and the porewater are considered to be nondeformable, and the impermeable flexible membrane possesses no membrane strength. If no volume change is allowed under an applied pressure P, the reaction in the test sample to the applied pressure will be demonstrated as a hydraulic pressure response, that is, $P = u$, where u is the porewater pressure. This is equivalent to applying a load or pressure to a piston-jack system with its rigid chamber containing water, with total load carried by the water in the chamber. In theory, there is no stress transfer between the soil grains. If water is allowed to escape from the impermeable membrane surrounding the soil grains, the porewater pressure u will be less than P, that is, $u < P$. The effective stress $\bar{\sigma}$ is the stress developed between contacting grains or particles (intergrain stress), and is obtained by the relationship $\bar{\sigma} = \sigma - u$, where σ is equal to P, that is, σ denotes the total stress applied to the system. This relationship is generally considered to be the underpinning of the *effective stress principle,* that is, the effective stress is equal to the total stress minus the porewater pressure. If all the water under stress is allowed to escape from the system, u becomes zero and the effective stress $\bar{\sigma}$ is the total applied stress σ.

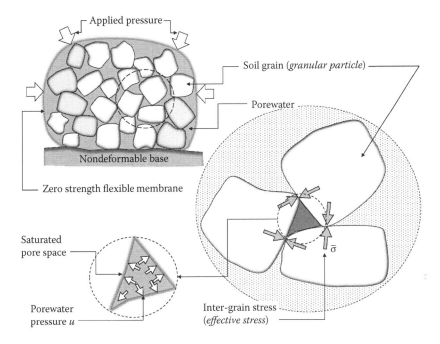

FIGURE 6.4
Schematic illustration of the concept of effective stress using a discrete granular system as the model for soil compression under applied load. The soil grains are contained by an imperme-able membrane with zero membrane strength; $\bar{\sigma}$ denotes effective stress.

6.3.2 Physical Concept in Relation to Clays

6.3.2.1 Clays with Adsorbed Water Layers

For soils that have discrete grains that can establish physical contact between grains, as illustrated in Figure 6.4, the meaning of the effective stress prin-ciple can be readily and easily visualized. However, in the case of soil–water systems with active soil particles such as clay mineral particles, the physical reality of inter-particle contact is not readily achieved. Figure 6.5 is an illus-tration of the interaction of the microstructural units formed by layer-lattice clay mineral particles. The role of porewater pressure in this type of soil–water system is particularly interesting—given that one needs to consider not only the adsorbed water layers associated with individual particles, but also the interlayer water in each particle.

For soil–water systems such as the one portrayed in Figure 6.5, physi-cal contact between microstructural units will be between their respective adsorbed water layers. Stress transfer between microstructural units will be actuated through these adsorbed water layers—illustrated by sketch (C) shown in the bottom right-hand corner of the figure. The porewater pressure developed in the macropores u will react against the microstructural units.

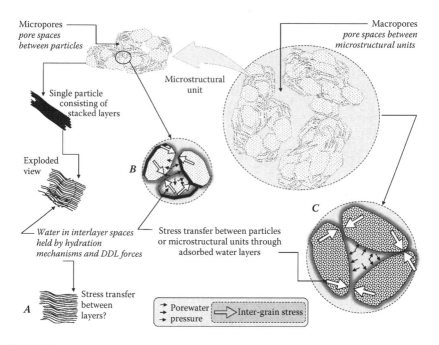

FIGURE 6.5
Interaction of clay microstructural units formed by layer-lattice clay mineral particles. Macroscopically, lower right-hand sketch (C) shows stress transfer between microstructural units will be through adsorbed water layers surrounding the units. Middle sketch (B) shows interactions between particles in a microstructural unit without necessarily having physical contact between particles. Lower left-hand sketch (A) shows the interactions that need to occur in the interlayers of individual particles, raising the question of what stress transfer between the layers means.

Within the microstructural units themselves, in addition to the micropores between particles, hydration water layers (adsorbed water) surround the individual clay particles constituting the microstructural units. Contact between particles will be through the adsorbed water layers—meaning that stress transfer between particles will be through these water layers, as depicted in the illustration shown in the middle sketch (B) in Figure 6.5. The porewater pressures developed in the micropores u_{mic} can be, or will be, different from those developed in the macropores.

In view of the clay mineral particles constituting typical clays, and accepting that contact between particles or microstructural units will be through adsorbed water layers, the physical concept of effective stress as physical intergrain or intergranular stress transfer requires some adjustment. One needs to accept that physical inter-grain stress transfer will be accomplished through adsorbed water layers. At equilibrium, since the porewater is a continuous phase liquid in the clay, $u = u_{mic}$. With proper measurements of the porewater pressure u, the effective stress for a fully saturated clay–water system is still $\bar{\sigma} = \sigma - u$.

6.3.2.2 High-Swelling Clays

In high-swelling clays where layer-lattice clay minerals are dominant constituents, the role of porewater pressure becomes an important issue. Interlayer water and the pressures developed therein are dependent on the exchangeable ions populating the interlayers and the chemistry of the water in the macropores. The meaning of porewater pressure, in relation to interlayer water, takes on a different form. Interlayer porewater pressure must account for the interactions in the diffuse ion layers. The osmotic potential ψ_π associated with interactions in the interlayers is manifested as a negative porewater pressure, that is, a negative water pressure is needed to keep water from being drawn into the interlayers in response to the osmotic potential. In soil engineering terminology, this phenomenon is referred to as *osmotic suction* S_π, with the corresponding porewater pressure denoted as u_π, which is a negative quantity.

In a situation where water is made available to a high-swelling clay, the amount of pressure required to keep the clay from swelling, that is, to maintain a constant volume, is the amount of force or pressure required to keep the water from being drawn into the clay. In a high-swelling clay under confining pressures required only to maintain constant volume, the effective stress will be $\bar{\sigma} = \sigma - (-u_\pi) = \sigma + u_\pi$. If the applied confining pressure is less than u_π, a negative porewater pressure will be measured. If the clay is allowed to take in water, it will swell until the measured porewater pressure u is zero. If on the other hand, a constant volume is maintained and if the external pressure applied is greater than the swelling pressure, a net porewater pressure will be generated. This net porewater pressure will be the difference between the applied pressure and the swelling pressure.

Trying to reconcile the different types of porewater pressures in a complex clay–water system with microstructural units and reactive clay mineral particles requires laboratory measurement capabilities that are not readily available. In theory, for a clay or any other type of soil under external loading, if one accepts that the water phase in soil is continuous, and if equilibrium is established, $u = u_{mic} = u_\pi$. The imperative here is to be able to determine when equilibrium in the porewater has been reached, and to be able to physically measure u. The physical concept of effective stress in high-swelling clays takes the same form as the previous case of normal clays—meaning that actual physical contact between mineral particles will not occur. Instead, contact between mineral particles will be between the adsorbed or hydration water layers surrounding the particles. Hence, effective stress transfer will be between particles through adsorbed water layers.

6.4 Shear Strength of Soils

Determination of the shear strength of soils is most often conducted in the laboratory using a combination of compression-shear devices. The procedures for testing using these devices, together with field measurement devices and techniques, are well documented and detailed as specifications. As stated in Section 6.2, testing standards have been issued by the relevant testing standards agencies, national professional bodies, national licensing bodies, and regulatory agencies. The discussion in this and the next section focuses on the mechanistic aspects of shearing in soils under load application, and interpretations relative to soil shear strength and failure.

6.4.1 Laboratory Determinations

The most common technique used for determining the shearing resistance and strength of soils is the triaxial test. To a large extent, this is because it allows one to determine the roles that volume change and porewater pressure can play in the resistance of the soil to shearing forces. The essence of the laboratory test system is shown in Figure 6.6 for uniaxial compression and triaxial (axisymmetric) tests. The common techniques used for determination of soil strength are (a) unconfined compression uniaxial (UC) tests,

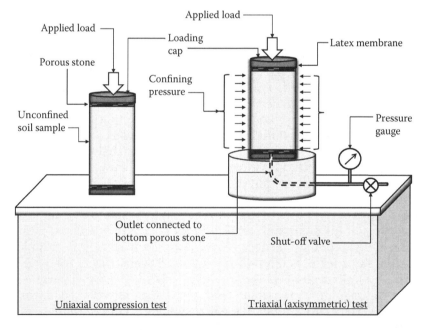

FIGURE 6.6
Uniaxial compression and triaxial test systems showing the basic elements of the test.

generally used for cohesive soils (fine-grained soils), but not for pure cohesionless soils (soils with zero cohesion), (b) and triaxial axisymmetric tests ($\sigma_2 = \sigma_3$) on samples in three possible states, such as unconsolidated undrained (UU), triaxially consolidated undrained (CU), and triaxially consolidated drained (CD). The choice of type of test depends on what one is seeking to determine, and also on the type of sample being tested.

In the unconfined compression uniaxial (UC) test shown in the left-hand sketch in Figure 6.6, no lateral or confining support is given to the sample under test. This means that granular soils possessing zero cohesion cannot be tested with this scheme because of their inability to stand upright as a coherent sample. UC testing of samples seek to determine the unconfined shearing strength of sample, without loss of porewater. The unconsolidated undrained (UU) test conducted with the triaxial set-up shown in the right-hand sketch of the figure is essentially similar in its intent with the UC test. The only difference between the UU and the UC tests is the confinement afforded the test sample. The UU and UC tests are sometimes referred to as constant-volume shear tests.

6.4.1.1 Triaxial Test System

The major differences in test objectives come with the consolidated undrained (CU) and consolidated drained (CD) triaxial tests. These tests permit one to determine the role of porewater pressure in the development of shear resistance of the soil under load. In the CU test, the sample is allowed to consolidate isotropically under a prescribed load, that is, to reduce in volume whilst allowing porewater to escape under the prescribed isotropic load. By allowing the sample to consolidate, the test sample no longer retains the same void ratio or density as the undisturbed field sample. At the end of the consolidation process, the porewater pressure in the sample is zero. Figure 6.7 shows the stress plane for the axisymmetric triaxial test, and the point denoted by σ_c is shown on the isotropic line (space diagonal) where $\sigma_1 = \sigma_2 = \sigma_3$. The shearing process of the consolidated sample is obtained from load application, which develops a stress increase of $\Delta\sigma$ on the specimen, as shown in the figure. In the undrained test condition, no porewater is allowed to escape. Since the porewater pressure acts uniformly in all directions, its line of action is parallel to the isotropic line. Applying the effective stress principle $\bar{\sigma} = \sigma - u$, one obtains the effective stress path shown in the figure. Assuming that the Mohr–Coulomb model for failure applies to the test sample, the various effective stress paths should terminate when they reach the Mohr–Coulomb envelope. The parameter c^* shown in the figure is a function of the cohesion parameter c in the Mohr–Coulomb model.

6.4.1.2 Other Strength Test Systems

The direct shear test predates the popular triaxial test system, and is perhaps the oldest form of shearing test devised to determine the shear resistance of

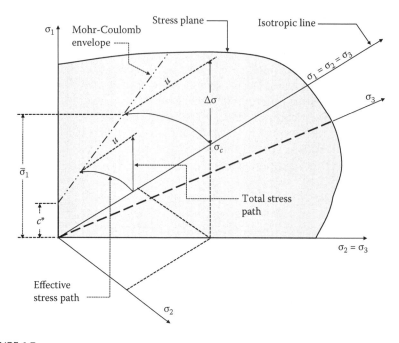

FIGURE 6.7
Stress plane for conventional axisymmetric triaxial test with porewater pressure measurements obtained to permit determination of effective stress.

soils. This test is still widely used because of its simplicity. The essence of this test is shown in the schematic illustration shown at the left-hand side of Figure 6.8. The other two types of shear testing shown in the figure are the plane-strain shear device, and the double ring-shear test device. The plane-strain shear apparatus was devised to counter the problems of shear distortion and progressive shear attributed to the direct shear device. As with the double ring-shear device, these two types of shearing tests still do not provide one with the capability for measurement of porewater pressure.

The following are amongst some of the drawbacks or limitations of the direct shear test and its measurements:

- Since the shearing plane, illustrated in the blow-up shown at the bottom left-hand portion of the figure, is essentially a "forced" shearing plane, it is not clear that this is the weakest plane in the soil sample.

- Because of the nature of the test equipment, there is no ability for one to undertake porewater pressure measurements, that is, tests are drained shear tests.

- The nature of the test itself, where the top portion is forced to displace laterally under an applied load or force, results in progressive failure which begins at the forcing-boundary interface.

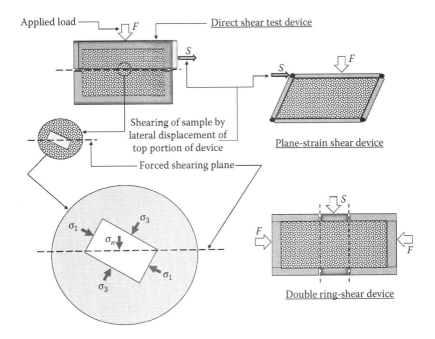

FIGURE 6.8
Schematic illustration of shearing intent of some soil shearing test devices.

- The "softer" the sample is, the greater is the strain distortion from progressive shear failure.

In interpreting or plotting the results from the direct shear test, Hill (1950) has noted that for plane strain, that is, no strain in the third ($i = 3$) direction, "The present modification of Mohr's statement of the yield criterion does not seem to be widely known to workers in that field, who generally regard the shear-box test as giving a direct measure of points on the envelope. Since the deformation is so constrained as to be effectively a simple shear in a narrow zone, it seems that the test really gives a direct measure, not of E [a point on the failure envelope in the Mohr diagram] but of F, for if the material is isotropic, the directions of the maximum shear stress and maximum shear strain-rate must coincide." $F(p, \tau_m) = 0$, where p is the mean compressive stress $= -(1/2)(\sigma_1 + \sigma_2)$, and τ_m is the algebraic maximum shear stress $= (1/2)(\sigma_2 - \sigma_1)$.

6.4.2 Yield and Failure Theories

6.4.2.1 Principal Stress Space

Whilst the concept that *yield* refers to the onset of plastic deformation, whereas *failure* signifies collapse of the material (under test) as a terminal stage of the process of yielding is commonly applied to many materials, it is not always consistent with observed behaviour of many soils. A useful method for examination of the state of stress resulting in yield or failure in a soil is to plot the principal stress components σ_1, σ_2, and σ_3 at yield or failure in principal stress space (Figure 6.9). The bounding surface shown in the figure identifies the limiting surface, meaning that all stresses at yield or failure are assumed to lie on this surface, depending on whether the condition of yield or failure is sought in the test measurement. The surface normal to the isotropic line (line where $\sigma_1 = \sigma_2 = \sigma_3$) is the octahedral plane, often referred to as the π plane. If one is seeking to determine yielding of the test specimen, the limit stresses will be defined by the measurements at the yield point, that is, points A, B, and C, assuming that this point has been properly identified. The bounding surface will be the yield surface, and the shape of the surface on the π plane will identify the yield theory. If on the other hand, the limit stresses sought are the failure stresses, the bounding surface will be the failure surface, and a corresponding failure theory or model will be identified from the shape shown on the π plane.

In order for yield and failure in soils to be properly accounted for, it is necessary for both experimental and analytical techniques for assessment and analysis to be consistent and realistic in their evaluation of soil behaviour

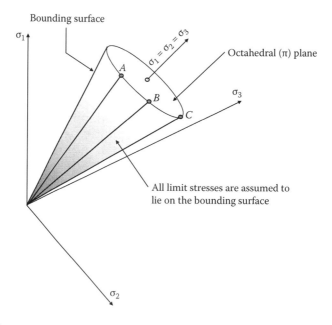

FIGURE 6.9
Principal-stress space showing principal stresses at time of limiting stresses. Limit stresses may be yield or failure stresses, depending on choice of stress measurement during shear strength test. If yield is chosen as limit stress, the bounding surface will be a yield surface and the shape of the surface on the octahedral plane will identify the yield theory.

under stress. For this to happen, the phenomenological coefficients and parameters must properly represent material properties, and soil behaviour under stress must be accurately portrayed.

6.4.2.2 Yield Criteria

The earliest criterion of yield and failure is the Rankine criterion, a *maximum-stress criterion* which postulates that the maximum principal stress in the test material determines failure, regardless of the magnitude and sense of the other two principal stresses. By this criterion, the onset of yielding in a material under stress occurs when the absolute value of the maximum stress reaches the yield point of stress of the material, in simple tension or compression. This criterion has some limitations, as for example, for isotropic materials under a hydrostatic stress-state $\sigma_1 = \sigma_2 = \sigma_3$, yield or failure will not occur. However, for strongly anisotropic materials such as layered materials, where a pronounced difference in strength properties exists in different directions, there could be some validity in applying this yield criterion. In general, the criterion does not fit well with known performance of soils under stress.

There are three versions of the *maximum shear-stress criterion*. In the *Tresca criterion*, yielding in a material begins when the maximum shear stress in the material is equal to the maximum shear stress at the yield point in simple tension $\sigma_1 - \sigma_3 = \sigma_y$, where σ_y is the absolute value of the yield stress in tension or compression. In principal stress space, the Tresca yield surface would be a right hexagonal cylinder. In the Drucker modification of the Tresca criterion, generally recognized as the *extended Tresca yield criterion*, the yield surface in principal stress space is a right hexagonal pyramid instead of a right hexagonal cylinder. This contrasts with the irregular hexagonal pyramid of Coulomb. The yield surfaces for the three yield criteria (extended Tresca, Coulomb, and extended von Mises) are shown in Figure 6.10 in the octahedral (π) plane.

The constant *elastic strain-energy-of-distortion theory* of yielding, which is also known as the constant *octahedral shearing-stress criterion*, is variously attributed to Huber, Hencky, and von Mises. The criterion wherein onset of plastic yielding occurs is when the strain energy of distortion given by W_D:

$$W_D = \frac{1+\mu}{6E}\left[\left(\sigma_1 - \sigma_2\right)^2 + \left(\sigma_2 - \sigma_3\right)^2 + \left(\sigma_3 - \sigma_1\right)^2\right] \tag{6.8}$$

reaches a critical value. In the relationship given as Equation (6.8), μ is Poisson's ratio, and E is Young's modulus. Drucker and Prager (1952) proposed a modification of the von Mises criterion in which the principal stress space is a circular cone inclined to the principal axes with the isotropic line passing through the middle of the cone (e.g., Figure 6.9). The yield function

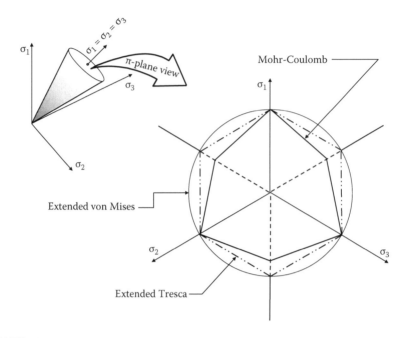

FIGURE 6.10
Octahedral (π)-plane view of Coulomb, extended Tresca, and extended von Mises yield surfaces.

for this modification, which is known as the extended von Mises criterion, is given as:

$$\alpha \frac{\left(\sigma_1 + \sigma_2 + \sigma_3\right)}{3} + \sqrt{\frac{1}{6}\left[\left(\sigma_1 - \sigma_2\right)^2 + \left(\sigma_2 - \sigma_3\right)^2 + \left(\sigma_3 - \sigma_1\right)^2\right]} = \chi \qquad (6.9)$$

where α and χ are material constants.

The relationship $\tau = c + \sigma \tan \phi$, which was first proposed by Coulomb (1776) as a yield criterion for soils, remains as one of the cornerstones in soil engineering. In the relationship, τ represents the shear stress, c is the cohesion stress, σ is the normal (compressive) stress, and ϕ is the angle of internal friction of the soil. The Mohr representation of the principal stresses (σ_1, σ_2, and σ_3) acting on a sample is shown in Figure 6.11. The normal and shear stresses acting on the any section, such as the one shown in the diagram in the top left corner of the figure, are contained in the largest circle defined by σ_1 and σ_3. Yielding or failure occurs when the combination of principal stresses acting on a sample produces the σ_1–σ_3 circle that touches the Mohr envelope, as shown in Figure 6.12.

The Coulomb relationship represents a special case of the Mohr theory of strength in which the Mohr envelope is a straight line inclined to the normal stress at an angle of ϕ. The combination of the Coulomb relationship with the Mohr diagram and the envelope is referred to as the Mohr–Coulomb theory.

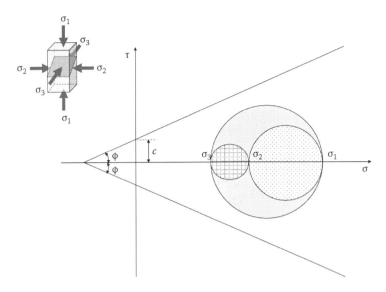

FIGURE 6.11
Mohr representation of stresses in three-dimensional system, together with Coulomb yield criterion.

In the Navier extension of the Coulomb theory (known as the Coulomb–Navier theory) account is given to the stress normal to the failure plane.

6.4.3 Mohr–Coulomb Theory

The failure theory, which was proposed by Mohr (1900) following the earlier work of Coulomb and Navier, considers failure by both yielding and fracture of the material under consideration. The functional relationship between normal and shear stresses on the failure plane is stated as: $\tau = f(\sigma)$, where τ is the shearing stress along the failure plane, and σ is the normal stress across the failure plane (see top left corner drawings in Figures 6.11 and 6.12). The functional relationship of the two-parameter theory can be plotted on the τ-σ plane, as shown in Figure 6.12. The obtained (shown as the dashed line in the figure) is called the *Mohr rupture envelope*. This represents the locus of all points that define the limiting values of both τ and σ in the slip planes under the different states of stress σ_1, σ_2, and σ_3.

Mohr's hypothesis states that failure depends upon the stresses on the slip planes, and that failure will occur when the obliquity of the resultant stress exceeds a certain maximum value. Additionally, the elastic limit and the ultimate strength of materials are dependent on the stresses acting on the slip planes. The Mohr envelope reflects a property of the material which is independent of the stresses imposed on it, and: (a) there is no requirement for the material to obey Hooke's law, (b) no requirement for Poisson's

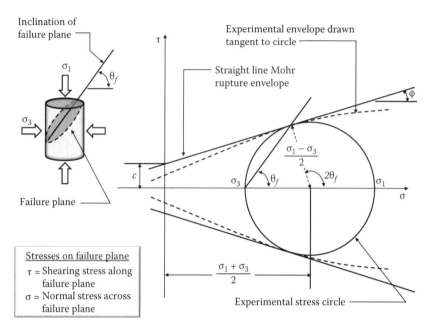

FIGURE 6.12
Mohr's failure theory plotted in the $\tau - \sigma$ plane. The major principal stress acting on the cylindrical sample shown at the top left drawing is σ_1, and the cylindrical stresses ($\sigma_2 = \sigma_3$) acting on the sides of the cylinder are σ_3.

ratio be constant, and (c) the strength and stiffness of the material in tension and compression need not be equal. The straight line envelope shown in Figure 6.12 represents the Coulomb relationship $\tau = c + \sigma \tan \phi$ referred to in the previous section. This is a special case of the Mohr theory of strength in which the Mohr envelope is a straight line inclined to the normal stress axis at an angle of ϕ, and the use of the Coulomb relationship to represent the Mohr envelope in the Mohr diagram is called the *Mohr–Coulomb theory*.

In the Mohr representation of stresses acting on the three principal planes shown in Figure 6.11, the stresses on any plane in the test material must lie within the shaded area of the large circle defined by ($\sigma_1 - \sigma_3$). The slope of any line which joins the origin and any point on the circumference of the large circle establishes the obliquity of stress, with the maximum inclination defined by angle ϕ being tangent to the large circle. Failure occurs when this maximum obliquity is attained on planes where stresses are determined at the points on the large circle where the tangent is obtained (Figure 6.12). Since the stresses act on planes that are parallel to the diameter of the principal stress, the diameter of the largest Mohr circle and the magnitude of stresses located on any point on the circumference of the large circle are independent of the intermediate principal stress σ_2. The two parallel sets of slip planes developed when an isotropic specimen is stressed slightly beyond the plastic limit by a state of homogeneous stress are symmetrically inclined with respect to the directions in which the major and minor principal stresses act. Since the two planes intersect each other along the direction in which the intermediate principal stress acts, Mohr assumed that the intermediate principal stress is without influence on the failure of a material. Accordingly, this means that some point on the circumference of the large circle in Figure 6.11 must represent the limiting condition, as shown by the tangent point in Figure 6.12.

The Mohr theory affords one with a relatively simple means for arriving at a failure theory for soils through appropriate strength tests. The triaxial test technique shown in Figure 6.6 is one such test system. The various Mohr's stress circles are plotted for the limiting states of stress and the unique failure stress on the failure plane for each test is taken as the point of common tangent between a smooth limiting curve or envelope and the various $(\sigma_1 - \sigma_3)$ circles. Whilst actual Mohr rupture envelopes are often curves, the use of straight-line envelopes for soils does not introduce meaningful errors or deviations because the curvature is relatively insignificant—over the range of normal stress encountered in practice. Identifying the radius of the Mohr's circle shown in Figure 6.13 as R, the following relationship is obtained:

$$R = \frac{\sigma_1 - \sigma_3}{2} = c \cos\phi + \left(\frac{\sigma_1 + \sigma_3}{2}\right)\sin\phi \qquad (6.10)$$

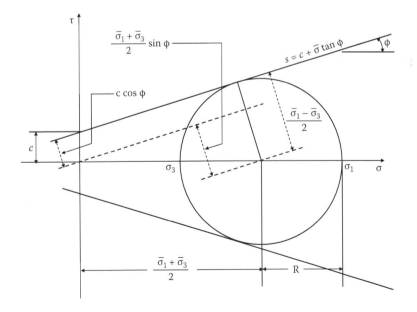

FIGURE 6.13
Properties of a straight-line Mohr rupture envelope. The stresses shown in the figure are limiting effective stresses at failure.

In the Coulomb relationship, c and ϕ denote cohesion and angle of internal friction, respectively. As shown in Figures 6.12 and 6.13, when one plots the data obtained from the tests in the Mohr diagram using the tangent points on the Mohr circle as the limit condition, the very natural consequences of this technique are the c and ϕ parameters determined from the diagram. These are, by virtue of the procedure of determination, analytical parameters obtained as a direct consequence of the application of the Mohr–Coulomb theory. They may or may not bear any physical semblance to the real material properties.

6.4.4 Yield/Failure Models

6.4.4.1 One-Parameter Models

Examples of the one-parameter model can be seen in the von Mises and Tresca yield criteria. The von Mises criterion, which identifies the occurrence of failure as the point when the octahedral shear stress τ_{oct} reaches a limiting value, and which has been stated in Equation (6.8), can be expressed in terms of the second invariant of the deviator stress J_2 or the octahedral shear stress τ_{oct} as follows:

$$f\left(J_2\right) = J_2 - k^2 = 0, \ \text{ or } \ f\left(\tau_{oct}\right) = \tau_{oct} - \frac{\sqrt{2}}{3}k = 0$$

where

$J_2 = \sigma'_{ij}\sigma'_{ji} = 1/3(I_1^2 - 3I_2)$, and $\sigma'_{ij} = \sigma_{ij} - 1/3\sigma_{ii}\delta_{ij}$,
I_1 and I_2 are the first and second stress invariants: $I_1 = \sigma_1 + \sigma_2 + \sigma_3$; $I_2 = \sigma_1\sigma_2 + \sigma_2\sigma_3 + \sigma_3\sigma_1$

$$\tau_{oct} = \frac{\sqrt{2}}{3}\left(I_1^2 - 3I_2\right)^{1/2}$$

In pure shear, k is the failure/yield stress, and δ_{ij} is the Kronecker delta (See Section 6.2.3).

In terms of J_2 and the *angle of similarity* θ, the Tresca criterion, which is a maximum shear stress criterion, can be written as:

$$\frac{\sigma_1 - \sigma_3}{2} = \frac{1}{\sqrt{3}}\sqrt{J_2}\left[\cos\theta - \cos\left(\theta + \frac{2}{3}\pi\right)\right] = k \tag{6.11}$$

where k is an experimentally determined material parameter representing the yield stress in pure shear. The angle of similarity θ, which defines the direction of the octahedral shear stress τ_{oct}, varies in the range of $0 \le \theta \le \pi/3$.

It is related to the second (J_2) and third (J_3) invariants of the deviator stress through the following relationship:

$$\cos 3\theta = \frac{3\sqrt{3}}{2} \frac{J_3}{J_2^{3/2}} \tag{6.12}$$

where $J_3 = \sigma_1' \sigma_2' \sigma_3'$.

In theory, the intermediate principal stress does not play a role in the Tresca criterion, but is considered on an equal basis with the other two principal (major and minor) stresses in the von Mises criterion. To reconcile these two separate views of the role of the intermediate principal stress, Lode (1926) introduced a *stress parameter* μ^* designed to differentiate between the two criteria. Taking the condition where $\sigma_1 \geq \sigma_2 \geq \sigma_3$, Lode expressed the Tresca yield criterion as: $(\sigma_1 - \sigma_2)/\sigma_0 = 1$, where σ_0 is the yield stress. The stress parameter μ^* is given as:

$$\mu^* = \frac{2\sigma_2 - \sigma_3 - \sigma_1}{\sigma_1 - \sigma_3} = -\frac{\sigma_2 - \dfrac{(\sigma_1 + \sigma_3)}{2}}{\dfrac{(\sigma_1 - \sigma_3)}{2}} \tag{6.13}$$

As shown in the right-hand side of the above equation, μ^* is the ratio of the difference between the intermediate principal stress σ_2 and the average of the major and minor principal stresses σ_1 and σ_3, respectively, and half the difference between σ_1 and σ_3. The von Mises criterion stated by Lode in the same format as the Tresca criterion is given as:

$$\frac{\sigma_1 - \sigma_3}{\sigma_0} = \frac{2}{\left(3 + \mu^{*2}\right)^{1/2}} \tag{6.14}$$

It is seen that if $\mu^* = -1$, Equation (6.14) becomes the Tresca criterion.

6.4.4.2 Two-Parameter Models

The two common two-parameter models are: (a) the Mohr–Coulomb, described in the previous section and (b) the Drucker–Prager, otherwise known as the extended von Mises failure criterion. The generalized relationship for the Drucker–Prager criterion is expressed as: $f(I_1, J_2) = \sqrt{J_2} - \alpha I_1 - k = 0$, where I_1 is the first invariant of the stress tensor, J_2 is the second invariant of the deviator stress, and α and k are in pure shear Mohr–Coulomb plot. The intercept of the failure line on the ordinate defines the value of k, and the angle of the slope is measured as α. In terms of the Mohr–Coulomb internal friction angle ϕ and cohesion c obtained from conventional triaxial tests, α and k are obtained as follows:

$$\alpha = \frac{2\sin\phi}{\sqrt{3}\,(3-\sin\phi)} \qquad \text{and} \qquad k = \frac{6c\cos\phi}{\sqrt{3}\,(3-\sin\phi)} \qquad (6.15)$$

6.4.4.3 Plasticity Models

The common plasticity models are the *Cam Clay model* and the group known as *cap models*. These deserve mention in this discussion because they bring concepts that differ somewhat from the ones previously discussed. In the Cam Clay model, the concept of the critical state or critical void ratio of the soil is used in establishing the state boundary surface. This is an important concept that describes the yielding behaviour of soils. Whilst cap models also describe the continuous yielding behaviour of soils, they differ somewhat in the manner in which the caps are portrayed. The reader is advised to consult the specialized textbooks dealing with stress-strain and yielding of soils.

6.5 Mechanisms in Granular Soil Strength

The key factors in determining the kinds of mechanisms involved in the resistance of a soil to shearing forces, other than the manner of soil shearing, include (a) the nature of the various soil fractions, (b) the proportions and distribution of the soil fractions, (c) its origin and maturation, and (d) its water content. The shear strength parameters such as c and ϕ, determined from a combination of laboratory tests and application of specific failure theories such as the Mohr–Coulomb theory, are mechanistically different between cohesionless and cohesive soils. Even within each of these two groups of soils, the mechanisms involved in shear resistance differ with individual soils in each of these groups, depending on soil structure, density, water content, compositional features, and so forth. Their differences will be manifested in how they resist shearing forces, and consequently on the shear strength parameters.

6.5.1 Frictional Resistance

The frictional resistance of soils to shearing forces is developed differently between cohesionless (granular soils) and cohesive soils. To a large extent, this is due to the control of shear resistance by mechanisms associated with gravitational forces for the former, and dominance of surface forces over gravitational forces in shear resistance for the latter.

Granular soils consist of discrete interacting particles that are not readily influenced by surface forces. Their behaviour is governed by gravitational and mechanical forces, and resistance to shear deformation is related directly by particle contact and packing. The frictional property of granular soils is the main physical component involved in development of shearing resistance (and hence shear strength) of such soils. In the shearing process, displacement of particles occurs along the shear plane. In this process, the shear resistance is comprised of two primary mechanisms: (a) sliding friction, and (b) interlocking "friction."

- *Sliding friction* consists of microscopic resistance of particles due to surface roughness of contacting particle surfaces. No significant volume dilatation is associated with sliding friction. The corresponding parameter is denoted as ϕ_s, the angle of friction developed from sliding resistance. The analogy of a solid block resting on another is used to describe this phenomenon. Resistance to displacement is a function of the surface roughness of the contacting particles.

- *Interlocking "friction"* consists of interlocking of particles, requiring physical displacement, overriding of particles, and sliding friction of the overriding particles (Figure 6.14). The overriding of adjacent particles along the shear plane requires physical displacement of the obstructing particles, resulting in a positive volume change (dilatation). The greater the density of the granular soil being sheared, the greater is volume dilatation, as illustrated in Figure 6.2. The parameter ϕ_i is used to denote the "friction angle" developed as a result of particle packing.

In simplistic terms, without much semblance of physical reality, and assuming that linear superposition is permitted, one obtains $\phi_s + \phi_i = \phi_a$, where ϕ_a is the apparent friction angle. The term *apparent friction angle* is used to indicate that the parameter ϕ_a is composed of a parameter ϕ_i which is not directly measured or computed, and which is not really an "angle." The apparent friction angle is obtained as a direct measure of the angle of the slope of the Mohr envelope as shown in Figure 6.15. On the other hand, the parameter ϕ_i is obtained by deduction and analysis through an evaluation of the developed shear strengths and corresponding particle packing (densities) of the soils under consideration.

To illustrate the role of particle packing in development of the parameter ϕ_i, we revisit the graphical presentation of the three different types of axial stress–strain curves and their corresponding changes in volume, previously shown as Figure 6.2. Consider the energy applied to a test sample in a conventional axisymmetric triaxial test ($\sigma_2 = \sigma_3$) such as that shown in the right-hand portion of Figure 6.6. At around the peak stress point, shown in Figure 6.2, it is reasonable to assume that $\Delta\sigma_1/\Delta\varepsilon_1 = 0$. Considering the strains

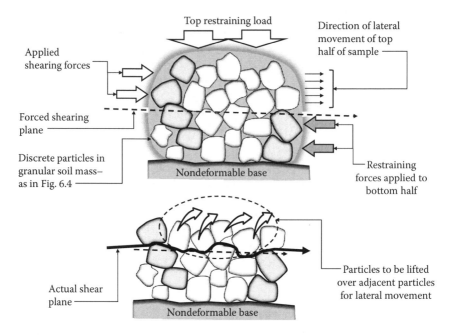

FIGURE 6.14
Schematic illustration of microscopic interlocking of soil particles (grains) requiring over-riding of adjacent particles along shearing plane. Top drawing shows the granular soil mass under shearing forces and the shearing plane forced by the actions of the shearing forces. The bottom drawing shows the actual shear plane and the required over-riding action of the particles along the actual shear plane.

ε_1, ε_2, and ε_3 to be in the same directions as the stresses σ_1, σ_2, and σ_3, respectively, the total work input in application of the shearing force must be the sum of (a) the work required to overcome the internal shear strength w_i, and (b) the work associated with volume change. This means:

$$\sigma_1\Delta\varepsilon_1 = w_i + \sigma_3(-2\Delta\varepsilon_3) \tag{6.16}$$

where the negative sign associated with $\Delta\sigma_3$ indicates extensional strain. Since the applied axial stress $\sigma_1 = \sigma_3 + \Delta\sigma$, substituting σ_1 in Equation (6.16) and specifying the following:

$$\Delta v = -(\Delta\varepsilon_1 + 2\Delta\varepsilon_3), \tag{6.17}$$

one obtains

$$\Delta\sigma\Delta\varepsilon_1 = w_i + \sigma_3\Delta v \tag{6.18}$$

where Δv is refers to the differential volumetric strain, and the negative sign associated with Δv indicates volume decrease.

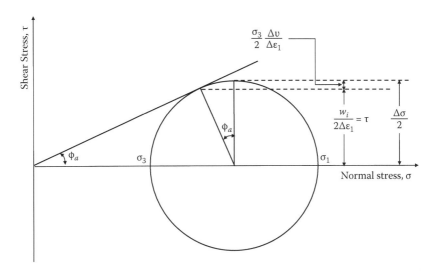

FIGURE 6.15
Mohr diagram for drained axisymmetric test on sand. Since the porewater pressure $u = 0$ for the drained test, the stresses shown in the diagram are effective stresses.

This relationship (Equation [6.18]) can be interpreted as follows:

- $\Delta\sigma \cdot \Delta\varepsilon_1$ = work put into the test sample.
- w_i = intrinsic energy of the sample. This represents the conversion of work into heat energy because of friction resistance across the rupture plane, that is, an entropy increase.
- $\sigma_3 \Delta v$ = work done by the sample, that is, mechanical work of the system (no entropy increase).

Considering the applicability of the Mohr–Coulomb relationship for the drained test on sands, as shown by the Mohr diagram in Figure 6.15, if the shear stress τ is assumed to act on the rupture surface developed under shear, we will obtain $\tau = w_i/2\Delta\varepsilon_1$, and the stress associated with work output will be given by $\Delta\sigma/2 = \tau + (\sigma_3/2)(\Delta v/\Delta\varepsilon_1)$. The assignment of $w_i/2\Delta\varepsilon_1$ to τ assumes that mechanistically, one can separate the energy associated with volume change (or no volume change) from the total work without influencing normal relationships. From Figure 6.15, one obtains: $\sigma_3(\Delta v/\Delta\varepsilon_1) = \Delta\sigma(1 - \cos\phi_a)$. Since $\Delta\sigma = \sigma_1 - \sigma_3$, this can be substituted in the relationship to give us: $\Delta v/\Delta\varepsilon_1 = (\sigma_1/(\sigma_3 - 1))(1 - \cos\phi_a)$. From Mohr's circle: $\sigma_1/\sigma_3 = 1 + \sin\phi_a/1 - \sin\phi_a$. Making this substitution, the following relationship is obtained:

$$\Delta v = 2\sin\phi_a \left(\frac{1 - \cos\phi_a}{1 - \sin\phi_a} \right) \Delta\varepsilon_1 = \Delta\varepsilon_1 f(\phi_a) \tag{6.19}$$

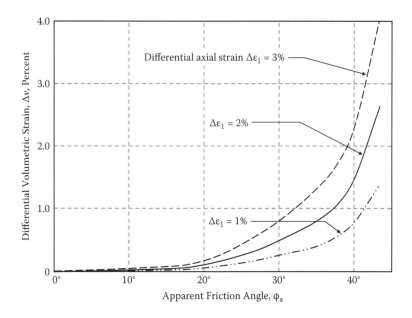

FIGURE 6.16
Relationship between differential volumetric strain and apparent friction angle ϕ_a for varying values of differential axial strain $\Delta\varepsilon_1$ at failure.

From Equation (6.19), it will be noted that $\Delta v/\Delta\varepsilon_1$ is completely describable in terms of the apparent friction angle ϕ_a, and that the important and necessary conditions are that failure is occurring through the test sample, and that the relationships that have been derived are limited to the failure state of the test sample. Figure 6.16 gives an example of the volume change characteristics that can be predicted if ϕ_a is known.

6.5.2 Mohr Failure Envelopes—Sliding and Interlocking Friction

In the stress–strain curves shown in Figure 6.2, two stress points on the individual curves are worthy of note. These are (a) peak stress and (b) residual stress. Residual stresses are obtained after straining beyond the peak stress point, as shown by the three stress–strain curves in the figure. Note that at the residual stress points, the corresponding differential volumetric strains would be zero. If one plotted the peak stress points of test samples on a Mohr diagram, the angle described by the failure envelope would be different from a corresponding failure envelope described by the residual stress points. Identifying the peak stress as σ_{pk} and residual stress as σ_r, the corresponding failure envelope angles as ϕ_{pk} and ϕ_r, one would observe that $\phi_{pk} > \phi_r$.

With no incremental volumetric strain at the residual stress point and beyond, it can be assumed that sliding friction is the dominant shear resistance mechanism, and that interlocking does not contribute much to the resistance effort. Accordingly, the friction parameter ϕ_r can be taken to be equal to the sliding friction parameter ϕ_s. From a very simplistic determination, and from a qualitative viewpoint, one can see the influence of interlocking friction on the shear resistance of granular soils—taking the difference between ϕ_{pk} and ϕ_s to be equal to ϕ_i, the interlocking parameter. There are many caveats associated with this simplistic generalization, not the least of which is the fact that (a) the ϕ_i parameter cannot realistically be an angle subtended by some interlocking failure envelope, and (b) that linear superposition of the ϕ parameters brings with it constraints regarding work expended by the shearing force exclusively on the failure plane. That being said, this mechanistic view of the contributions from interlocking of particles to the shear resistance of granular soils provides one with a perspective of soil structure and density effects on soil strength.

6.5.3 Volumetric Strain

Analytical determination of volumetric strain of granular soils such as sands and other discrete particle systems can be modelled using an assemblage of discrete particles. In the simplest case of two equal spherical particles in contact, the Hertz theory predicts that the contact (tangent) plane developed between the two particles under load F will be circular, with a radius of r (left-hand sketch of Figure 6.17). If d is the diameter of each sphere, the decrease in distance between the centres of the two spheres, denoted as δd, in contact under the normal load F is given as

$$\delta d = \sqrt[3]{\frac{9\pi^2 k^2 F^2}{d}} \tag{6.20}$$

where k is the Hertzian contact stiffness that is equal to $(1 - \mu^2/\pi E)$, and where μ and E are Poisson's ratio and Young's modulus respectively. The elastic macroscopic volumetric strain v_e of an assemblage of spheres can be obtained as

$$v_e = 3\left\langle \frac{\delta d}{d} \right\rangle = 3\frac{\left(9\pi^2 k^2\right)^{\frac{1}{3}}}{d^{\frac{4}{3}}}\left\langle F^{\frac{2}{3}} \right\rangle, \tag{6.21}$$

where carats $< >$ denote the mathematical expectation of the random variable contained between the carats.

For a random array of spheres such as that shown in the right-hand sketch of Figure 6.17, one can construct a space structure with members defining the forces at the contact points, including magnitudes and directions. A section cut through the space structure, as shown in the sketch in Figure 6.17, will allow for analysis to be conducted using the method of sections. The

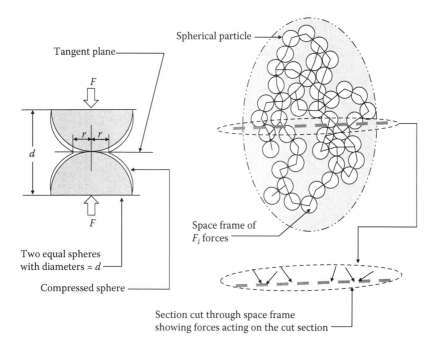

FIGURE 6.17
Schematic illustration of equal spheres in contact and under load. Left-hand sketch shows two equal spheres in normal contact under compression force F—with resultant compression. Right-hand sketch shows a simulated force space structure obtained from forces acting between adjacent equal spheres. The bottom cut-out shows the forces acting on the cut section in the space frame.

relationship reported by Yong and Wong (1972) for the elastic macroscopic strain v_e using such a method of analysis is given as follows:

$$v_e = \frac{GR}{\left(1-n_0\right)^{\frac{2}{3}} a_t^{\frac{2}{3}}} \left[\sigma + \frac{E^{\frac{2}{3}}}{m} \log_e \frac{a_t - \left(a_t - a_i\right) e^{-(m\sigma^{\frac{2}{3}})/E^{\frac{2}{3}}}}{a_i} \right] \qquad (6.22)$$

where σ is the elastic macroscopic stress, n_0 is the initial porosity, a_i and a_t are the average coordinate numbers of the assemblage of particles at hydrostatic pressure $\sigma = 0$ and $\sigma = \infty$ respectively, m, G, R, a_t, and a_i are material parameters

$$G = 2^{\frac{2}{3}} 3 \left(81\pi^2 k^2\right)^{\frac{1}{3}} \left(\frac{\pi}{6}\right)^{\frac{2}{3}}$$

and

$$R = \frac{<F^{\frac{2}{3}}>}{<F>^{\frac{2}{3}}}.$$

The material parameters m, a_i, and a_t determine the average coordinate number and its variation with stress. m is a packing factor that describes the rate of increase of contact points in an assemblage of particles in plane strain. The comparison of predicted volumetric strain relative to applied pressure for Ottawa sand, reported by Yong and Wong (1972), showed good correlation for a void ratio of 0.55 and about 15% deviation (higher experimental values) at a higher void ratio (0.65).

6.6 Cohesive Soil Strength

Cohesive soils are called *c*-φ soils since they have physical properties of *cohesion* (*c'* and *friction* (φ, *the angle of internal friction*) that have been called the basic shear strength parameters of cohesive soils. Strictly speaking, these should be called analytical properties or properties determined from application of the Mohr–Coulomb theory of failure. To avoid confusion, we will refer to them as the shear strength parameters *c* and φ. The links between the shear strength parameters with physical reality range from good (actual) to highly tenuous. Other than soil composition, the roles of water content and porewater pressure feature significantly in "what is measured" and "what they mean."

6.6.1 Analytical and Physical Strength Parameters

Determination of the shear strength of cohesive soils is generally achieved via appropriate laboratory strength test techniques such as the triaxial test system discussed in Section 6.4.1. Data analysis can be conducted along two different lines: (a) using a yield or failure theory (model) that will provide one with the analytical shear strength parameters associated with the theory chosen to analyze the data obtained, and/or (b) analyzing the test data with a view to determine the mechanisms involved in shear resistance of the soil under test, and deduce the kinds of activities, and hence the parameters, involved in mounting the resistance to shear failure of the soil. The former approach to data reduction is akin to deterministic analysis and the latter approach is more along the lines of a mechanistic analysis.

6.6.2 Analytical Shear Strength Parameters

The two different approaches to data analysis provide one with the opportunity to gain information on the physical behaviour of soil in relation to applied stresses, and to determine how the behaviour is characterized analytically. In the case of data reduction using a yield/failure model, the shear strength parameters obtained are a direct consequence of the model. As has been indicated previously, to avoid confusion in how the data are handled, these strength parameters should be identified as analytical shear strength parameters. The crucial element in all of this is the admissibility or viability of the yield/failure model used to analyze the test data. This means that one needs to be confident that the manner in which the test soil sample responds to the applied stresses is properly modelled.

The model fitting the Mohr–Coulomb theory of failure is one where the specimen under stress remains rigid until the point of failure. At failure, the rigid sample is bisected into two parts by an inclined rupture plane. The theory essentially models the stress response of a rigid-plastic material. All the work performed by the applied stress is expended along a definite rupture (failure) plane, meaning that no work is expended in deforming any of the two rigid halves separated by the failure plane. Other than inattention to (a) the effects of the intermediate principal stress σ_2 (referred to in Section 6.4.3), and (b) inertia terms, the requirement of no volume-change rigid-plastic behaviour under shearing stresses is perhaps the most restrictive element of the model.

There are other yield/failure models that are in common use, as noted in the discussion in the earlier sections. They bring with them different sets of restrictions and conditions. The common thread running through these models is the fact that they are designed to produce specimen behaviour under stress that mimics the stress–strain relationship obtained in laboratory tests, with particular attention to conditions at failure. The shear strength properties reported by users of these models must be recognized as analytical shear strength parameters. This does not mean that these "properties" are faulty or inappropriate. On the contrary, these have been found to be very useful in deterministic analyses of situations related to field practice.

6.6.3 Mechanistic Analyses

There is no hard and fast rule that governs the structure of a mechanistic analysis. What is required is a good understanding of soil behaviour in relation to compositional features and stressing situations. One begins by examining the interactions between soil constituents, that is, interparticle action, with a knowledge of the various kinds of forces of interaction available with the type of soil fractions involved. In essence, one is required to examine the interaction between soil and water in the development of the physical mechanism for yield or failure of the soil under test. Unlike many other kinds of materials, the strength of cohesive soils is determined by microscopic, mesoscopic, and macroscopic elements of the soils: layer-lattice minerals, soil particle morphology and soil structural units, and macroscopic soil structure. How all these elements are put together, and how they interact and respond to the chemical nature of the porewater will determine the response of the soil to stress input. The nature and form of particle and soil structure interaction makes it difficult to establish a definite strength quantity for the soil—except perhaps in terms of certain governing conditions and limitations.

6.6.4 c and φ Parameters and Properties

We have established that the c and ϕ shear strength parameters are the direct outcome of application of the Mohr–Coulomb failure theory, as seen in Figures 6.12, 6.13, and 6.15. Other than describing c as the intercept on the ordinate of a Mohr diagram, the concept of cohesion c as an actual physical property is not easily grasped. Explanations of "stickiness" or "glue-like effect" have been advanced in previous times—to no avail. What is clear, however, is the fact that a cohesive soil sample will stand upright as a test cube or test cylinder without any lateral support, whereas a cohesionless soils such as sand will not. The fundamental questions have always been, "What is holding all those microscopic particles together?" and "If it is water, why is water not leaking out from the test sample?"

It has been found that one can consider cohesive soils to be somewhat similar to colloidal systems, especially in respect to interface phenomena as witness the diffuse ion-layer type of behaviour of the layer-lattice minerals. Treatment of the shearing strength of cohesive soils as colloidal systems does not easily lead to the production of c and ϕ and mechanistic parameters or properties. The discussion in Section 6.3.2 in relation to effective stresses makes the point that contact between clay particles or microstructural units is effectively between adsorbed (hydrated) water layers, as demonstrated by the schematic shown in Figure 6.5. Given that picture of the physical interaction between the soil particles of a cohesive soil, one can advance the concept that cohesion in a cohesive soil is a direct measure of the resistance developed in relation to the interactions between adsorbed water layers of contacting particles and microstructural units.

Figure 6.18 shows the interacting adsorbed (hydration) water layers surrounding individual particles and microstructural units. Resistance to displacement of the particles or displacement/distortion of the microstructural units will be offered by the interacting adsorbed water layers. The greater the degree or level of interactions between the water layers, and the greater the water layers are held by their respective particles or units, the greater will be the resistance to shear displacement or distortion. In essence, cohesion c is a physicochemical property derived from interatomic, intermolecular, and interparticle actions.

Resistance to shear displacement by interacting adsorbed water layers will explain the nature of the cohesive c resistance. The physical concept of friction between the particles or microstructural units accepts the fact that actual physical contact between solid particles will occur for soils that are not composed exclusively of layer-lattice clay minerals. Typical cohesive soils consist of a mixture of clay minerals, and other fine-grained particles that are not clay minerals. There will be contact between the solids that do not

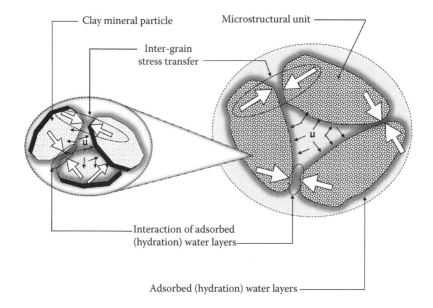

FIGURE 6.18

Interaction of clay microstructural units formed by layer-lattice clay mineral particles. Right-hand sketch shows stress transfer between microstructural units through adsorbed water layers surrounding the units, and interaction of adsorbed water layers. Left-hand sketch shows interactions between particles in a microstructural unit with their individual adsorbed water layer interacting with each other. Resistance to displacement of particles or displacement/distortion of the microstructural units will be resisted by the interacting adsorbed water layers.

have pronounced adsorbed water layers, or between primary mineral solids. The effect of initial bonding between particles and microstructural units will be seen as a pseudo frictional phenomenon.

6.7 Porewater Pressure

Porewater pressure in cohesive soils plays an important role in the mechanisms involved in resisting applied shearing forces, more so than granular soils because of the role of the physicochemical forces in cohesive soils. Unlike granular soils where application of an external pressure will evoke an almost instantaneous response in equilibrating porewater pressure in the soil, the equilibrating response time in cohesive soils is slower. The term *equilibrating porewater pressure* is used to mean the porewater pressures at equilibrium under the applied external load. The more active the clay minerals are in the cohesive soil, the slower will be the equilibrating porewater pressure response time. The discussion in Section 6.3.2 regarding stress transfer amongst the soil particles and microstructural units (see Figures 6.5 and 6.18) made reference to porewater pressures u in the macropores and u_{mic} in the micropores. For soils containing swelling clay minerals, there exists another porewater pressure component u_π which is the porewater pressure due to osmotic effects from diffuse ion-layer phenomena in interlayers.

6.7.1 Partly Saturated Soil Porewater Pressures

Measurements of porewater pressures in soils are generally conducted with laboratory triaxial tests using a null-point technique for soils that are fully water saturated, meaning that no volume change is required for generated porewater pressures to be measured. This is a significant requirement. Problems arise in seeking accurate portrayal of porewater pressures when soils are not fully water-saturated because of the presence of air and vapour in the pore spaces of the partly-saturated soils. To determine the effective stresses in such soils when subject to applied pressures or loads, one needs to obtain an accurate picture and measurements of the various forces or pressures that contribute to the total resistance offered by the soil to the applied loads.

The subject of effective stress determination in partly saturated soil has been, and continues to be, vigorously studied in soil mechanics and soil engineering, to a large extent because of the challenges offered in determining the physics of the phenomena involved when confronted with soils that contain layer-lattice clay minerals. The discussion in this section focuses on the elements of the problem vis-à-vis interactions between soil particles and water and air/vapour in the pores spaces.

6.7.1.1 Pore Pressures in Granular Medium

We use the term *pore pressures* instead of *porewater pressure* for the obvious reason that the pressures generated in the soil pore spaces of a partly saturated soil will consist of both pore air and porewater pressures. The left-hand sketch in Figure 6.19 shows an idealized representation of two contacting spheres to illustrate the simplest of the basic elements involved in pore pressure development.

The left side of the left-hand sketch in Figure 6.19 shows the conventional concept of porewater u_w and pore air u_a pressures in the pore space between the two contacting spheres. The right side of the left-hand sketch shows the equivalent phenomena as a capillary potential ψ_c, which in this case is equal to the matric potential ψ_m. In soil mechanics usage, the effective stress for partly saturated soils takes its cue from Bishop's pioneering studies (Bishop, 1959; Bishop, 1960; Bishop et al., 1960), where the expression for the effective stress was proposed as

$$\bar{\sigma} = \sigma - u_a + \chi\left(u_a - u_w\right) \tag{6.23}$$

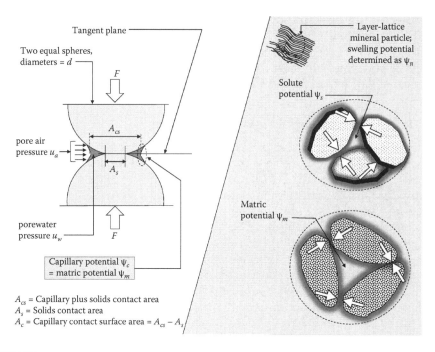

FIGURE 6.19

Physical and physicochemical representation of forces of interaction in the void spaces in partly saturated soils. Left-hand sketch shows interaction of two ideal contacting spheres with a capillary fringe. The porewater pressure shown is in the water phase within the capillary whereas the pore air pressure is defined by the capillary configuration. The equivalent physicochemical mechanism is shown as the capillary potential ψ_c which is equal to the matric potential ψ_m. The right-hand sketch shows the physicochemical phenomena demonstrated in three different components of a clay. Top right shows the osmotic potential ψ_π developed as a result of interlayer DDL mechanisms. Centre-right shows the effect of dissolved solutes in interactions between particles and bottom-right shows soil–water interaction in terms of the matric potential ψ_m.

where u_a refers to the pore air pressure, u_w refers to the porewater pressure, χ refers to the capillary contact surface area per unit gross area of soil over which the porewater pressure acts, represented as A_c in Figure 6.19. This means that the pore air pressure would act over an area of $(1 - \chi)$. It has been pointed out by Aitchison and Donald (1956) that strictly speaking, χ is not an area. However, as noted by Skempton (1960), "This assumption (of χ as an area) leads to a simple model of the problem and a correct form of the expression for equivalent pore pressure." Present usage of χ considers it as a numerical parameter dependent on the saturation level of the soil.

Using Equation (6.23) as the building block, the maximum shearing resistance τ obtained via the Mohr–Coulomb failure model provides one with the basic expression upon which considerable research and study continues. The expression of τ in terms of effective stresses is given as follows:

$$\tau = c' + \left[\sigma - u_a + \chi\left(u_a - u_w\right)\right]\tan\phi' \tag{6.24}$$

where the superscripted primes associated with c and ϕ are the apparent cohesion and apparent angle of friction. In conventional soil mechanics, $(u_a - u_w)$ is considered to be the matric suction S_m. In terms of soil–water potentials, this would mean that $(u_a - u_w)$ is equivalent to ψ_m.

6.7.1.2 Pore Pressures in Clays

The right-hand sketches in Figure 6.19 show the various soil–water potential components active in partly saturated clays. Measurement of any of these as separate quantities is not easily accomplished. The osmotic potential ψ_π developed as a result of interactions within the interlayers (top right corner of figure) can be taken to be equal to the solute potential ψ_s in the interparticle spaces at equilibrium conditions, that is, $(\psi_\pi = \psi_s)$. Note, however, that when shearing or dynamic conditions exist, this assumption will not hold because of the different equilibrating response times of these potentials—similar in principle to the equilibrating response time of the porewater pressure comparison between granular and cohesive soils referred to at the beginning of this section.

At equilibrium, the matric potential ψ_m that is active in the macropores of the soil should be equal to the solute potential ψ_s. Again, if shearing forces are active resulting in nonequilibrium conditions, this equality will not hold for the same reasons cited previously regarding differences in equilibrating response times. In practical terms, this means that if one desires to obtain an accurate picture of the resisting forces in the pore spaces, one needs to ensure that equilibrium conditions are attained after each applied load or increment of load, assuming that one desires to use Equation (6.24) or its equivalent. The important element in the implementation of relationships for effective stress $\bar{\sigma}$, or shearing stress τ of partly saturated clays is a reliable and accurate technique for measurement of (a) the negative porewater pressure u_w, and/or (b) the pore air pressure u_a, and (c) the matric suction S_m and/or matric potential ψ_m.

6.7.2 Porewater Pressure Coefficients

The characterization of developed porewater pressures in terms of *pore pressure coefficients* A and B, first proposed by Skempton (1954), has proven to be a useful tool in the assessment of partly saturated soil performance under shearing stresses. To be consistent with the development of the A and B coefficients, usage of the term *pore pressure* is maintained. The derivation for both the coefficients A and B considers an element of soil at equilibrium under some initial isotropic effective pressure p, and porewater pressure u at zero.

Consider an axisymmetric triaxial test where increases in pressures $\Delta\sigma_1$ and $\Delta\sigma_3$ occur in two stages. The corresponding porewater pressure changes to pressure increases of $\Delta\sigma_1$ and $\Delta\sigma_3$ will be Δu_1 and Δu_3, respectively. Note that the subscripts "1" and "3" associated with the porewater pressure u refer to the increments of pore pressure corresponding to the respective pressure increases ($\Delta\sigma_1$ and $\Delta\sigma_3$)—that is, they do not refer to the major and minor principal axes. The pore pressure coefficient B is given as follows. The corresponding increases in effective stress $\Delta\bar{\sigma}_3$ is given by $\Delta\bar{\sigma}_3 = \Delta\sigma_3 - \Delta u_3$ and the corresponding change in the total volume of the sample is determined as $\Delta V_c = -C_c \cdot V(\Delta\sigma_3 - \Delta u_3)$, where C_c is the compressibility of the soil structure. The change in volume of the pore spaces is obtained as $\Delta V_v = -C_v n V \cdot \Delta u_3$, where C_v is the compressibility of the fluid and air, n is the porosity of the soil, and V_v is the volume of the pore space. These two volume changes must be identical, that is, $\Delta V_c = \Delta V_v$ since the solid particles of the soil structure are assumed to be incompressible. Hence, one defines the pore pressure coefficient B as

$$B = \frac{\Delta u_3}{\Delta\sigma_3} = \frac{1}{1 + \dfrac{nC_v}{C_c}} \tag{6.25}$$

Note that the pore pressure coefficient B is a measure of the level of saturation of a soil. It can be determined in laboratory consolidated undrained axisymmetric triaxial tests by measuring the porewater pressure increase Δu_3 as one increases σ_3 by $\Delta\sigma_3$. In saturated soils, an increase in σ_3 should be matched by an equal increase in u_3, that is, $(\Delta u_3 = \Delta\sigma_3)$. This means that $B = 1$. This same value for B can also be obtained if one considers the compressibility coefficients C_v and C_c. Since water is less compressible than the mineral structure, $(C_v/C_c = 0)$. At the other end of the scale, for dry soils, one can consider C_v to approach infinity. Hence $B = 0$. The same result will be obtained if one considers this in terms of pore pressure increase, that is, there will be no porewater increase with an increase in σ_3.

The pore pressure coefficient A, which is a measure of the shearing stress and compressibility of the soil structure, that is, soil skeleton, can be characterized as follows: The application of a stress difference in triaxial testing $(\Delta\sigma_1 - \Delta\sigma_3)$ gives one $\Delta\bar{\sigma}_1 = (\Delta\sigma_1 - \Delta\sigma_3) - \Delta u_1$ and $\Delta\bar{\sigma}_3 = -\Delta u_1$. Assuming, as a first approximation, that the soil skeleton is elastic, one obtains from elastic theory the following relationship:

$$\Delta V_c = -C_c V \frac{1}{3}\left(\Delta\bar{\sigma}_1 + 2\Delta\bar{\sigma}_3\right),$$

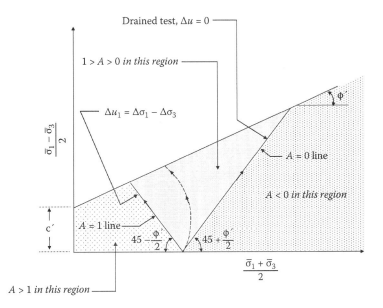

FIGURE 6.20
Mohr–Coulomb plot with effective stresses, showing stress paths and pore pressure coefficient *A* domains. For normally consolidated clays, A is expected to be between 0.5 and 1, meaning that the stress paths will be in the region defined by $0.5 \leq A \leq 1$. The stress path shown in the grey area in the diagram is an illustration of the stress behaviour of a normally consolidated clay.

where V_c is the volume of the soil skeleton. The corresponding change in pore space in the soil will be obtained as $\Delta V_v = -C_v n V \Delta u_1$.

Since the two volume changes must be the same, that is, $\Delta V_c = \Delta V_v$, one can obtains $\Delta u_1 = B(1/3)(\Delta\sigma_1 - \Delta\sigma_3)$. Replacing the idealized assumption of an elastic skeleton with a nonelastic soil skeleton and expressing the 1/3 factor in terms of the pore pressure coefficient A, one obtains $\Delta u1 = B \cdot A(\Delta\sigma_1 - \Delta\sigma_3)$, and since $\Delta u = \Delta u_1 + \Delta u_3$, one will obtain

$$\Delta u = B\left[\Delta\sigma_3 + A\left(\Delta\sigma_1 - \Delta\sigma_3\right)\right]$$
$$= B\Delta\sigma_3 + \overline{A}\left(\Delta\sigma_1 - \Delta\sigma_3\right) \tag{6.26}$$

where $\overline{A} = B.A$. When $B = 1$, $\overline{A} = A$ and hence for a fully saturated soil, one will obtain

$$\Delta u = \Delta\sigma_3 + A\left(\Delta\sigma_1 - \Delta\sigma_3\right) \tag{6.27}$$

Similar to the laboratory experimental determination of the pore pressure coefficient B, using consolidated undrained axisymmetric triaxial testing, the determination of $\overline{A}$ can be made by measuring the porewater pressure generated under an application of axial pressure $\Delta\sigma_1$, (where $\Delta\sigma_1 = \sigma_1 - \sigma_3$). With a corresponding porewater pressure increase of Δu_1 that can be measured, one can express as a ratio:

$$\frac{\Delta u_1}{\Delta\sigma_1} = \frac{\Delta u_1}{(\sigma_1 - \sigma_3)} = \bar{A} \qquad (6.28)$$

Figure 6.20 shows the domain of the pore pressure coefficient A in relation to porewater pressures developed in undrained triaxial testing. In the undrained test on a saturated soil sample, if one assumes that $(\Delta u_1 = \Delta\sigma_1 - \Delta\sigma_3)$, then $A = 1$.

6.8 Shear Resistance Mechanisms

6.8.1 Gravity Controlled

As previously discussed (Section 6.5), shear resistance mechanisms in soils refer to the nature of the actions and interactions of soil particles as they resist relative displacement with each other. The resistance mechanisms have root in whether interparticle actions are controlled by gravitational forces or by intermolecular and interatomic forces. In gravity-controlled soil–water systems, macroscopic interlocking of particles require appreciable overriding and movement of particles relative to each other in the region of the failure plane, provoking dilatation in the test sample. The frictional resistance developed between contacting particle surfaces consists of: (a) microscopic interlocking due to surface roughness of the particles, and (b) adhesive bonding between the contacting particle surfaces, the strength of which is a function of the distance between atoms and molecules in the contacting surfaces and the composition of the soil particles. If one assumes that the spacing between oppositely charged atoms for a given material is constant, it follows that the strength of the bond must be proportional to the total contact area established between the interacting particles.

6.8.2 Bonding Effects

Bonding between particles and microstructural units (*msu*) is an important element in (a) the development of soil strength, and (b) structuring the response mechanism to shearing forces. Edge-to-edge (*ee*) and edge-to-surface (*es*) bonds can include (a) cementing bonds obtained from oxides of iron, silicon and aluminium, carbonates and organic matter precipitated between particles and *msu*, and (b) those resulting from interparticle forces.

Quantitative differentiating between *ee* and *es* bonding and their strengths is difficult. Qualitatively, however, deductions can be made on the basis of type of minerals and cementation agents present in the soil, and the specific charge of the active minerals. When cementing bonds are strongly developed, it follows that the soil structure (that is, structural framework) formed by the particles and *msu* will be strongly bonded. This is a characteristic of bonded clays which distinguishes it from unbonded or remoulded clays. An illustration of the kinds of stress–strain curves obtained from such clays is shown in Figure 6.21.

In the illustrative stress–strain curves shown in Figure 6.21, the contribution made by the strong bonds developed from cementation and carbonates is seen in terms of (a) the increase in peak strength, and (b) the steeper gradient in the portion of the stress–strain curve before failure. The latter phenomenon is due to the fact that cementation and carbonate bonds are brittle bonds, that is, they will break at low strains. In comparison, clay-organic bonds are generally thought to be semi-rigid or "plastic" in nature. In unbonded clays, the structural framework formed by the particles and *msu* is not tightly held at the juncture points normally formed by *ee* and *es* configurations. Cementation bonds are either poorly developed or absent. Remoulding of bonded clays destroys the bonds at the juncture points, with the result that shear resistance mechanisms will have to rely on particle and *msu* interactions that include their respective adsorbed water layers, as discussed previously in Section 6.6.

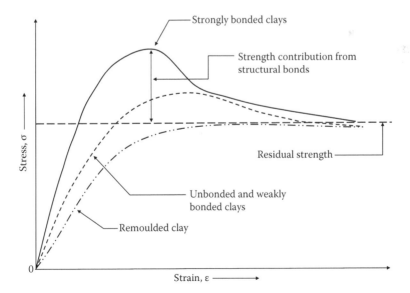

FIGURE 6.21
Schematic illustration of stress–strain curves for strongly bonded, weakly bonded, and remoulded clays.

In addition to the cementation bonds, interparticle bonds contribute to the characterization of stress–strain performance. The varied nature of interparticle bonds and their respective strengths, combined with the heterogeneity of the network of soil particles and microstructural units (i.e., heterogeneity of the macrostructure), means that the reaction of an element of soil to imposed stresses will be complex. As with any mechanical or physical structure, the integrity of the structure is compromised by the *weakest link*, meaning that the weakest bonds between particles and between microstructural units are most vulnerable. Under stress, the weakest hydrogen bonds will fail first, thereby transferring to stronger junctions represented by bridging cations and dipole-type attraction, and further to cementations bonds and primary valence bonds in mineral/mineral contacts. The sequence of bond breakage and the collapse of the total soil structure is the accumulation of the collapse of microstructural units, which can be observed in terms of deformation of the sample.

6.9 Compressibility and Consolidation

The *compressibility* of a soil refers to its response behaviour under compressive stresses or loads. The roles of pore air and porewater in a soil under compressive stresses, together with the nature of the applied compressive stresses σ (i.e., $\sigma_1 = \sigma_2 = \sigma_3$ or $\sigma_1 \neq \sigma_2 \neq \sigma_3$, etc.) are significant in determining how a soil responds to these imposed stresses (Figure 6.22). The sketches shown on the left side of the figure illustrate triaxial compression of cylindrical (upper left) or rectangular-sided (bottom left) samples. When the stresses applied to the samples are hydrostatic (i.e., $\sigma_1 = \sigma_2 = \sigma_3$), volumetric compression will result when porewater is allowed to escape under the stresses. This is the consolidation step in consolidated undrained triaxial tests discussed in the previous sections. When the three principal stresses are not equal, shear stresses are introduced in the sample. The result is sample volumetric compression accompanied by shearing and distortion occurs during porewater drainage.

The sketch on the right side of Figure 6.22 shows the elements of a one-dimensional consolidation test scheme. During the application of monotonically increasing loads, porewater is allowed to escape and equilibrium is established before addition of the next load increment, meaning that the pore pressure developed under the load must be allowed to fully dissipate before addition of the next incremental load. The graphical diagram shown at the bottom right of the figure shows the typical consolidation-time relationship under a single load.

Whilst the terms compressibility, consolidation, and creep have been used in the literature to indicate volume change of a soil under compressive stresses, it is not always clear that the mechanistic differences between these types of behaviour are fully recognized. From the mechanistic point of view:

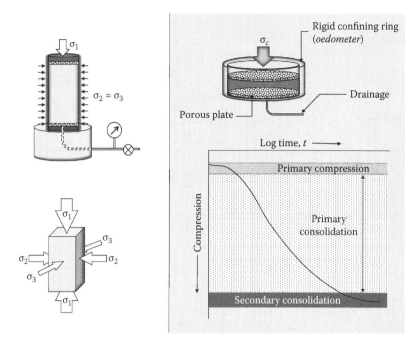

FIGURE 6.22
Compression of soil. Top left is the axisymmetric triaxial test system where $\sigma_2 = \sigma_3$. For axisymmetric compression/consolidation, $\sigma_2 = \sigma_3 = \sigma_1$. Bottom left is the triaxial test sample where all three principal stresses can be different or equal to each other. For triaxial compression/consolidation, all three principal stresses are equal to each other. The right-hand sketch shows the elements of a consolidation test system where one-dimensional consolidation loading σc is applied in monotonically increasing fashion. The bottom portion of the sketch shows the consolidation-time characteristics under a single load.

- *Compressibility* of a specimen refers its behaviour under compressive forces, and may be determined in terms of volumetric or axial strains in relation to total (undrained) or effective (drained or with pore pressure measurements) stresses. The stress–strain behaviour of a soil is a particular case of compressibility.

- *Consolidation* of a specimen is generally considered as a one-dimensional volume change event under consolidation loads, where the volume change is due to the expulsion of porewater under the incremental loads (pore pressure dissipation), and where the amount of volume change is a direct function of the amount of porewater extruded.

- *Creep* in soils is thought of as axial strain occurring over a prolonged period under a constant load without measurable porewater discharge. Because creep tests are not generally conducted in oedometers (consolidation devices), the results of creep tests are considered as descriptors of the rheological behaviour of a soil, tested to failure.

- *Secondary consolidation* is a deformation process that occurs after primary consolidation under a constant load, where the porewater discharge is vanishingly small, that is, no measurable pore pressures.

6.9.1 One-Dimensional Consolidation

Considering one-dimensional volume change under a consolidating load, the total change in volume of a fully-saturated soil can be stated as $\partial V / \partial t = \partial V_v / \partial t$, where V and V_v represent the total soil volume and volume of voids (in the soil) respectively. Assuming that (a) the soil particles are incompressible, (b) that the permeability coefficient k is constant, and (c) that volume decrement results solely from the extrusion of porewater from the saturated pore spaces, one can state that:

$$k \frac{\partial^2 h}{\partial x^2} + \frac{\partial V_v}{\partial t} = 0 \qquad (6.29)$$

where x is the spatial coordinate, and h is the total head due to the porewater pressure u, that is, $(h = u/\gamma_w)$ where γ_w is the density of water. Relating the change in volume of voids to the change in porewater pressure u through a modulus of volume change m_v, one obtains $m_v(\partial u/\partial t) = -\partial V_v/\partial t$. Substituting this in Equation (6.29) together with the relationship between h and u, and rearranging terms, one obtains:

$$\frac{\partial u}{\partial t} = \frac{k}{m_v \gamma_w} \frac{\partial^2 u}{\partial x^2} = C_v \frac{\partial^2 u}{\partial x^2} \qquad (6.30)$$

where the coefficient of consolidation C_v is given as $C_v = k/m_v \gamma_w$. The modulus of volume change m_v can be expressed in terms of the void ratio e as follows: $m_v = a_v/(1+e)$, where a_v is the coefficient of primary or theoretical compressibility: $a_v = de/d\bar{\sigma}$. The solution to Equation (6.30) depends on the boundary conditions used. For drainage of porewater at the top and bottom of the soil under consideration, and considering a soil stratum of depth $2H$, the following boundary conditions and initial condition can be stated:

$u = 0$ for top of soil layer, $x = 0$

$u = 0$ for bottom of soil layer, $x = 2H$

$u = u_o(x)$ for time, $t = 0$

The solution to Equation (6.30) for the stated boundary conditions is:

$$u = \sum_{n=1}^{n=\infty} \left(\frac{1}{H} \int_0^{2H} u_0(x) \sin \frac{n\pi x}{2H} dx \right) \left(\sin \frac{n\pi x}{2H} \right) e^{\frac{-n^2 \pi^2 C_v t}{4H^2}} \qquad (6.31)$$

The derivation of the consolidation relationship given as Equation (6.30) relies on certain assumptions, namely, (a) that there is no change in permeability during the consolidation process, (b) that temperature has no effect on the void ratio e-pressure p relationship, (c) that the load placed for compression is applied instantaneously, and (d) that the resultant compression or consolidation is a direct result of the instantaneous increase in applied load.

The change in hydraulic conductivity with applied pressure and with elapsed time is a reality, given that deformation (consolidation) of the sample under consolidating loads will decrease the void volume of the test sample. Correspondingly, there will be changes in the structure of the microstructural units, rearrangement of the soil particles, changes in the tortuosity, and hence changes in the permeability of the soil. Since the permeability coefficient k is related to porosity and other soil structural factors, if one assumes that for small deformations, the relationship between volume of voids and the excess porewater pressure is linear, the following relationship is obtained: $k = k_o - \beta(u_{max} - u)$, where β is the coefficient concerned with variation in soil permeability, u_{max} is the maximum porewater pressure, and k_o is the permeability coefficient at the beginning of the consolidation process. When $u = 0$, k becomes $k_{final} = k_o - \beta u_{max}$.

6.9.1.1 Laboratory Test Data

The general procedure for determination of the consolidation behaviour of soils employs a device similar in principle to the one shown in the right-hand sketch in Figure 6.22, that is, an oedometer. The choice of a one-dimensional consolidation test (a) provides for a simpler means for detection of volume change and axial deformations, and (b) avoids the complexities associated with both measurements of three-dimensional strains and mathematical analysis of three-dimensional consolidation behaviour. In the one-dimensional consolidation test procedure, there are at least three methods by which load is applied to the specimen in the oedometer:

- *Monotonically increasing loads (constant total-stress σ procedure)*, with each incremental load applied following complete porewater pressure dissipation under the previous load. The next load increment is commonly taken to be twice the previously applied load.

- *Controlled strain-rate procedure*, where the intent is to provide for a more rapid test through continuous load application. The rate of load application is dictated by the requirement that no appreciable distress occurs in the test sample through excessive shearing and high porewater pressure response. The specification of a particular strain-rate for load application is determined accordingly.

- *Controlled gradient procedure*, where the rate of load application is dictated by the requirement for controlling the generated porewater pressures. If one maintains a constant porewater pressure u, the

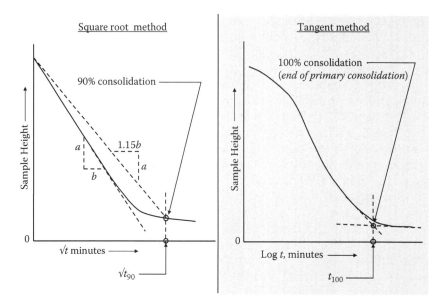

FIGURE 6.23
Data representation from oedometer tests. The left-hand sketch is the square root (Taylor) method for determining 90% primary consolidation; (a) and (b) represent the vertical and horizontal components of the slope. The right-hand sketch is the tangent (Casagrande) method for determining 100% primary consolidation, and the right-side graph is a typical *e-log P* curve from a consolidation test.

total stress σ will need to be increased continuously because if σ remained constant, u will decrease under a constant σ. With the increasing σ and a constant u, this means that the effective stress $\bar{\sigma}$ will be maintained continuously.

Determination of *end of primary consolidation* under an applied load most often requires a graphical technique or some rule-of-thumb procedure. The reasons for this can be seen in the nature of the time-deformation relationship obtained under an applied load (constant total stress procedure), as demonstrated in Figure 6.23, where the two commonly used graphical procedures for determination of 90% and 100% primary consolidation are shown. The square-root procedure shown on the left side of the figure is known as the Taylor method, and the tangent method shown on the right side of the figure is known as the Casagrande method. The asymptotic nature of the curves, as they reach equilibrium primary consolidation and vanishingly small pore-water pressures, requires one to set some type of criterion to establish when the next incremental load may be applied.

The information gathered from a series of load-deformation results is generally portrayed in terms of the applied loads expressed as pressures P, and deformations expressed as changes in the void ratio e of the test sample. The

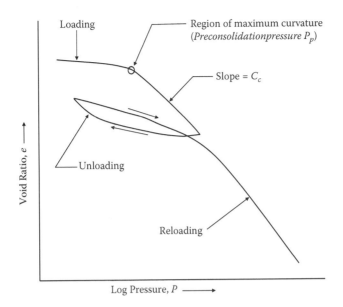

FIGURE 6.24
Typical *e-log P* curve obtained from a one-dimensional consolidation test. The slope *Cc* is known as the coefficient of compression.

typical *e-P* relationship shown in Figure 6.24, which is commonly referred to as the *e-logP* curve, allows one to use the slope of the loading curve (C_c, coefficient of compression) to formulate a simple working model for predicting the total compression of the soil stratum in the field in one-dimensional consolidation.

Assuming that the change in volume of a saturated soil ΔV under consolidating loads is equal to the change in volume of the voids in the soil ΔV_v, and assuming incompressibility of water and soil solids: $\Delta V = \Delta V_v = V_s \Delta e$, where V_s is the volume of soil solids. This means that $\Delta V = V \cdot \Delta e / (1+e)$, and from Figure 6.24, $\Delta e / (\log P_2 - \log P_1) = C_c$, where $P_2 > P_1$. Hence, $\Delta V = (1/(1+e)) V C_c \log P_2 / P_1$, and for one-dimensional consolidation, $\Delta V = A \Delta h$, where A is the cross-sectional area of the test sample, and h is its height. Recognizing that P_2 is $P_1 + \Delta P$, and making the necessary substitutions and rearrangement of terms, one obtains:

$$\Delta h = \frac{C_c}{1+e} h \log \frac{P_1 + \Delta P}{P_1} \tag{6.32}$$

If P_1 represents the effective stress $\bar{\sigma}_1$ at some initial point, and the increase in pressure ΔP produces and increase in effective stress $\Delta \bar{\sigma}$, Equation (6.32) can be written as:

$$\Delta h = \frac{C_c}{1+e_1} h \log \frac{\bar{\sigma}_1 + \Delta \bar{\sigma}}{\bar{\sigma}_1} \tag{6.33}$$

where e_1 is the void ratio at the point taken for $\bar{\sigma}_1$.

The standard procedure recommended for the monotonically increasing load technique, otherwise known as the constant total stress procedure, makes use of load-increment ratios of one, as mentioned previously. This means loads or pressures that are doubled for each loading: $\Delta P/P = 1$. There is convincing evidence to show that load or pressure-increment ratio is an important factor in characterizing the *e-logP* curve (left-hand sketch in Figure 6.25).

When the incremental loads are small, the shape of the *e-logP* curve will be characterized by readjustments of individual particles and microstructural units (Figure 6.26) in response to their interactions with each other to reattain and maintain equilibrium. If the readjustments of the microstructures and macrostructure are allowed to occur with minimal presence of porewater pressures, the resultant consolidation will be small. If on the other hand, the load increment is large, readjustment of the microstructure, and hence the macrostructure, will be affected by shear distortions, and a larger degree of deformation will occur.

Given the different ways in which consolidating loads can be applied to the test sample, it must be accepted that the results obtained

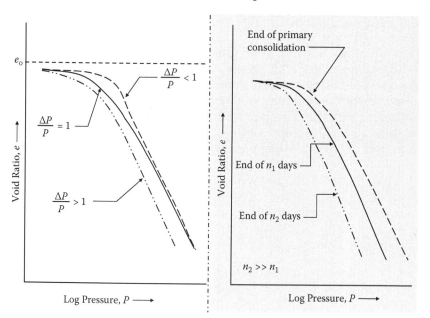

FIGURE 6.25

Typical *e-log P* curves obtained from one-dimensional consolidation tests showing the effect of different load increment ratios $\Delta P/P$ in the left-hand sketch. The common procedure of doubling loads is shown as $\Delta P/P = 1$. The right-hand sketch shows the effect of elapsed time after end of primary consolidation before application of the next incremental load on the position of the maximum curvature of the *e-log P* curve.

FIGURE 6.26
Scanning electron micrograph of microstructural units and soil particles in interaction with each other. The black band at the middle bottom of the picture represents 10 μm.

vis-à-vis coefficient of compression, and so forth are operationally defined. Experiments have shown that if one sets an arbitrary one-day, or one-week, or even a fortnight "rest" period after each load increment (before adding the next incremental load), the *e-logP* curve will move laterally towards the ordinate of an *e-logP* plot, as shown in the right-hand sketch in Figure 6.25. This means that the longer one waits before addition of the next incremental load in the constant total stress procedure, the lesser will be the preconsolidation pressure P_p determined from maximum curvature of the *e-log P* curve.

6.9.2 Consolidation of Swelling Clays

The *e-logP* curve of a *Na*-montmorillonite is shown in Figure 6.27 together with the measured swelling behaviour of the same montmorillonite prepared with parallel-particle orientation as the initial soil fabric. The sketches showing the parallel-particle orientation of the interlayer particles forming the microstructures at the end and beginning of the swelling pressure curve are deemed to be reasonable, based on the fact that the measured swelling pressures accord well with theoretical calculations based on the diffuse double-layer (DDL) model. At the beginning of the swelling test, interlayer

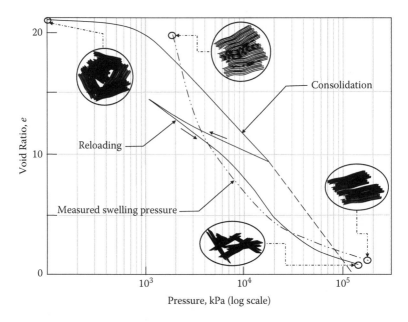

FIGURE 6.27
Comparison of consolidation and swelling pressure for sodium montmorillonite. Measured swelling pressure is from laboratory-prepared oriented particles at a salt concentration of 10–4 *M NaCl.*

expansion is at a minimum. The interlayers expand more as swelling against the prescribed pressure is allowed, to the point where maximum expansion is attained, consistent with DDL theory.

The sketches showing the possible particle orientations for the consolidation sample are postulated. For the bulk sample (consolidation sample), the microstructure depicted in the sketch shows a random arrangement of particles at the beginning of the consolidation process, with a degree of expansion in the interlayers. With load increments, the interlayer spaces are compressed and the random arrangement of the particles becomes more compact. Increased pressure on the sample will tend to reorient the particles and microstructures into a parallel configuration.

6.10 Creep Behaviour

Secondary consolidation in soils tested in the oedometer is essentially determined by empirical techniques—for example, graphical procedure defining end of primary consolidation (Figure 6.23). From a microstructural viewpoint, and also from a mechanistic perspective, the phenomenon

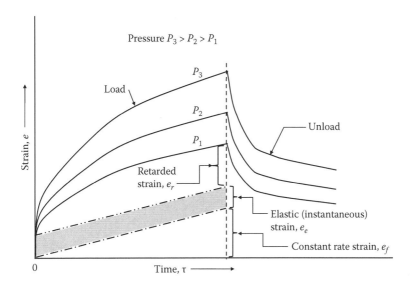

FIGURE 6.28
Load and unload-deformation (strain) relationships in creep testing of clays, showing the various categories of strain (deformation) developed under an applied constant load.

that is called secondary consolidation can be categorized as creep associated with drainage of interlayer water. In determination of creep performance, standard incremental load compression-time tests are conducted. Increments of load are introduced at the end of each creep, producing typical creep testing curves under constant load (Figure 6.28). The characteristic creep curve can be broken into three distinct performance characteristics as shown in the figure, instantaneous elastic deformation or strain e_e, retarded deformation or strain e_r and a constant rate deformation or strain e_f.

6.10.1 Creep in Clay: The Role of *Microstructure*

Since creep in soils pertains primarily to clays, the discussion in this section will focus its attention to the role of the microstructure of clays in the overall (macro) deformation or (creep) strain e. The basic macrostructure of a clay is considered to be composed of the collection of microstructural units such as those shown in Figure 6.27. The overall creep strain consists of the aggregation of strains of individual and groups of microstructural units. Designating the strain of a grouping of microstructural units as microstrain ε, one can make the case that ε is similar in characteristics to the macrostrain e. Accordingly, one obtains:

$$\varepsilon = \varepsilon_e + \varepsilon_r + \varepsilon_f = a_i \xi_i + \alpha f_i(t) \xi_i + \beta b_i \xi_i t \tag{6.34}$$

where ε_e, ε_r, and ε_f represent the elastic, retarded, and constant-rate strains of the microstructural units, respectively; a_i is the compliance of the elastic strain; $f_i(t)$ represents the creep compliance for the ith microstructural unit, b_i is the viscous coefficient responsible for continuous flow; ξ_i is the micro-stress acting on a basic grouping of microstructural units; and α and β are constants:

$\alpha = 0, \beta = 1$ for $\xi_i <$ yield point of a basic grouping of microstructural units,

$\alpha = 1, \beta = 0$ for $\xi_i >$ yield point of a basic grouping of microstructural units.

The macro or overall strain e will be given as:

$$e = \sum a_i \xi_i + \sum_1^m f_i(t)\xi_i + \sum_1^n \beta b_i \xi_i t \tag{6.35}$$

where m represents the number of basic groups of microstructural units whose yield strengths have not been exceeded, and n is the number of basic groups of microstructural units whose interaction strengths have been exceeded. If one considers the number of groups of microstructural units to be semi-infinite, that is, $(n + m) \to \infty$, it follows that if ξ_i exceeds the basic strength of the ith group, continuous flow of the clay under test will occur. In that instance, the previous equation (Equation 6.35) can be represented as a continuous function of the following form:

$$e = A\sigma + \int_{-\infty}^{t} f(t - \tau)\sigma d\tau + Bt\sigma \tag{6.36}$$

where A is an instantaneous compliance, B is a flow parameter, σ represents the total stress, t is the current time, and τ is a time variable.

One can consider that the energy state of a group of microstructural units defines its state of being. This means that the probability of occurrence P_i of a group of microstructural units at a particular state is a function of its energy state E_i, that is, $P_i = f(E_i)$. If each group of microstructural units with volume v constitutes an elemental system in a canonical ensemble of volume Ω, and if there are M elemental systems in the ensemble, the probability of occurrence of one system v in Ω in a state i is proportional to the number of states accessible to Ω. Accordingly,

$$P_i = c\omega(E^{(0)} - E_i) \tag{6.37}$$

where c is a proportional constant independent of i, $E^{(0)}$ is a energy state at a fixed ground state, and $\omega(E^{(0)} - E_i)$ represents the number of energy states accessible to the systems remaining in Ω. From a normalization procedure, where the summation includes all possible states of v, $\Sigma P_i = 1$, and with the use of a partition function: $1/C = \Sigma e^{-\lambda E_i}$ it can be shown that

$$P_i = \frac{e^{-\lambda E_i}}{\sum e^{-\lambda E_i}} \qquad (6.38)$$

and that the mathematical expectation of the total deformation strain $<\varepsilon(t)>$ can be obtained as follows:

$$< \varepsilon(t) >= A\sigma + \int_0^t \frac{P_i(\tau)}{g_i} f(t-\tau)\sigma \, d\tau + B\sigma t \qquad (6.39)$$

where g_i represents the number of groups of microstructural units at the same energy level. Since $\left[\partial \ln \omega / \partial E'\right]_0 = \lambda$ is a constant at a fixed energy level $E' = E^{(0)}$, one obtains $\lambda = 1/KT$. The parameter K represents a positive constant with dimensions of energy. T is the temperature, and E' is the energy of $m-1$ elementary systems remaining in Ω for the situation where any one system has energy E_i.

6.11 Concluding Remarks

6.11.1 Integrity and Environment Stressor Impacts

The integrity of a soil, in respect to mechanical properties, is defined by its ability to resist shearing and compression stresses. In terms of a given set of compositional features, the shearing strength and compressibility of a soil is determined by (a) *factor A*, initial cementation and precipitate bonds between particles and microstructural units, and (b) *factor B*, mechanical and physicochemical interactions between the particles and microstructural units that constitute macrostructure of the soil. A loss of integrity will mean a degraded capability to resist shearing and compression stresses. This means that if one desired to determine the impact from environment stressors, it is necessary to establish the stressor cause–effect relationship in terms of *factors A and B*.

The discussions in the previous sections have demonstrated that when one breaks the initial (formational) and/or precipitate (generally formed during soil ageing) bonds of a soil, the shearing and compressive strength of the soil is degraded—and also its compliance, that is, slope of the stress–strain curve. The kinds of stressors that will result in breakage of bonds include (a) events that create forces that result in relative displacement of particles, for example, earthquakes and tsunamis, and (b) events that result in dissolution of cementation and precipitate bonds over some finite time interval, for example, weathering forces such as those described in Chapter 2. The forces associated with weathering processes are also active in the ageing processes. The actions of these result in bond disruption and changes in the interparticle forces responsible for resistance to relative displacement of particles and microstructural units. They may, on the other hand, create cementation or other types of age-related bonding mechanisms. To the extent that any or all of the events, with their associated stressors and forces, may arise in the future, it is clear that there is a need to incorporate due consideration and analysis of the impact of these events on soil integrity as part of the subject of environmental soil behaviour. The challenges facing one relate to determination of how one mimics weathering and/or ageing processes in laboratory experiments in time frames that are considerably shorter than natural circumstances.

6.11.2 Soil Performance under Loading Stresses

There is one underlying message that needs to be clearly understood: Analyses and/or reporting of the rheological properties and performance of soils must consider the fact that by and large, testing results obtained in laboratory and field experiments are operationally defined. This means to say that whatever message one extracts from strength and compression experiments, these must take into account sample preparation procedures and the methods, techniques, test devices, data-reduction models, and so forth involved in producing the numbers and/or parameters used for subsequent analyses.

6.11.2.1 Yield and Failure Theories

Whilst the resistance to shearing forces can be "measured" in soils, determination or characterization of the internal mechanisms responsible for the "resisting shear forces" is not easily obtained. To a large extent, this is because of such extrinsic and intrinsic factors as type of test system and technique, temperature, soil structure, stress and strain history, bonding and skeletal strength, and others. By and large, yield and failure theories in common use are deterministic in nature, and are based on the performance of homogeneous and uniform-type materials. The popular Mohr–Coulomb criterion (model), for example, does not account for the effects of intermediate principal stresses, neither does it pay attention to inertia terms and the requirement for development of limit values. Furthermore, it is assumed that all the work expended in shearing the soil to failure is expended along the plane of shear failure, and that no work is expended in volume change or other aspects of volume distortion. This assumption is less than accurate because experience in soil testing has shown measurable volume changes and distortions occurring in compact granular soils and in clays ranging from over-consolidated to soft clays.

In granular soils, plastic yielding and flow constitute a small portion of the failure performance of the soil. By and large, failure occurs due to rupture, with accompanying volume changes in the soil except for soils at the critical void ratio. In the case of cohesive soils, failure occurs as a consequence of yielding or progressive flow, leading to large strains and the appearance of a surface of separation between particles, which is not to be confused with a rupture surface.

The interesting point arising from the application of such a criterion is the specification or determination of the cohesion c and ϕ parameters from graphical representation of test results. The question of how well these parameters represent physical reality is often begged, to a large extent because these parameters have been found to be useful, and that their use in analyses and yield/failure theories have provided some measure of success in predicting soil yield/failure.

Strength theories describing yield and/or failure of soils are useful because they provide the material parameters needed for design or prediction of soil performance. The c and ϕ parameters obtained from application of the Mohr–Coulomb model are analytical or mathematical parameters—to be distinguished from actual mechanistic properties of cohesion and friction. The mechanisms involved in "production" of the mechanistic properties are distinguished by the types of interparticle, intermolecular, and interatomic forces involved. An understanding of how these intrinsic forces interact to provide the mechanistic properties will provide one with a proper appreciation of the contributions from the various soil fractions that constitute the soil under consideration.

6.11.2.2 Porewater Pressure

The importance of porewater pressure in the response of a soil to shearing and compressive stresses cannot be overstated. It is important to have proper and accurate knowledge not only of the porewater pressures generated under the various stresses, but also of their status, that is, whether the pressures will dissipate and the nature of the dissipation process. Their relationship to the application to the effective stress concept is fundamental. The difficulties in respect to this important parameter/property lie in obtaining a proper measure of the pressures in the porewater in partly saturated soils and in high-swelling clays.

In partly saturated soils and high-swelling clays, the technical challenges are twofold: (a) availability of a porewater pressure sensing device that will provide measurements of porewater pressures in samples with voids that contain air, or in samples that have different types of porewater such as interlayer water, and (b) determining the meaning of "what is measured." The other challenges relate to the reconciliation of the mechanistic picture with the application of the effective stress principle.

References

Axelrad, D.R., and Yong, R.N., 1966, On an isothermal flow function for a heterogeneous medium, in S. Eskanazi (Ed.), *Modern Developments in the Mechanics of Continua*, Academic Press, New York, 1966, pp. 183–192.

Aitchison, G.D., and Donald, I.B., 1956, Effective stresses in unsaturated soils, *Proc. 2nd. Australia–New Zealand Conf. on Soil Mech.*, pp. 192–199.

Bishop, A.W., 1959, The principle of effective stress, *Teknisk Ukeblad*, 39:859–863.

Bishop, A.W., 1960, The measurement of pore pressure in the triaxial test, *Proc. of Conf. on Pore Pressures and Suction in Soils*, Butterworth, London, pp. 38–46.

Bishop, A.W., Alpan, I., Blight, G.E., and Donald, I.B., 1960, Factors controlling the strength of partly saturated cohesive soils, *Proc., ASCE Research Conf. on Shear Strength of Cohesive Soils*, Boulder, Colorado, pp. 503–532.

Coulomb, C.A., 1776, Essai sur une application des règles de maximis et minimis à quelques problem des statique relatifs à l'architecture, *Mem. Acad. R. Pres. Sav. Etr.* 7:343–382.

Drucker, D.C., and Prager, W., 1952, Soil mechanics and plastic analysis or limit design, *Q. Appl. Mech.*, 10:157–165.

Hill, R., 1950, *The Mathematical Theory of Plasticity*, Oxford Univ. Press, London, 355 pp.

Lode, W., 1926, Versuche ueber den Einfluss der mittleren Hauptspannung auf das Fliessen der Metalle Eisen Kupfer und Nickel, *Z. Physik.*, 36:913–939.

Mohr, O., 1900, Die elastizitätsgrenze und bruch eines materials, *Z. Ver. Dtsch. Ing.*, 44:1524.

Prigogine, I., 1961, *Introduction to Thermodynamics of Irreversible Processes*, 2nd ed., John Wiley & Sons, New York.

Skempton, A.W., 1954, The pore pressure coefficients A and B, *Geotechnique*, 19:77–101.

Skempton, A.W., 1960, Effective stress in soils, concrete and rocks, *Proc. of Conf. on Pore Pressures and Suction in Soils*, Butterworth, London, pp. 4–16.

Yong, R.N., and Wong, C.Y., 1972, Experimental studies of elastic deformation of sand, *Proc. 3rd. South East Asian Soils Conf.*, pp. 323–327.

7

Thermal and Hydraulic Properties

7.1 Introduction

The two frequently encountered phenomena in soils—thermal and hydraulic—involving movement of water in the soil in response to thermal- and hydraulic-driven processes, produce changes in soil water content and porewater chemistry. Correspondingly, there will also be changes in solute content and advective-diffusive transfer of solutes in the soil. In hydraulic-driven processes, water content increases and/or decreases are the direct result of movement of water from one location to another under water content gradients, or movement of water from a water source under hydraulic gradients. In thermal-driven processes, water content changes are due to two different mechanisms occurring simultaneously: (a) water transfer (liquid water and vapour) in response to temperature gradients, thereby creating water content gradients, and (b) water movement because of the water content gradients. These two mechanisms of water movement have been termed as *heat-mass simultaneous movement processes* or *heat-mass coupling phenomena* by de Vries (1958). In this chapter, we will be concerned with hydraulic properties in relation to hydraulic-driven processes, and thermal and hydraulic properties of soils in heat-mass coupling phenomena.

Confusion sometimes arises in the use of the terms *transfer of water, moisture transfer,* and *movement of water* in soils. Whilst *transfer of water* and *moisture transfer* are often used to refer to the collective movement of vapour transport and liquid water (Section 3.5 in Chapter 3), and *movement of water* in soils is used to refer to the movement of liquid water, the literature shows a slight preference for *movement of water* as the inclusive term, that is, water transfer by both vapour transport and liquid water movement. In discussions in this chapter on the thermal and hydraulic properties of soils in relation to heat-mass coupling flow kinetics, the term *movement of water* will be used to refer to (a) the movement of liquid water in the saturated state, and (b) the movement of both liquid water and vapour in the unsaturated state. (Unsaturated flow kinetics has been covered in Chapter 3.)

7.2 Thermal Properties

Thermal effects on water movement in soils can be very complicated because of the development of thermal and hydraulic gradients, and electrical and osmotic potentials, resulting in the transport of heat, electricity, water, and solutes. Water movement in soils due to imposed thermal gradients is a combination of vapour and liquid phases. Evaporation of water at the hot end and condensation at the cold end provides the driving mechanism for water movement in soils. Thermal gradients are most important in relatively dry soils, for example, 1°C/cm can be equivalent to 10^3 cm water head/cm. Where thermal transport is high, vapour movement is the larger component. Vapour transfer has been discussed in detail in Section 3.5.1 (Chapter 3). Diurnal and seasonal changes in soil temperature contribute to the *up and down* movement of water in a soil profile.

In unsaturated soils, in addition to film transport (that is, transport in the film boundary or adsorbed water layer surrounding soil particles) in response to thermal gradients, water flow due to thermo-capillary and advection processes occurs. This latter flow, which is also known as Bénard–Marangoni advection, is the result of surface tension gradients that are temperature dependent. In the pore spaces of unsaturated soils, evaporation causes increasing curvature of the liquid films at air–water interfaces, and condensation results in decreasing curvature of the liquid films. It should be noted that because of the nonuniformity in sizes of apertures of the pore spaces, surface tension gradients also exist under isothermal conditions. In this instance, liquid water movement resulting from isothermal surface tension gradients is known as the Gibbs–Marangoni effect.

7.2.1 Heat-Mass Simultaneous Flow

In general, water movement in both the liquid and vapour phases due to a thermal gradient occurs from hot to cold regions. In unsaturated soils, the adsorbed water that constitutes the water film surrounding particles and assemblages of particles or microstructural units will be affected by changes in temperatures. The result will be liquid water movement along film layers in response to thermal gradients, generally known as *film transport*. Application of a thermal gradient will produce induced osmotic gradients and will also give rise to electric potentials, thereby resulting in the production of an electro-osmotic flow of water.

Viewed from the perspective of the thermodynamics of irreversible processes, the flux relationships for the coupled phenomena in the water phase, using the rate of internal entropy production to determine equations for the driving forces, can be written as follows (Katscalsky and Curran 1967):

$$J_h = -L_{hh}\left(\frac{\Delta T}{T^2}\right) - L_{hs}\Delta\left(\frac{\mu_s}{T}\right) - L_{hw}\Delta\left(\frac{\mu_w}{T}\right)$$

$$J_w = -L_{wh}\left(\frac{\Delta T}{T^2}\right) - L_{ws}\Delta\left(\frac{\mu_s}{T}\right) - L_{ww}\Delta\left(\frac{\mu_w}{T}\right), \tag{7.1}$$

$$J_s = -L_{sh}\left(\frac{\Delta T}{T^2}\right) - L_{ss}\Delta\left(\frac{\mu_s}{T}\right) - L_{sw}\Delta\left(\frac{\mu_w}{T}\right)$$

where J and L are the fluxes and phenomenological coefficients; and the subscripts h, w, and s represent the heat, water, and solutes, respectively, where water refers to liquid water or vapour. T refers to the temperature, and the chemical potential of the solutes and the electro-chemical potential of water are represented by μ_s and μ_w, respectively. The coupling coefficients are normally assumed to be equal, meaning that $L_{wh} = L_{hw}$, $L_{ws} = L_{sw}$ and $L_{sh} = L_{hs}$. These are known as the Onsager's reciprocal relations. The phenomenological coefficients L_{ww} and L_{hh} are related to the hydraulic conductivity and the heat conductivity, respectively. Heat transfer occurs in liquid phase and/or in the vapour phase.

Applying the relationships

$$grad\left(\frac{-\mu}{T}\right) = \frac{1}{T}grad(-\mu) - \mu\,grad\left(\frac{1}{T}\right), \text{ and } grad\left(\frac{1}{T}\right) = -\left(\frac{1}{T^2}\right)grad\ T,$$

Equation (7.1) can now be written as

$$J_h = -\{L_{hh} - L_{hs}\mu_s - L_{hw}\mu_w\}\frac{1}{T^2}grad\,T - L_{hs}\frac{1}{T}grad\,\mu_s - L_{hw}\frac{1}{T}grad\,\mu_w$$

$$J_w = -\{L_{wh} - L_{ws}\mu_s - L_{ww}\mu_w\}\frac{1}{T^2}grad\,T - L_{ws}\frac{1}{T}grad\,\mu_s - L_{ww}\frac{1}{T}grad\,\mu_w \tag{7.2}$$

$$J_s = -\{L_{sh} - L_{ss}\mu_s - L_{sw}\mu_w\}\frac{1}{T^2}grad\,T - L_{ss}\frac{1}{T}grad\,\mu_s - L_{sw}\frac{1}{T}grad\,\mu_w$$

Equation (7.2) shows that a conjugate coefficient of heat flux J_h due to the thermal gradient is affected by the chemical potential μ_s of the solutes and the chemical potential μ_w of water. Additionally, it is in inverse proportion to the square of temperature (i.e., inverse to T^2). As for water and solute movement, the conjugate coefficient for mass flux J_w and J_s due to the thermal gradient also shows the same inverse proportional relationship to the temperature.

7.2.2 Fourier-Type Heat Flow

There are three principal modes of heat flow in soils: radiation, convection, and conduction, with radiation occupying a minor role. In the initial stages of water uptake in a partly saturated soil, vapour transfer occurs in the continuous pore spaces and heat transfer is considered to be principally by conduction and convection. Subsequently, that is, when vapour transfer becomes vanishingly small, the primary mode of heat transfer will be by conduction—with heat flow through soil solids, liquid water and soil air—and secondarily through convection.

Applying Fourier's empirical law to conductive heat flow, the total heat flow q_h in a soil can be expressed as

$$q_h = -k_h \frac{\partial T}{\partial x} + \left(c_l q_l + c_v q_v\right)\Delta T, \tag{7.3}$$

where k_h is the thermal conductivity, c_l and c_v are the specific heat capacity of liquid water and vapour, respectively, q_l and q_v refer to the advection flux for liquid water and vapour, respectively, and ΔT is the temperature difference between the soil volume under consideration and the external region. The first term on the right-hand side of the equation refers to heat flow by conduction, and the second term refers to heat flow by conduction.

The continuity condition for a soil volume subject to a thermal flux states that an amount of heat that flows into or out of the volume is equivalent to the product of the changes of temperature and heat capacity of the soil. This is expressed in terms of the heat conservation law as follows:

$$c\rho \frac{\partial T}{\partial t} = -\frac{\partial q_h}{\partial x} + L, \tag{7.4}$$

where T is the temperature, q_h is the flux of heat, L is a heat source or heat sink, c is the specific heat capacity, and ρ is the mass of the soil. The product $c\rho$ is the heat capacity C.

Substituting Equation (7.4) into Equation (7.3), one obtains

$$C\frac{\partial T}{\partial t} = \frac{\partial}{\partial x}\left(k_h \frac{\partial T}{\partial x}\right) + \left(c_l q_l + c_v q_v\right)\frac{\partial T}{\partial x} \pm L_T \rho_w I, \tag{7.5}$$

where L_T refers to the latent heat of water at evaporation or condensation, ρ_w is the density of water, and I refers to rate of phase change. Evaporation of liquid water provides a positive latent heat flux whereas condensation provides a negative latent heat flux, hence, the $\pm$ sign before the last term in Equation (7.5). Because of the different stages of evaporation and the highly unsteady nature of vapour diffusion, it is likely that there will be intermediate locations that will exhibit local condensation. This explains why we need to express the last term as a function of spatial distance and time. The second term on the right-hand side of the equation that refers to convective flow can usually be considered to be vanishingly small in comparison to conduction (first term on right-hand side of Equation [7.5]). Convection will be the primary mode of heat flow in desert soils and in soils in forest fires.

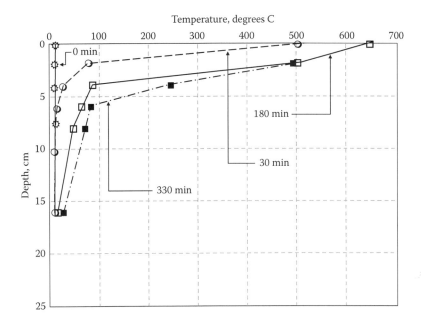

FIGURE 7.1
Time–temperature profiles of a surface soil in the field subject to a high surface temperature of 700–800°C (estimated) imposed by burning charcoal on the surface. All the processes associated with heat transfer are involved in the development of the temperature profiles. These include combustion of organic matter, and changes in chemical composition and soil minerals. No temperature changes appear below the depth of 25 cm. (Results communicated by Professor Taku Nishimura, University of Tokyo.)

Figure 7.1 shows a range of temperature profiles in a surface soil in the field with burning charcoal imposing a high temperature on the soil surface. The temperature profiles at various time intervals are the result of heat transfer due to all the processes associated with heat flow operation, and changing soil properties obtained as a result of combustion of organic matter, chemical composition, and soil minerals. The situation portrayed by the test results are analogous with time–temperature–depth profiles obtained in surface soils under conditions of forest fire and shifting cultivation, that is, allowing a land to remain fallow, after crop harvesting, for a period of a few years so as to regain soil fertility.

7.2.3 Thermal Properties

The thermal properties of concern include (a) thermal conductivity, (b) thermal diffusivity, and (c) heat capacity. Thermal conductivity is a coefficient introduced in heat flow equations and can be considered to be an impedance factor to heat flow in soil. Thermal diffusivity is a coefficient defined by both thermal conductivity and heat capacity of soil. The heat capacity of a soil is a function of such fundamental properties as soil composition, water content, and soil structure.

7.2.3.1 Thermal Conductivity

The first relationship in Equation (7.2) shows that the impedance factor to heat flow in a soil due to a temperature gradient is $(L_{hh}-L_{hs}\mu_s-L_{hw}\mu_w)/T^2$. The heat-mass simultaneous flow of liquid water or vapour in the soil pores assumes that the chemical potential gradients of solute and water are vanishingly small and can be ignored. It should be noted that the thermal conductivity in this relationship is defined for the water phase only, and is a function of three phenomenological coefficients, that is, chemical potentials of solute, water, and temperature.

In Fourier-type heat flow, the *thermal conductivity* of a soil is defined as the quantity of heat passing through a unit area of a soil mass in a unit time under a unit temperature gradient through soil solids, water, and soil air. It is very sensitive to soil structure and the relative proportions of solids, water, and gas—in addition to soil type, water content, and density. In partly saturated soils, thermal conductivity is very sensitive to the existing water content. Heat transfer by conduction involves heat transfer between molecules—that is, from one molecule to another. In comparison to conduction in the gaseous and liquid phases, conduction in and between the solid fractions is most efficient and dominant because of the tight configuration of molecules in the solids. The heterogeneity of microstructural units and their distribution in a soil mass result in thermal conductivities that are highly variable because of the spatial- and time-variable proportions and distributions of solids, liquid, and gaseous phases.

Campbell and Norman (1988) have proposed a relationship for thermal conductivity as follows:

$$k_c = \frac{\theta \alpha k_w + v_s \beta k_s + v_g \phi k_g}{\theta \alpha + v_s \beta + v_g \phi} \tag{7.6}$$

where k_c refers to the *thermal conductivity*; the subscripts w, s, and g associated with the thermal conductivity k refer to the thermal conductivity of water, solids, and gas, respectively; v_s and v_g are the volume fractions of solids and gas; and α, β, and ϕ are weighting factors accounting for heat associated with evaporation and condensation—that is, latent heat transfer. The thermal conductivities for most primary minerals including clay minerals at a temperature of 20°C are about 6 to 9 mcal/cm.s.K, depending on the specific gravity of the mineral under consideration; and the thermal conductivities for liquid water and air at the same 20°C temperature are 1.38 and 0.06 mcal/cm.s.K, respectively. Changes or fluctuations in the temperature will result in corresponding changes in the thermal conductivity. This is because changes in other heat transfer factors will impact on the thermal conductivity.

Measurement of the thermal conductivity in a soil requires the establishment of a constant temperature gradient across the soil. Experimental procedures for determination of thermal conductivity in soils include (a) the steady-state heat flux method of Manose et al. (2008), and (b) the twin transient-state cylindrical probe method that is based on the heat flow analysis in the unsteady state suggested by Kasubuchi (1977).

7.2.3.2 Thermal Diffusivity

The *thermal diffusivity* D_T, which is a measure of transient heat flow, is defined as the thermal conductivity k_h divided by the heat capacity C. Whilst the thermal conductivity k_h of a soil is a measure of how much heat will flow through the soil, the thermal diffusivity D_T tells us how rapidly the heat will flow through the soil. Methods of measurement of thermal diffusivity D_T can be difficult when thermal gradients are unsteady, and particularly in situations where both time-variable and point-variable properties characterize the soil. Under such circumstances, a dynamic measurement of thermal diffusivity is required, using, for example, a forced Rayleigh scattering method (an optical technique) to measure the thermal diffusivity of solids and liquids (Motosuke et al., 2003), or a transient pulse technique which allows one to simultaneously determine the thermal conductivity (Osako and Ito, 1997). The thermal diffusivity D_T can also be determined by measuring the temperature response in a soil under a thermal pulse using the following relationship, assuming that advection and latent heat flow can be ignored:

$$\frac{\partial T}{\partial t} = D_T \frac{\partial^2 T}{\partial x^2} \tag{7.7}$$

7.2.3.3 Heat Capacity

The heat capacity C of a soil can be expressed as specific heat capacity per unit volume (volumetric heat capacity c_{soil}). The *specific heat capacity* of a soil c_{soil} is defined as the change in heat content of a unit mass of soil per unit change in temperature. The product of c_{soil} and ρ, the density of the soil, gives us $\rho c_{soil} = C$, the *specific heat capacity per unit volume*, or the *volumetric heat capacity* of a soil. To account for the different heat capacities of each of the three phases in a soil—solids, water, and gas—a weighting procedure is used to obtain the overall heat capacity of a soil. Whilst this procedure is not precise, it will nevertheless provide one with an acceptable estimate of the heat capacity of a soil. To determine the volumetric heat capacity C of a soil, one will need to know the volume fractions of each of the three phases as follows: v_s for volume fraction of solids, θ for volumetric water content, and v_g for volume fraction of gas. These are combined with their respective specific heat capacities and densities to obtain C as follows:

$$C = v_s \rho_s c_s + \theta \rho_w c_w + v_g \rho_g c_g \tag{7.8}$$

where ρ_s, ρ_w, and ρ_g refer to the densities of solids, water, and gas, respectively; and the subscripts s, w, and g, associated with the specific heat capacity c refer to the specific heat capacities of solids, water, and gas, respectively. The volumetric heat capacities for most primary and clay minerals can be taken to be about 0.45 to 0.5 cal/m³K, depending on the specific gravity of the mineral under consideration, and the volumetric heat capacities for liquid water and air are 1.0 and 0.003 cal/m³K, respectively.

7.3 Water Movement under Thermal Gradient

Imposition of a temperature gradient, as a thermal stressor, on a soil–water system results in a combined heat and thermally affected water transfer in the soil. The thermally affected water transfer processes are in part related to the water content of the soil. For a partly saturated soil at low moisture contents, thermally affected vapour transfer is a significant factor, and because of rapid evaporation and condensation of water in pore space, it can be from four to five times larger than would be predicted from diffusion calculations using a Fickian model. In addition to vapour transfer, the processes associated with liquid water transfer in a soil volume element includes the effects of evaporation and condensation of water in the pore spaces.

7.3.1 Thermally Induced Vapour Movement

The Fickian model for vapour diffusion can be stated as: $q_v = -D_v(\partial c_v / \partial x)$, where q_v is the vapour flux, D_v is the diffusion coefficient, c_v is the concentration of vapour, and x is the spatial coordinate (Section 3.5.1, Chapter 3). Assuming vapour to be an ideal gas, in combination with the equation of state: $c_v = p_v / RT$, where p_v is the partial pressure of vapour, R is the gas constant and T is the temperature. Introducing a mass flow factor γ defined as $\gamma = P/(P - p_v)$, where P is the total pressure of the gas phase in soils, the Fickian equation for vapour flux in a pore space is given as follows:

$$q_v = -\gamma \frac{D_v}{RT} \frac{dp_v}{dx} \tag{7.9}$$

Introducing the Clapeyron equation $d\ln p_s / dT = H/RT^2$ and volumetric air content and tortuosity factor to Equation (7.9), the following approximate relationship that describes vapour flux in soils is obtained:

$$q_v = -a\gamma\tau \frac{D_v}{R^2 T^2} p_v \left(\frac{L_T}{T} \frac{dT}{dx} + \frac{\partial \psi_w}{\partial x} \right) \tag{7.10}$$

where a is the volumetric air content, τ is the tortuosity, and ψ_w is the potential of water (Cary, 1966; Kay and Groenevelt, 1974; Nakano and Miyazaki,

1979). Equation (7.10) shows that the Fickian equation includes two terms: (a) one term expressing vapour flux due to a temperature gradient, and (b) the other term showing vapour flux due to the chemical potential gradient of water, as shown by the second relationship of Equation (7.2) in Section 7.2.1. The effect by temperature by itself is expressed in Equation (7.10).

7.3.2 Thermally Induced Liquid Water Movement

In partly saturated soils, the primary driving force for liquid water movement is the matric potential gradient of water, taking note of the fact that the matric potential is a temperature-dependent characteristic. The matric potential is a function of the surface tension of water, and the curvature of the meniscus established at the soil pore air–water interface: $\psi_m = -2\sigma/r\rho_w$, where ψ_m is the matric potential of water, σ is the surface tension, r is the curvature radius of meniscus, and ρ_w is the density of water. The curvature radius of meniscus is considered to be equivalent to radius of tube-shaped pores (Figure 3.1, Section 3.2.1, in Chapter 3). Assuming that the temperature dependency of the matric potential is due primarily to the surface tension of water since the temperature dependency of the matric potential on the density of water is negligible, one obtains Equation (7.11) by invoking the chain rule $\partial\psi_m/\partial T = (\partial\sigma/\partial T)(\partial\psi_m/\partial\sigma)$ and $\sigma = 75.636 - 0.1409T - 0.0003T^2$ and from least square approximation for $0 \leq T \leq 90°C$ to the Darcy equation $q_l = -k(\partial\psi_m/\partial x) = -k(\partial\psi_m/\partial T)(\partial T/\partial x)$ for a partly saturated soil at low moisture content, in which the gravitational term is negligible, and where k refers to the hydraulic conductivity.

$$q_{lT} = -\frac{2k}{r\rho_w}(0.0006T + 0.1409)\frac{\partial T}{\partial x}, \tag{7.11}$$

where q_{lT} refers to liquid water flow expressed explicitly in terms of temperature T and its gradient, according to Darcy's empirical law. For a partly saturated soil at high water content, liquid water flow in response to the matric potential is minimally influenced by the temperature gradient. For swelling clays, in which the major portion of the liquid water is interlayer water, that is, water in the interlayer space of clay minerals, the chemical potential of the interlayer water will be the driving force for liquid water, instead of the matric potential ψ_m used in the Darcy-typed flow equation.

The relationship between the soil–water potential (chemical potential) and temperature can be calculated using the DDL (diffuse double-layer) model, using the expression for the midplane potential ψ_c as follows (Section 3.3.2 in Chapter 3):

$$\psi = \psi_c = \frac{2\kappa T}{z_i\varepsilon}\ln\left[\frac{Kd\sqrt{S}}{\pi}\right], \tag{7.12}$$

where $K = \sqrt{8\pi z_i^2 e^2 n_o / \varepsilon \kappa T}$; and κ, S, n_o, z_i, ε, e, and d refer to the Boltzmann constant, coefficient of effective dialysate concentration, number of ions per unit volume in the bulk solution, valency of the ith species of ion, dielectric constant, electronic charge, and half distance between interacting layers, respectively. The influence of temperature T on the dielectric constant ε is given as follows: $\varepsilon = (62445/(T + 120)) - 70.91$.

The results from an experimental study on the relationship between the soil–water potential (determined from psychrometer measurements) and temperature for a 50-50 sand–sodium bentonite mixture are shown in Figure 7.2. The results show that the higher the water content of the test sample, the lesser the influence of temperature on the soil–water potential. The factors contributing to the dependence on temperature include (a) viscosity of porewater η, (b) surface tension of porewater, σ, (c) the dielectric constant ε, and (d) activation energy of the water molecules and ions. For measurements made with psychrometers, we need to include the effect of temperature on the electromotive force since this increases with temperature because of the increase in evaporation of water between the reference junction and the wet junction of the psychrometer.

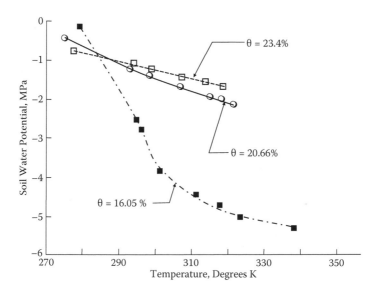

FIGURE 7.2
Variation of soil–water potential (obtained from psychrometric measurements) with temperature for a 50-50 sodium bentonite–silica sand mixture. The θ values shown in the graph are the volumetric water contents of the samples tested. (Adapted from data reported by Mohamed, A.M.O. et al., 1992, *Geotech. Testing J.*, 15:330–339.)

7.3.3 Thermal Gradients and Water Content Profiles

The result of all forms of water transfer in soils—vapour and/or liquid water—in response to all kinds of driving forces and mechanisms (e.g., heat and mass transfer), can be seen in changes in the spatial water content profiles of the affected soils. Analyses and/or predictions of changes in water contents require knowledge of the various physical and chemical constants, and coefficients and parameters describing the flow mechanisms initiated by the various driving forces and mechanisms. It goes without saying that whilst complex conceptual models of the various phenomena can be developed, these would require a host of information, parameters, and data that are not easily secured or even available. One should therefore look towards the use of less complex models with known phenomenological relationships and readily determined properties and characteristic coefficients. The following set of analyses provides an example of the kinds of relationships and phenomenological or characteristic coefficients used in determining changes in the water content profile of a particular soil.

Consider a partly saturated soil column with a heat source at one end and a constant cool temperature at the opposite end. For the analysis, Yong and Xu (1988) considered the unsaturated column of length L to be fixed at both ends with $x = 0$ at the left and $x = L$ at the right end as shown in Figure 7.3. The temperature at $x = 0$ is T_2 and at $x = L$ it is T_1. Both ends of the partly saturated soil column are impermeable to water but not for heat transfer. The initial water content is assumed to be uniformly distributed along the length of the column, $\theta(x,0) = \theta_0$.

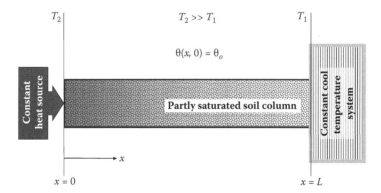

FIGURE 7.3
Partly saturated soil column subject to constant heat source at $x = 0$ and with constant cool temperature T_1 at $x = L$. Initial water content of soil column is constant, $\theta(x,0) = \theta_0$, and the boundaries at both ends are impermeable.

From the second postulate of irreversible thermodynamics, the governing sets of relationships pertaining to the heat and mass transfer phenomena for the experimental set-up shown in Figure 7.3 are

$$
\left\{ \begin{matrix} Q_w \\ Q_T \end{matrix} \right\} = \left[\begin{matrix} L_{ww} & L_{wT} \\ L_{Tw} & L_{TT} \end{matrix} \right] \left\{ \begin{matrix} \dfrac{\partial \theta}{\partial x} \\ \dfrac{\partial T}{\partial x} \end{matrix} \right\},
\tag{7.13}
$$

where

Q_w is the fluid flux, that is, fluid flow through per unit area per unit time;

Q_T is the heat flux, that is, heat flow through per unit area per unit time;

$(\partial \theta / \partial x)$ is the thermodynamic force due to water content gradient;

$(\partial T / \partial x)$ is the thermodynamic force due to temperature gradient;

θ is the volumetric water content, $\theta = \theta(x,t)$;

L_{ww} is the diffusion coefficient for fluid flow due to gradient of θ. The total accumulation of moisture due to the combined movements of vapour and liquid is considered as equivalent to liquid water.

L_{TT} is thermal conductivity coefficient for heat transfer due to the gradient of T;

L_{wT}, L_{Tw} are coupling coefficients;

T represents the temperature, $T = T(x,t)$;

t represents time.

From mass and energy conservation considerations, the following relationships are obtained:

$$
\frac{\partial \theta}{\partial t} = \frac{-\partial Q_w}{\partial x}
$$
$$
C \frac{\partial T}{\partial t} = \frac{-\partial Q_T}{\partial x}
\tag{7.14}
$$

where C represents the specific heat capacity. From Equations (7.13) and (7.14), the governing relationships are obtained as follows:

$$
\left\{ \begin{matrix} \dfrac{\partial \theta}{\partial t} \\ \dfrac{\partial T}{\partial t} \end{matrix} \right\} = \left[\begin{matrix} 1 & 0 \\ 0 & 1/C \end{matrix} \right] \frac{\partial}{\partial x} \left\{ \left[\begin{matrix} L_{ww} & L_{wT} \\ L_{Tw} & L_{TT} \end{matrix} \right] \left\{ \begin{matrix} \dfrac{\partial \theta}{\partial x} \\ \dfrac{\partial T}{\partial x} \end{matrix} \right\} \right\}
\tag{7.15}
$$

The boundary conditions for T and θ are:

$$T(0,t) = T_2 = \text{constant}; \; T(L,t) = T_1 = \text{constant}; \; T_2 > T_1 \tag{7.16}$$

$$Q_w \bigg|_{x=0} = \left(L_{ww} \frac{\partial \theta}{\partial x} + L_{wT} \frac{\partial T}{\partial x} \right)\bigg|_{x=0} = 0$$

$$\tag{7.17}$$

$$Q_w \bigg|_{x=L} = \left(L_{ww} \frac{\partial \theta}{\partial x} + L_{wT} \frac{\partial T}{\partial x} \right)\bigg|_{x=L} = 0$$

and the initial conditions for T and θ are given as:

$$T(x,0) = \begin{cases} T_2 & at \; x = 0 \\ T_1 & at \; x = L \end{cases} = T_2 + (T_1 - T_2)u(x)$$

$$\tag{7.18}$$

$$\theta(x,0) = \theta_0$$

where $u(x)$ is a step function.

To determine the phenomenological coefficients, Yong and Xu (1988) used trial functions which were assumed to be certain functions of θ and T. The functional forms of the trial functions were determined on the basis of previous experiments, and the constants associated with the functional relationships were evaluated using the *identification technique*. Figure 7.4 shows the computed water content–distance (θ-x) relationship for a partly saturated clay column test with a heater temperature of 100°C at one end and a constant temperature of 10°C at the other end. Since the computed diffusion coefficients used to calculate the θ-x curve are valid only for the chosen temperature gradient and a specified water content, separate calculations for other diffusion coefficients will be necessary when other temperature gradients and water contents are considered. The experimental data obtained from the set-up shown in Figure 7.3 are shown in Figure 7.4.

7.4 Hydraulic Properties

Movement of water in soils, otherwise known as hydraulic flow, occurs in response to the actions of one or more sets of forces on the soil mass or on the

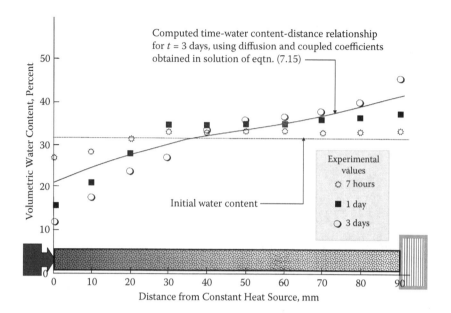

FIGURE 7.4

Comparison of computed water content-distance relationship using phenomenological coefficients (diffusion and coupled coefficients) computed from solution of Equation (7.15) at $t = 3$ days, with experimental values. The calculated coefficients are for one imposed temperature gradient and one volumetric water content. (Adapted from results reported by Yong, R.N., et al., 1992, *Int. J. Num. Analyt. Meth. Geomech.*, 16:233–246.)

porewater itself. These forces can be internal forces, such as those generated by chemical potential and thermal gradients, or externally applied loads. The soil property that is directly responsible for controlling the rate of hydraulic flow is the *hydraulic conductivity*. This soil property is one of the many *transmission properties* of soils that include heat transfer properties, and properties associated with application of electromagnetic geophysical methods such as time-domain reflectometry (TDR) and frequency domain reflection and transmission (FDRT).

The main factors and parameters of concern in determining the hydraulic conductivity of soils fall into five groups:

- Factors and parameters associated with the permeating fluid (permeant)—viscosity η, pressure p, density γ, and chemistry of permeant.
- Factors and parameters associated with the physical properties of the soil—tortuosity, dry density, porosity, pore-size distribution, particle size, and soil structure.
- Factors and parameters associated with the physicochemical properties of the soil—soil–water potential, heat of wetting, ionic concentration, soil functional groups, and thickness of Stern layer of particles.

- Water content of soils—whether soil is partly saturated or fully saturated with fluid.
- Factors and parameters associated with the type of permeameters used, and especially with test techniques and protocols.

The importance of each of these groups, relative to the others, cannot be determined in absolute terms—to a large extent because of the interdependencies between various factors and parameters in all the groups, and the fact that the levels of contribution of one factor on another cannot be fully assessed. For example, if one keeps the viscosity η constant, hydraulic conductivity will increase if the density γ is increased. On the other hand, if density γ is kept constant and viscosity η is increased, hydraulic conductivity will decrease.

Hydraulic conductivity is termed *unsaturated hydraulic conductivity* for partly saturated soils and termed *permeability* for fully saturated soil. The factors and parameters influential in both hydraulic conductivities are in similarity. In this section, we will discuss the influence of factors and parameters to hydraulic conductivity in term of *permeability* to water

7.4.1 Water Permeability

7.4.1.1 Darcy Model

Water permeability is normally defined for flow of water in a saturated soil and is simply called *permeability* in this section. Permeability is considered to be an impedance factor to flow of water, and is a function of dry density, particle size, porosity, the volumetric pore density, pore size distribution, shape and connectivity of pores, tortuosity, and characteristics of water such as viscosity and density. The conventional procedure for determining permeability of soils in the water-saturated state is to conduct an experiment where the rate of water flow (flux v) through a laterally confined sample is measured in relation to the hydraulic gradient.

Figure 7.5 is a schematic illustration portraying the essence of the experimental procedure employed by Darcy (1856) in proposing a methodology for determining the water permeability of a soil. The important pieces of data include the cross-sectional area of the sample A, the hydraulic gradient $i = h/L$, where h is the constant height of hydraulic head, L is the length of the test sample, and Q is the quantity of water collected at the outlet end at specified time interval. From the test information, the Darcy model for data reduction is used to obtain the coefficient k ("*un coefficient dependant de la perméabilité de la couche de sable*") as follows: $Q/A = -k(h/L) = -ki$, where v refers to the flux of water flow through the test sample and k is the Darcy coefficient of permeability. Common practice now omits the term "Darcy" and identifies k as the *permeability coefficient*.

By virtue of the measurement technique, this discharge flux v is a macroscopic value. The actual porewater velocity v_{pw}, which is the velocity of the

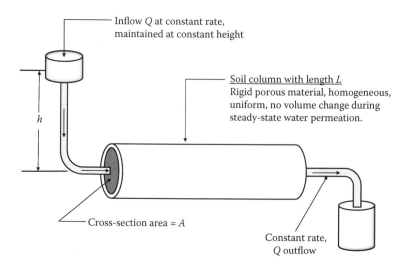

FIGURE 7.5

Schematic illustration of the Darcy flow model for determination of *"un coefficient dependant de la perméabilité de la couche de sable"* $k = v/i$. Note that the *soil* in the horizontal column shown in the drawing in the Darcy experiment is *sand*.

fluid in the pore spaces is obtained also as an average quantity as $v_{pw} = v/n$, where n is the porosity of the test sample. The porewater velocity, which is also often called the *advective velocity*, is used in mathematical formulations tracing the transport and fate of solutes in soil. In partly saturated soil, the average porewater velocity v_{pw} is defined by $v_{pw} = v/\theta$ using the volumetric water content θ, in place of porosity n.

7.4.1.2 Darcy Permeability Coefficient

The relationship $v = -ki$ is a phenomenological relationship where the rate of flow v is related to the applied hydraulic gradient i through a compliance factor k. In that sense, it is similar to the other phenomenological relationships such as Fourier's law for heat transfer, and Fick's diffusion law. It is interesting to note that not much has changed since the time of Henry Darcy's experiments in respect to present-day laboratory experiments conducted to determine soil permeability—except for the fact that Darcy-type experiments are being conducted on granular soils and clays, including swelling clays.

It is useful to determine how valid or how useful the k values are for the various kinds of soils tested using Darcy-type experimental systems. To do so, it is necessary to recognize the restraints, conditions, test requirements, and limits of applicability of the Darcy model. First and foremost, it is important to know that the coefficient k is not a measured parameter, that is, the values obtained for k are not directly measured. The k values are computed from measurements of fluid flow rate v under a specified hydraulic gradient i. In

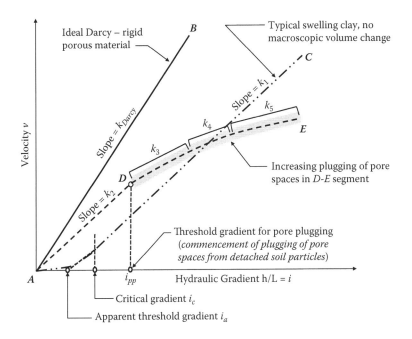

FIGURE 7.6
Illustration of critical i_c, apparent threshold i_a, and pore space plugging i_{pp} threshold gradients obtained in hydraulic conductivity tests. Line *AB* is the standard Darcy model results. Curve *AC* represents the results for tests on a swelling clay with no macroscopic volume change. Curve *ADE* portrays the results for tests where particle detachment occurs when applied hydraulic gradients become excessive.

other words, the Darcy-type experiment only provides the relevant information allowing one to calculate the coefficient k—"the coefficient dependent on the permeability of the layer of soil (*un coefficient dependant de la perméabilité de la couche de sable*)." This is the slope k_{Darcy} of the straight line *AB* in Figure 7.6.

From the report of the experiments conducted by Darcy, one obtains the following as controlling experimental factors for what is now called the *Darcy model*: (a) granular medium, (b) laminar flow through the test sample, (c) steady-state fluid flow, and (d) no macroscopic volume change during water permeation. In present-day water-permeation experiments on soils, it appears that the "granular medium" factor is not a critical factor so long as the conditions and requirements of the other three factors are satisfied. The *Darcy model* is considered to be valid for low velocity flow. In other words, it is best applied for cases with Reynold's number R lower than 1. In any event it should not be applied to cases where R exceeds 10. R is defined as $R = \rho v d / \mu$, where ρ is the density of water, v is the velocity of water in pores, μ is the viscosity of water, and d is the average diameter of soil pores or the average particle diameter.

The effect of particle interactions such as those demonstrated by swelling soils is shown as curve *AC* in Figure 7.6 for the condition where no measurable macroscopic volume change occurs during hydraulic permeation. The initial nonlinear *v-i* relationship for curve *AC* is the result of resistance to flow offered by the energies of interaction between particles and ion-particle interactions. This means to say that the nonlinearity between zero and the critical gradient is due to the impedance offered by the interlayer DDL forces. Evidence concerning the lack of proportionality between flow velocities and hydraulic gradients in recent historical experimental studies of flow in saturated clays reported for example by Philip (1957), Lutz and Kemper (1959), Hansbo (1960), and Yong and Warkentin (1966) show that at low water contents, linear proportionality between flow velocity and hydraulic gradient is not established until some critical gradient is reached.

For situations where the clay has a high swelling potential or high swelling pressure upon wetting, the standard procedures need to be modified or even changed drastically. Because of the high swelling pressures that can develop, rigid confinement is required if the cross-sectional area and macroscopic sample volume presented to the permeating fluid are to be preserved. Linearity in the *v-i* relationship is not obtained until the critical gradient i_c is reached, so long as the macroscopic volume of the test sample remains unchanged (Figure 7.6). The greater the swelling potential of the clay is, the shallower the slope (k_1) will be.

7.4.1.3 Effect of High Hydraulic Gradients

The question of effect of magnitude of hydraulic gradients applied in tests for hydraulic conductivity of soils has often been raised. The two most important conditions for linearity to be achieved between *v* and *i* are (a) test sample is a rigid porous medium, meaning that no internal restructuring of the particles and microstructures occurs, that no re-ordering or re-arrangement of pore geometries and distributions occurs, and that the volume of the sample remains constant during permeation; and (b) laminar fluid flow through the test sample.

Departures from any part of these conditions will lead to calculated values of *k* that may not be representative of real hydraulic conductivity of the soil. This is especially true since most tests are conducted with one specifically prescribed hydraulic gradient—that is, the *k* value for the test soil is obtained from only one set of test conditions. What is required is a series of tests with different values of *i*. Curve *ADE* in Figure 7.6 shows the results of tests using different various hydraulic gradients on the *v-i* relationship for a test sample, with little-to-no macroscopic volume change. At the lower range of *i*, the portion *AD* shows linearity between *v* and *i*—meaning that the value of k_2 is independent of the applied hydraulic gradient. However, when the hydraulic gradients exceed the threshold value for particle detachment (point *D* on the curve), onset of pore space plugging occurs, and succeeding tests with increasing hydraulic gradients will produce a nonlinear relationship between v and i. This phenomenon of pore space plugging has been called *densification* of the sample in the literature, and has been identified as occurring at the outflow end of the test sample. The schematic illustration in Figure 7.6 shows that if one uses tangents for the DE portion of the v-i curve to calculate the appropriate k value, one would obtain at k_n values, which would not be equal to k_2. With increased pore space plugging, the computed k_n values will be lower than the k_2 value computed for the initial linear portion of the curve.

7.4.1.4 Darcy Model with Permeant Properties

To accommodate the influence of viscosity η and density γ of the permeant in data reduction techniques using the Darcy model, one could express the Darcy relationship as follows: $v = k^+(\gamma/\eta i)$, where k^+ is defined as the intrinsic permeability, dependent on the characteristics of soil pore structure only, i is the hydraulic gradient, and the properties of the permeant are expressed as the ratio of its density to its viscosity (i.e., γ/η). The viscosity and density of water in a pore space changes with distance from the surface of a solid. It is not constant over the entire pore space occupied with fluids. Whilst there is some debate about the extent of the changes, there is agreement that these changes affect the actual porewater velocity distribution in pore space, as shown in the idealized sketch in Figure 7.7.

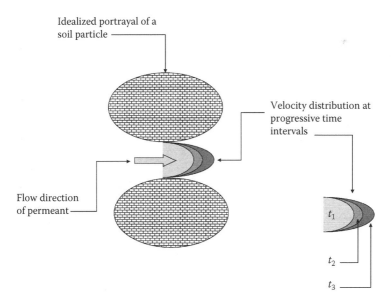

FIGURE 7.7
Velocity distribution of permeant in pore space between two highly idealized portrayal of soil particles.

The influence of the profiles of viscosity and density of water in a pore space on permeability can be estimated by solving the Navier–Stokes relationship, which describes the motion of fluids under the condition that viscosity and density are dependent on the position of the water. Such a solution has been reported by Ichikawa et al. (1999), who used a perturbation theory (homogenization analysis) in their analysis of water in simple pore spaces formed between parallel planes in an infinitely small volume of clay.

7.4.2 Permeability and Soil Structure

The Darcy model for determination of the Darcy coefficient k does not directly consider the impact of the properties of the permeant on permeability, nor does it take into account the structure of the soil (micro- and macrostructure). The scanning electron microscopic (SEM) picture of a Kanto loam shown in Figure 7.8 is an example of the kinds of pore channels through which fluid is expected to flow at varying velocities, depending on the size of the pore spaces, connectivity of pore spaces, and tortuosity of the pore channel paths.

7.4.2.1 Kozeny–Carman Permeability Coefficient

For fully saturated soils, the *Kozeny–Carman* (*K-C*) relationship for permeability includes the influence of soil structure on permeability through

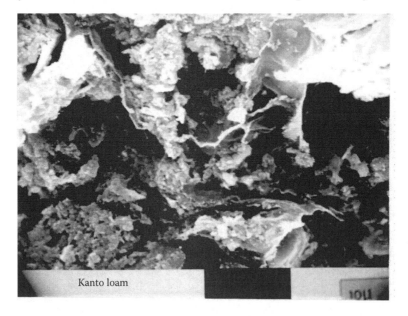

FIGURE 7.8
SEM picture of Kanto loam illustrating differences in pore spaces and sizes of microstructural units. The lateral size of the black bar at the bottom of the picture is 10µm in length.

consideration of tortousity, surface area of soil particles, and porosity of the soil, in addition to the viscosity and density of the permeant. The Poiseuille relationship for viscous flow through narrow tubes serves as a starting point for development of the analysis, that is,

$$v^* = \frac{\gamma r^2}{8\eta} \frac{\Delta\psi}{\Delta l}$$

(7.19)

where v^* is the mean effective flow velocity through a tube of radius r and length Δl, γ and η are the density and viscosity of the fluid, and $\Delta\psi$ is the potential difference between the ends of the tube. To accommodate the features of a porous medium, including path length and porosity n, a shape factor C_s can be introduced into Equation (7.19) in place of the fraction $\frac{1}{8}$, as follows:

$$v_{pw} = \frac{v}{n} = C_s \frac{\gamma r^2}{\eta} \frac{\Delta\psi}{\Delta l}$$

(7.20)

where v_{pw} is the porewater velocity and v is the flux of water through unit area per unit time, as described in a previous section. The other steps in the development of the K–C relationship for its permeability parameter include

- Defining S_w as the wetted surface area per unit volume of the porous medium.
- Replacing the r term in the Poiseuille relationship with n/S_w, taking advantage of the fact that the shape factor C_s accounts for the noncircular shape of the pore channels.
- Applying $S_w = S(1 - n)$, $r = n/S(1 - n)$.
- Recognizing that the water flow path through the soil is not a straight-line direct path (length of Poiseuille flow tube) Δl, an effective flow path Δl_e is introduced, meaning that $\Delta l_e > \Delta l$.
- Defining a tortuosity parameter, $\tau = \Delta l_e / \Delta l$.
- Incorporating the tortuosity parameter in the mean effective velocity as: $v_{pw} = v/n \, \Delta l_e / \Delta l$.
- Using the effective path length Δl_e instead of the length of the flow tube Δl to describe the potential gradient used in Equation (7.20).

With the above steps, the following is obtained for flux of water through the unit area:

$$v = \frac{nv_{pw}}{\tau} = \frac{n^3}{S^2(1-n)^2} \cdot \frac{C_s}{\tau} \frac{\gamma}{\eta} \frac{\Delta\psi}{\Delta l_e}$$

$$= \frac{C_s\gamma}{\eta\tau^2 S^2} \frac{n^3}{(1-n)^2} \frac{\Delta\psi}{\Delta l}$$

(7.21)

Designating k_{KC} as the permeability coefficient obtained from this equation to distinguish it from the Darcy coefficient of permeability k, one obtains from the relationship just shown,

$$k_{KC} = \frac{C_s \gamma}{\eta \tau^2 S^2} \frac{n^3}{(1-n)^2}$$

(7.22)

The graph in Figure 7.9 shows the relationship between the Kozeny–Carman permeability coefficient k_{KC} and the amount of surface area wetted S_w, calculated on the basis of (a) properties of water at 20°C, (b) a tortuosity τ value of $\sqrt{2}$, and (c) a shape factor $C_s = 0.4$.

7.4.3 Soil Strata and Hydraulic Properties

Macroscopically, soils consist of several differing layers (of soils) that are the result of differences in the processes of soil genesis, that is, weathering of fragmented rock, sedimentation or accumulation of composition, biotic actions, and anthropogenic actions such as agricultural land use. The soil properties of each layer are different in grain size, water retention, and permeability (Chapter 1), and hence the transport of water in and through layered soils does not follow the same patterns as in homogeneous uniform soils.

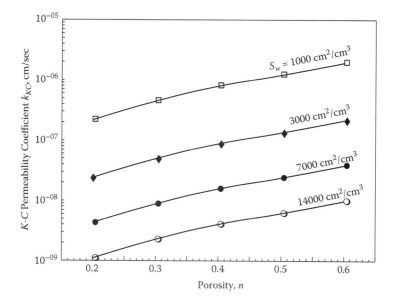

FIGURE 7.9
Variation of permeability coefficient k_{KC} with calculated porosity, in relation to wetted surface area S_w.

In the case of a soil layered-structure consisting of (a) a coarse-grained soil layer overlying a fine-grained soil layer, water infiltration in the upper soil layer migrates quickly into the underlying fine-grained soil when the wetting front reaches the boundary between the two layers. On the other hand, in the case of (b), a fine-grained soil layer overlying a coarse-grained soil layer, storage of infiltration water occurs in the upper fine-grained soil layer because of difference in pore size, that is, capillary force, between fine-grained and coarse-grained soils. The stored water does not migrate easily into the lower coarse-grained soil because of the high retention capacity in the upper fine-grained soil. The boundary between both soil layers is a barrier obstructing the downward flow of water into the lower soil layer. This is called the *capillary barrier*—a characteristic feature that is applied to prevent water entry into underground disposal facilities of hazardous wastes, and to prevent deep percolation of rain and irrigation water for water harvesting in dried agricultural land, especially in desert areas. When the total potential of the overlying fine-grained soil increases to values greater than that of the underlying coarse-grained soil at the boundary between both layers, water stored in the fine-grained soil will migrate into the underlying coarse-grained soil. When water flow occurs uniformly in both soil layers in both cases (a) and (b), even though the volumetric water contents are different between the upper and lower soil layers, the water potential profile is continuous at the boundary (Hanks and Bowers, 1962). For analyses of uniform water transfer through layered soil, one needs to use Equation (3.27) in Chapter 3, where transfer is expressed in terms of the potential energy of water.

7.4.3.1 Finger Flow

When both soils are initially in the air-dried state, for the case of (b) where a fine-grained soil layer overlies a coarse-grained soil, water uptake into the underlying coarse-grained soil occurs as a finger flow—that is, water moves through a particular water passage in the underlying soil (Figure 7.10a), generated by local disparities in pore sizes at the boundary between the two layers that favour finger-flow penetration. Applying the capillary model to soils (see Chapter 3), finger flow will be considered to occur at a location where the following condition is met: the location is a random selection determined by the result of slight differences in pore size along surfaces of both layers, and a random combination between the upper and lower pores at the boundary.

$$\gamma_w g H_U \geq \frac{2T \cos \alpha_U}{r_U} - \frac{2T \cos \alpha_L}{r_L} \tag{7.23}$$

where H_U is the height of water held by capillary action at the bottom of the upper soil layer, r_U and r_L are the mean pore radii of the upper and lower soil layers, respectively, T is the surface tension of water, α_U and α_L are the contact angles of the upper and lower soil solids, respectively, γ_w is the density of water, and g is the acceleration due to gravity. The term on the left-hand side expresses the action of gravity on soil water, and the first and second terms on the right-hand side of the equation refer to the actions of capillary forces in the upper and lower soil layers, respectively.

Finger flow occurs in monolayer soils composed of dried, coarse grains in the process of redistribution of water after water uptake ceases at the surface (Wang et al., 2003). The reason is because matric potential decreases behind the wetting front formed at the initial stage of water uptake, and water is drawn into the smaller pores at the wetting front by capillary force and flows downward. The water content in the finger is in a partially saturated state. The growth of the finger flow is attributed to two factors: (a) water is continuously applied to the upper surface of the overlying fine-grained soil at a high rate, and (b) a horizontal flow at the bottom of the upper layer occurs toward the location where finger flow occurs, resulting in the growth of the finger downward with expansion of its width. When the permeability k_L of the lower coarse-grained soil is larger than the water uptake rate q_{suf} of water at surface of the upper fine-grained soil, that is, when $k_L > q_{suf}$ is obtained, the finger will be stable and water will be transported into the groundwater zone with a considerably high water velocity (Iwata et al., 1995; Jury and Horton, 2004).

For the sloping layered soil structure shown in Figure 7.10b, water stored at the bottom of the upper layer moves toward the lower location along the

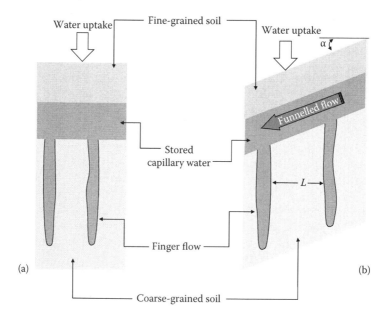

FIGURE 7.10
Schematic illustration of finger flow in a layered soil consisting of a fine-grained soil overlying a coarse-grained soil. Left-hand sketch (a) shows flow pattern in a level (horizontal) surface situation, and right-hand sketch (b) shows the finger flow pattern in a sloped surface situation.

boundary by gravitational action. This lateral flow is termed *funnelled flow*. Finger flow occurs and grows at a location where the condition expressed by Equation (7.23) is satisfied, in the context of funnelled flow. Steenhuis et al. (1991) have reported that the interval between fingers could be determined experimentally as follows:

$$L \le \frac{k_S}{q_{suf}} \tan \alpha \cdot \left[\frac{1}{b} + \left(h_U - h_L \right) \right]$$

(7.24)

where L is the distance between fingers, referred to as the *diversion length*, k_S is the fully saturated hydraulic conductivity of the overlying fine-grained soil, α is the angle of the slope, and h_U and h_L are the air-entry head of the upper and lower soil layers, respectively. The coefficient b is given as $k = k_S \cdot \exp(b\psi_m)$, where k is the unsaturated hydraulic conductivity of the overlying fine-grained soil layer, and ψ_m is the matric potential. The maximum diversion length is considered to be in range of 5 to 50 m.

Finger flow transfers local contaminants into the underlying coarse-grained soil, which in turn allows for rapid transport of the contaminants to the groundwater. Because of the rapidity of contaminant transport to the groundwater, the surrounding soil remains uncontaminated. This phenomenon causes difficulty in detecting the source of contamination and the processes of leading to groundwater contamination from soil analyses of deep soils, because contaminated soil samples are not readily or easily obtained in vast fields.

7.4.3.2 Water Flow in Layered Soils

Calculation or estimation of water flow in layered soils requires one to take into account the different permeabilities of the different soil layers that constitute the soil profile of interest. Assuming complete saturation and water flow capability through the soil layers, calculations for the least permeable layer—that is, the least Darcy coefficient k value, as the control for water flow—will yield flow values that would be too low. For a soil profile of depth H, with layers of different thickness H_i, the Swartzendruber (1960) analysis shows that the Darcy relationship can be used to define an equivalent permeability coefficient k_e for the total profile depth H as follows:

$$\frac{H}{k_e} = \sum_{i=1}^{i=n} \frac{H_j}{k_j}$$

(7.25)

The results are independent of the order of summation, meaning that this is a general relationship that is valid for any pattern of layering.

7.4.4 Void Space and Hydraulic–Thermal Properties

There is a likelihood that the vadose zone, because of wetting-drying cycles, will have macropores, cracks, and fissures in the zone. The void spaces of these macropores, cracks, and fissures constitute (a) channels of water flow in soils, and (b) conduits for sensible heat flow in soils from water flux and the heat capacity of water, that is, $c_l q_l \Delta T$ in Equation (7.3), where c_l is the heat capacity of liquid water, q_l is the flux of liquid water, and T is the temperature (Section 7.2.2). Water flow from soil surface to underground water is, more or less, affected by flow through these channels, in addition to the normal subsurface flow that is governed by potential gradients in the soil matrix. Since flow in the macropore channels is under the influence of gravitational forces, the velocity of flow in deep percolation will be considerably higher than that of the uniform subsurface flow due to potential gradients in the soil matrix. This means that deep percolation of water from macropore channel flow will constitute preferential flow in a soil mass where such conditions exist. This means that water and contaminants migrating from the soil surface will reach deeper locations in the vadose zone quickly, to the extent that this will happen before the soil matrix can fully adsorb both water and contaminants in the upper regions (Charbeneau, 1984). Contaminant flow in soils is discussed in Section 9.3.2 in Chapter 9. According to analyses by the *kinematic wave approximation* technique, it is possible to estimate such quick flow properties or deep percolation flow properties as flux and location of the wetting front, and even water content profiles in average on cross section, based on the principles of conservation of water in channels or soil cross sections.

7.4.4.1 Flow-Through Channels

When water migrates into soils from the soil surface, the flux of water migrating in channels formed by roots, soil animals, drying, and various external forces is assumed to be a function of the volumetric water content in the channels. This is because water is considered to flow in channels that are both partially and fully occupied. According to Germann and Beven (1985), $q_{ch} = b\theta_{ch}^a$, where q_{ch} and θ_{ch} are the flux and volumetric water content in channels, respectively, and a and b are constants. Applying this relation to the conservation law, with a sink term S representing water in the channels, one obtains $\partial\theta_{ch}/\partial t = -(\partial q_{ch}/\partial z) - S$, where z is the spatial coordinate in a vertical direction, and t represents time. Because flux is assumed to decrease because of water uptake into the surrounding soil matrix from channel wall, the following is obtained:

$$\frac{\partial q_{ch}}{\partial t} = -C_{ch}\frac{\partial q_{ch}}{\partial z} - C_{ch}S \tag{7.26}$$

C_{ch} is the *kinematic wave velocity* that is given by $C_{ch} = \partial q_{ch}/\partial\theta_{ch} = ab\theta_{ch}^{a-1}$. Assuming that S, the sink term is given by changes of θ_{ch} per unit time as follows: $S = d\theta_{ch}/dt = (\theta_{ch}/aq_{ch}) \cdot (dq_{ch}/dt)$, Equation (7.26) can be expressed as a first order partial differential equation of q_{ch} with a solution that would give a flux profile that is observed in channels opening on the soil surface after a rainfall. Germann and Beven (1985) and their coworkers have obtained a solution for location of wetting front and draining front using ingenious techniques for the condition for continuation of rain during time t, and ceasing thereafter.

7.4.4.2 Deep Percolation of Water

Flow of water migrating into the vadose zone from the soil surface, that is, deep percolation of water, can be estimated using the mean values on both volumetric water content and flux of water in a soil section. Assuming that migrating water flow from the soil surface is due only to the gravitational force—since water potential gradients will be negligible in comparison to gravitational potential gradient (that is, the water potential in soil is almost in equilibrium with the gravitational potential at the deeper locations)—the mean water flux q will be given by the gravitational potential gradient of 1 in the Darcy model. The mean water conductivity in a soil section will be $q = -k(\theta)$, where $k(\theta)$ is the mean water conductivity in a soil cross section. Accordingly, the water conservation equation without source or sink term $\partial\theta/\partial t = -\partial q/\partial z$ can be restated as a simple first order partial differential equation using the mean volumetric water content θ as follows (cf. Equation (7.26)):

$$\frac{\partial\theta}{\partial t} = \frac{dk}{d\theta}\frac{\partial\theta}{\partial z} \tag{7.27}$$

The characteristic differential equation of Equation (7.27) is given by $dt/1 = -dz/(dk/d\theta)$, and dz/dt, the *kinematic wave velocity* is given as $-(dk/d\theta)$. This shows that the mean volumetric water content moves downward in soils as a wave propagation front. Assuming a relation of $k(\theta) = \alpha\theta^P$, where α and P are constants, one obtains:

$$\frac{dz}{dt} = -\alpha P\theta^{P-1} = -k(\theta)\frac{P}{\theta} = \frac{qP}{\theta} \tag{7.28}$$

where q is the flux of water through a cross section of soil. Since the mean velocity v of water in soils is given as $v = q/\theta$, the kinematic wave velocity dz/dt of water content is given as Pv. This means that the kinematic wave velocity of water content is P times larger than the mean velocity v of water in soils. Values of P have been observed in range of 3 for sand to 12 for clay in literature (Campbell, 1974).

7.5 Swelling Clay Hydraulic Conductivity

7.5.1 Hydraulic Conductivity and Clay Structure

Whilst the macroscopic mechanistic hydraulic conductivity performance of swelling clays is closely similar to that of the nonswelling clays, there are some significant differences in respect to porewater conductivity. The schematic diagram shown in Figure 7.11 illustrates the main elements of the differences encountered, most of which are due to the nature of the clay and the diffuse ion layers that characterize the swelling clay (smectite). The figure illustrates the differences in proportions of micropore porosity and macropore porosity. The kaolinite used as an example of a nonswelling clay has platelets as particles, and these do not have interlayers that "explode" upon exposure to water, as is the case of the smectite.

Bulk measurements of the velocity of fluid flow through the types of clay shown in the diagram are reflective of flow through the macropores. The contribution of fluid flow from the interlayers and micropores of the swelling

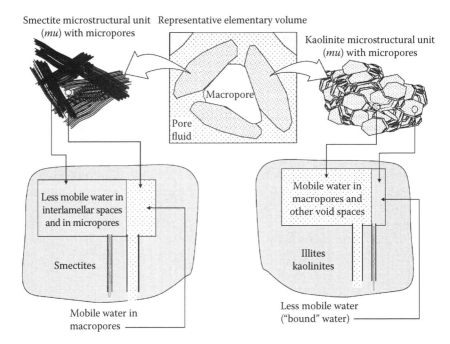

FIGURE 7.11
Schematic illustration of the proportions of water held in interlamellar spaces, micropores and macropores of smecties and kaolinites. The two bottom diagrams show the respective proportions of mobile water associated with the two clays. Water held in the micropores and interlamellar spaces is significantly less mobile than water held in the macropores and other void spaces.

clay (smectite) to the macroscopic fluid flow velocity is more in terms of a boundary-layer drag. Because of the considerably higher proportioning of micropores to macropores, fluid flow will not only be extremely slower, but the effect of restructuring of the microstructural units will be greater with the swelling clays. Changes in wetted surface areas would be significant if and when restructuring of the microstructures occurs.

7.5.2 Laboratory Test for Permeability

Permeameters are most often used in laboratory experiments using either constant head (that is, constant hydraulic gradient) conditions, as shown in the Darcy-type experiment in Figure 7.5, or using falling head conditions. The procedures for permeability testing are well documented in the testing standards of most countries (e.g., ASTM, EC standards). Instead of standard-size permeameters, oedometer-type and triaxial confinement devices have been used as test systems for determination of the permeability of swelling and nonswelling clays. The advantage of these types of systems is their ability to confine lateral and axial expansion of the test samples during permeation of water. The triaxial-type system is used for most clays, whereas the oedometer-type system is preferred for high swelling clays because of its robust confinement against the swelling pressure developed during fluid permeation of the test sample. The principles of the two systems are shown in Figure 7.12.

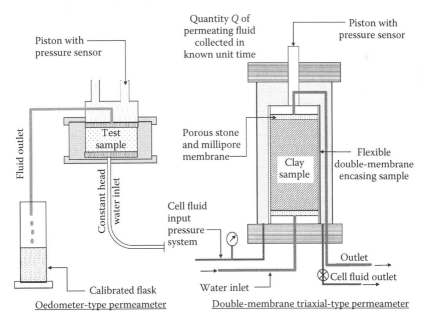

FIGURE 7.12
Oedometer-type (left) and double-membrane triaxial type (right) for determination of permeability of clays.

Although there are no particular standards for the dimensions of samples used for testing in the two kinds of cells shown in the figure, there are some simple guidelines to follow. It is necessary to ensure that (a) the cross-sectional area of the test sample must be sufficiently large so that one has a representative portion of the sample, and (b) the height of the sample should be sufficient to permit proper hydraulic transport of the permeating fluid. Sample sizes used in the oedometer-type and triaxial-type permeameters have been varied, depending on whether the devices used were adapted from standard oedometers and triaxial test equipment. By and large, deviations from these standard devices have tended to follow the same dimensional ratios for sample sizes. For tests on high-swelling clays with the oedometer-type device, the research literature shows that sample heights of from one-half to three-quarters of the sample diameter have been used. When such is the case, it is necessary to ensure that steady-state flow is achieved, and that duplicate tests with different hydraulic gradients are conducted to ensure test validity.

7.5.2.1 Oedometer-Type Permeameter

The robust oedometer-type permeameter shown in left-side portion of Figure 7.12 is designed to provide radial rigidity, that is, no radial expansion of the confined sample. The piston-pressure cell system at the top of the cell allows one to record swelling pressures generated axially in the permeability tests. The Darcy model is used to determine the permeability coefficient. The permeating fluid is applied under a specified constant head to the test sample, and the fluid quantity Q accumulated within a unit time period is collected at the outlet end in a calibrated flask. Since the cross-sectional area A of the sample is known, the permeating flux $v = Q/A$ for the unit time period can be related to the test-specified gradient i via the impedance to water flow k, known as the Darcy coefficient of permeability, as given in the Darcy model. For swelling clays, the hydraulic gradient applied must be greater than the swelling pressure. Defining the *excess hydraulic gradient* (EHG) to be the hydraulic gradient above that required to equalize the swelling pressure, experience has shown that excess hydraulic gradients above 60 or so applied to dense swelling clays would begin to cause local channelling and piping. For less compact swelling clays, the upper limit of the EHG will be lower. For example, if a hydraulic gradient of 30 is required to counter the swelling pressure developed in a saturated swelling clay sample contained in the oedometer-type device, and if one desires to apply an EHG of 50 to the test sample, this means that a total hydraulic gradient of 80 would be required to implement the test plan.

7.5.2.2 Triaxial-Type Permeameter

The triaxial-type permeameter shown in the right-side portion of Figure 7.12 has been used for permeability tests in sand-bentonite mixtures and also for medium-swelling clays. The primary difference in sample performance between this and the oedometer-type device is the ability to exercise three-dimensional confinement or lateral confinement through control of the cell pressure. Monitoring of the confining sample volume and control of the vertical piston permits one to exercise a no-volume change condition on the sample being permeated. In addition to the control of volume change, the triaxial-type permeameter allows one to apply a back-pressure in the test sample to counter the swelling pressure. The test protocol for this device is similar to that of the oedometer-type device.

7.6 Vapour and Gas Conductivities

Vapour, air, and chemical gases in the pore spaces of soil move under concentration and pressure gradients, in association with temperature gradients, in partly saturated soils. Movements associated with concentration gradients have been discussed in Section 3.5.1 in Chapter 3 and Section 7.3.1 in this chapter. In this instance, a diffusion coefficient or a transformed diffusion coefficient, with several relevant factors, characterizes the gas conductivity in soils. When external pressures are imposed on the *gas* in soils—where the term *gas* is used as a generic name for vapour and other chemical gases and air that occupy pore spaces in soils—flow under the resultant pressure gradients occurs as aerodynamic fluid flow in the soil under consideration. The rate of flow is considered to obey Darcy's law, and conductivity is defined as a coefficient that relates the flow rates with their respective pressure gradients. This gas conductivity is usually termed *gas permeability*. Since gas is a compressible fluid, its flow through a soil is accompanied by changes in volume. Gas permeability can be experimentally measured with oedometer-type and triaxial-type permeameters (Figure 7.12) where measurements of gas flow are made for samples subject to pressure gradients generated by pressure imposed on opposites sides of the sample.

The two types of expression that can be used to calculate the mean gas permeability over a certain thickness of samples are (a) steady-state flow, and (b) unsteady-state flow. Development for both steady-state and unsteady-state relations assumes validity of the Darcy model $v = -ki$, and the equation of state for an ideal gas in respect to gas flow in a soil sample, that is, $pV = nRT$, where p is the pressure, V is the volume, n is the number of moles, R is the gas constant, and T is the temperature.

7.6.1 Steady-State Experiments

The permeability at a certain pressure p in the samples is obtained by directly combining Darcy's law with the equation of state for gas (Horseman et al., 1999; Tanai and Yamamoto, 2003) as follows:

$$k = \frac{2\mu qpL}{P_1^2 - P_2^2} = \frac{2\mu QLp}{A\left(P_1^2 - P_2^2\right)} \tag{7.29}$$

where P_1 and P_2 are the external pressures applied at the opposite sides of the sample, p is the pressure at a certain point in the sample, q is the gas flux, Q is the outflow of gas, L is the length of samples, A is the cross sectional area of samples, and μ is the viscosity of the gas. Applying the mean pressure of samples, for example, $p = \bar{p} = (1/2)(P_1 + P_2)$ to p, we obtain:

$$k = \frac{\mu \bar{Q} L}{A\left(P_1 - P_2\right)} = \frac{\mu \bar{q} L}{P_1 - P_2} \tag{7.30}$$

where the overbars associated with Q and q are the mean values.

7.6.2 Unsteady-State Experiments

Gas chambers are set at the opposite sides of samples, with pressures p_{VO} and p_{SO} ($p_{VO} > p_{SO}$) applied to both chambers at the beginning. The pressure in the chamber that applies gas to the samples is decreased gradually as gas migrates into the samples, with pressure maintained constant in the chamber that captures gas outflow (Yoshimi and Osterberg, 1963). Unsteady gas flow in the samples can be determined, and the relationship for gas in the chamber that applies gas to the samples is obtained using mass conservation principles as: $d(\rho V)/dt = -\rho q A$, where ρ is the density of the gas in the chamber, V is the volume of the chamber, q is the flux defined by Darcy's law, and A is the cross-sectional area of test samples. Applying both the state equation for gas and Darcy's law to this relationship, integrating the resultant relationship and rearranging, we obtain the gas permeability at a mean pressure: $p = (p_{vt} + pso)/2$ in the samples that follow:

$$\bar{k} = \frac{VL\mu}{Ap_{SO}\,t} \ln \frac{1}{C} \left| \frac{p_{Vt} + p_{SO}}{2\left(p_{Vt} - p_{SO}\right)} \right| \tag{7.31}$$

and

$$C = \left| \frac{p_{VO} + p_{SO}}{2\left(p_{VO} - p_{SO}\right)} \right| \tag{7.32}$$

where $\bar{k}$ is the mean gas permeability with time t that is defined as $\bar{k} = (1/t)\int_0^t k\,dt$, μ is the viscosity of the gas, p_{Vt} is the pressure of accumulated gas in the chamber used to apply gas to the sample at a certain time

t, p_{VO} is the pressure of accumulated gas in the chamber at $t = 0$, p_{SO} is the pressure in the chamber capturing gas outflow, L is the length of the sample, V is the volume of the chamber, and A is the cross sectional area of the sample. One should note that the gas permeability given by Equation (7.31) includes the effects of changes in pore spaces, solid density, and water content with time resulting from expansion and shrinkage of samples during permeation of gas.

7.7 Concluding Remarks

The thermal and hydraulic properties of soils discussed in this chapter have been focussed on the conductivity and permeability of soils in respect to fluid movement, and the movement of water as part of the heat-mass transfer process. We have also included *gas permeability* as a conductivity property because most water movement in soils in nature occurs as "unsaturated flow."

The water content of a soil is probably one of the most important properties of a soil. Its presence in a soil allows for interparticle actions realized through various internal forces developed from the surface properties of the soil particles and the chemistry of the porewater. To that end, the focus in this chapter has been on two particular types of environment stressors, thermal and hydraulic. Together with chemical stressors, they are most often the main "drivers" for water movement in soils. The coefficient of permeability, which is the soil property that accounts for hydraulic conductivity, is not measured directly. It is a *computed parameter*. This means to say that one does not obtain a direct measurement of the parameter.

It is important to bear in mind that the Darcy model, which is used to analyze permeability test measurements, will yield a coefficient of permeability that is consistent with the various test conditions encountered by Darcy in his own experiments. It is also very important to appreciate that the coefficient of permeability k computed from a Darcy-type experiment is *operationally defined*. What this means is that the computed k value is sensitive to procedures, techniques, methodology, and so forth used in experimentation, from sample preparation to final conduct of test. The discussions in this chapter have pointed out that the properties of the permeant and the soil being tested have not been factored directly into the Darcy-type analysis, and that there are procedures that have been developed to provide some accommodation for these properties. These are not perfect, nor do they necessarily provide the total set of corrections. They are nevertheless a step in the right direction.

References

Cary, J.W., 1966, Soil moisture transport due to thermal gradient, *Soil Sci., Am., Proc.,* 30:428–433.

Campbell, G.S., 1974, A simple method for determining unsaturated conductivity from moisture retention data, *Soil Sci.,* 117:311–314.

Campbell, G.S., and Norman, J.M., 1988, *An Introduction to Environmental Biophysics,* 2nd ed., Springer-Verlag, Berlin.

Charbeneau, R.J., 1984, Kinetic models for soil moisture and solute transport, *Water Resour. Res.,* 20:699–706.

Darcy, H., 1856, Les fontaines publiques de la ville de Dijon: Exposition et application des principles à suivre et des formules à employer, in V. Dalmont (Ed.), *Histoire les Flntaines Publiques Dijon, Libraire des Corps Imperiaux des Ponts et Chaussees et Des Mines,* Paris, pp. 305–3011.

de Vries, D.A., 1958, Simaltaneous transfer of heat and moisture in porous media, *Trans. Amer. Geophys. Union,* 39(5):909–919.

Germann, P.F., and Beven, K., 1985, Kinematic wave approximation to infiltration into soils with sorbing macropores, *Water Resour. Res.,* 21:990–996.

Hanks, R.J., and Bowers, S.A., 1962, Numerical solution of the moisture flow equation for infiltration into layered soil, *Soil. Sci. Soc. Am. Proc.,* 26:530–534.

Hansbo, S., 1960, Consolidation of clay, with special reference to influence of vertical sand drains, Swedish Geotechnical Institute, Proc. No. 18, pp. 41–61.

Horseman, S.T., Harrington, J.F., and Sellon, P., 1999, Gas migration in clay barriers, *Eng. Geol.,* 54:139–149.

Ichikawa,Y., Kawamura, K., Nakano, M., Kitayama, K., and Kawamura, H., 1999, Unified molecular dynamics and homogenization analysis for bentonite behaviour: current results and future possibilities, *Eng. Geol.,* 54:21–31.

Iwata, S., Tabuchi, T., and Warkentin, B.P., 1995, *Soil-Water Interaction,* Marcel Dekker, New York, 440 pp.

Jury, W.A., and Horton, R., 2004, *Soil Physics,* 6th ed., John Wiley & Sons, New York, 384 pp.

Kasubuchi, T., 1977, Twin transient-state cylindrical-probe method for the determination of thermal conductivity of soil, *Soil Sci.,* 124:255–258.

Katchalsky, A., and Curran, P.F., 1967, *Non-equilibrium Thermodynamics in Biophysics,* Harvard Books in Biophysics, 1, Harvard University press, Cambridge, Mass, 38–41.

Lutz, J.F., and Kemper, W.D., 1959, Intrinsic permeability of clay as affected by clay-water interactions, *Soil Sci.,* 88:83–90.

Mohamed, A.M.O., Yong, R.N., and Cheung, S.C.H., 1992, Temperature dependence of soil water potential, *Geotech. Testing J.,* 15:330–339.

Manose, T., Sakaguchi, I., and Kasubuchi, T., 2008, Development of an apparatus for measuring one-dimensional steady state heat flux of soil under reduced air-pressure, *Eur. J. Soil Sci.,* 59:982–989.

Motosuke, M., Nagasaka, Y., and Nagashima, A., 2003, Measurement of dynamically changing thermal diffusivity by the Forced Rayleigh Scattering method (measurement of the gelation process), Paper presented to 15th Symposium on Thermophysical Properties, June, Boulder, Colorado.

Nakano, M., and Miyazaki, T., 1979, The diffusion and non-equilibrium, thermodynamic equations of water vapour in soils under temperature gradients, *Soil Sci.*, 128(3):184–188.

Osako, M., and Ito, E., 1997, Simultaneous thermal diffusivity and thermal conductivity measurements of mantle materials up to 10 GPa, Technical Report of ISEI (Institute for Study of the Earth's Interior), Ser.A. No.67.

Philip, J.R., 1957, The physical principles of soil water movement during the irrigation cycle, *Congr. Inter. Comm., Irrigation and Drainage*, 8:125–154.

Pusch, R., Moreno, L., and Neretnieks, I., 2001. Microstructural modelling of transport in smectite clay buffer, in K. Adachi and M Fukue (Eds.), *Clay Science for Engineering*, A. A Balkema, Rotterdam/Brookfield, pp. 47–51.

Swartzendruber, D., 1960, Water flow through a soil profile as affected by the least permeable layer, *J. Geophys. Res.*, 65:4037–4042.

Tanai, K., and Yamamoto, M., 2003, Experimental and modelling studies on gas migration in Kunigel V1 bentonite, JNC TH8400, 2003-024, JAEA.

Wang, Z., Turi, A., and Jury, W.A., 2003, Unstable flow during redistribution in homogenous soil, *Vedose Zone J.*, 2:52–60.

Yong, R.N., and Warkentin, B.P., 1975, *Soil Properties and Behaviour*, Elsevier, Amsterdam, 449 pp.

Yong, R.N., Xu, D.M., Mohamed, A.M.O., and Cheung, S.C.H., 1992, An analytical technique for evaluation of coupled heat and mass flow coefficients in unsaturated soil, *Int. J. Numer. Anal. Methods in Geomechanics*, 16:233–246.

Yong, R.N., and Warkentin, B.P., 1966, Introduction to *Soil Behaviour*, Macmillan Co., New York, N.Y., 451 pp.

Yoshimi, Y., and Osterberg, J.O., 1963, Compression of partially saturated cohesive soils, *J. Soil Mechanics, ASCE, SM*, pp. 1–24.

8

Sorption Properties and Mechanisms

8.1 Introduction

The term *sorption* is used to refer to the many types of adsorption processes and mechanisms that result in the partitioning of dissolved substances in the porewater of a soil onto the surfaces of the soil particles. The sorption properties of soils are particularly important since the type and the amount of substances sorbed by a soil will change the nature of the soil, and hence affect its functionality. It can be said that these are probably the most important environmental soil properties. This can have significant impact in geoenvironmental engineering projects, groundwater management, and agriculture. The discussions in this chapter will focus on the nature of the interactions and mechanisms involved in soil sorption. Since the types of interactions and mechanisms involved in soil sorption processes are dependent not only on the surface properties of the soil particles, but as much on the nature and surface properties of the substances involved, the discussions will pay attention to (a) the surface properties of the interacting parties (soil particles and substances), and (b) the mechanism leading to soil sorption.

8.2 Solutes, Contaminants, and Pollutants

We pay attention to presence of contaminants in soils (a) as solutes that are not part of the natural composition of the soil mass of interest, (b) as an environmental issue of importance because of their ability to alter the properties and behaviour of soils, and (c) because they present threats, by their very presence in the soil, to the health of biotic receptors and the environment. Figure 8.1 gives an example of the effect on the axial strain of a soil sample because of the penetration of a contaminant leachate into the sample.

Contaminants find their way into a soil mass:

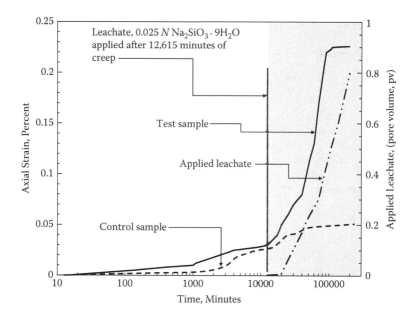

FIGURE 8.1

Effect of exposure of a natural lacustrian clay consisting of illite, chlorite, kaolinite, and other nonclay minerals, under a creep load (4.8 kPa) to a leachate (0.025N $Na_2SiO_3.9H_2O$) after 12,615 minutes of creep. The control sample of the same clay, which was not exposed to the leachate, is shown by the dashed line. (Adapted from data presented by Yong, R.N., et al., 1985, *Eng. Geol.*, 21:279–299.)

- Naturally, (a) as solutes leached or transported due to percolating rainfall, snowmelt, floods, and so forth from other parts of a soil mass to the portion of soil mass of concern, and (b) from atmospheric deposition of particulates.

- As a consequence of anthropogenic activities via illegal dumping, intensive agricultural practice, landfills and tailings ponds, mining activities, and so forth.

Soil water content and porewater chemistry are primary factors involved in establishment of soil integrity. Discussions in the previous chapters have shown that soil water content and the ionic strength and pH of soil–water (porewater) have considerable influence on the development of soil properties. This is because most soil properties are derived from the results of interparticle and interaggregate actions. The term *aggregate* is used to include microstructural units (*msu*) such as peds, clusters, and domains. The transport of contaminants in soils changes the porewater chemistry at the localities wherein transport occurs. The contaminants can be naturally occurring alkaline metals and alkaline-earth metals (groups I and II in the periodic table) as part of the soil formation process, and/or contaminants originating from external and anthropogenic sources. In the latter case, the contaminants may be inorganic in nature or may consist of organic chemicals. Examples of potential acute and chronic effects due to exposure to some of these contaminants are shown in Table 8.1.

Strictly speaking, *contaminants* in soils are substances (solutes, chemicals, etc.) that are not part of the initial composition of the soil. In general, most laboratory methods or techniques used to determine the concentration of contaminants in soils do not always distinguish between the amounts of contaminants and naturally occurring solutes measured. There is little need to distinguish between contaminants and naturally occurring solutes when it comes to examination and discussion on transport processes and impacts on soil properties and behaviour.

In discussions in this chapter, we use the term *contaminants* deliberately to call attention to the soil environmental problem of discharge and transport of contaminants in the ground via illegal dumping of waste, from waste landfills and other sources such as herbicides and fertilizers used in agro-industry. It is not uncommon, in the literature, to refer to contaminants originating from anthropogenic sources and/or activities as *pollutants*. Pollutants are contaminants that have been identified as threats to human health and the environment because of their nature, as opposed to their concentration. These pollutants are generally heavy metals and toxic chemicals or compounds listed in the Priority Listing of many jurisdictions (e.g., U.S. Environmental Protection Agency [EPA] Priority Pollutants List) designated as pollutants. The discussion on pollutants and contaminants in soils in respect to their fate, control, health hazard, containment, and mitigation can be found in books written specifically to address these problems (e.g., Fetter, 1993; Knox et al., 1993; Yong, 2001a; Yong and Mulligan, 2004). The attention to contaminants in this book is directed towards their role and effect on the properties and behaviour of soils. From this perspective, the two immediate sets of issues revolve around (a) the principal agents of transport of water and contaminants, and (b) the processes and properties involved in transport of contaminants. These form the basis for the discussions in this chapter.

8.2.1 Transport of Contaminants

Transport of contaminants in soils cannot occur without the presence of porewater. This means to say that water serves as the carrier for contaminants. What are the agents responsible for hydraulic-driven and thermal-driven transport processes? Seasonal and weather-related events constituting naturally occurring agents include, for example, summer–winter seasons and wetting–drying events. Events and/or agents generating transport of water and/or contaminants in soils associated with anthropogenic activities include water impoundment structures, waste landfills, buried heat-emitting structures or facilities, and load-carrying structures. The transport

TABLE 8.1

Acute and Chronic Effects of Some Contaminants/Pollutants on Human Health

	Acute						Chronic		
Contaminant/Pollutant	Nervous System	Gastro System	Neuro System	Respiratory	Skin	Death	Carcino-Genic	Muta-Genic	Terto-Genic
Halogenated organic pesticide	H		H	H		H	A	A	A
Organophosphorous pesticide	H		H	H		H	A	A	A
Carbamate insecticide	H		H	H		H			
Halogenated organics	H		H		H		H	H	
Organic lead compounds	H	H	H				H		
Zn, Cu, Se, Cr, Ni		H		H	H		H		
Hg	H	H	H				H	H	H
Cd		H					H	H	

Source: Adapted from Information in Yong, R.N., et al., 1992, *Principles of Contaminant Transport in Soils*, Elsevier, Amsterdam, 327 pp.

Note: H, A, = statistically verifiable effects on humans and animals respectively.

processes associated with these agents and events, and the outcome of these processes, will vary in accordance with the agent and/or event, and with the nature of the soils involved. In respect to transport of water, other than the type of agent or event, the primary sets of control on water movement include (a) geometry and connectivity of pore paces in the soil, (b) tortuosity, (c) sizes and distribution of apertures, (d) properties of the water, and (e) temperature. The primary externally imposed activating mechanisms for movement are hydraulic pressure and thermal gradients.

8.3 Contaminant Interactions with Soil Particles

The types of interactions occurring between contaminants and soil particles are, to a very large extent, dependent on (a) the surface properties of both the soil particles and the contaminants themselves, and (b) the availability of surfaces of both particles and contaminants for interactions. The surface functional groups of soil particles, which have been described in Section 2.2 of Chapter 2, will react with the surface functional groups of the contaminants when these two parties are brought into contact with each other. The attachments formed between the contaminants and soil particles can be strong or weak, dependent on the types of functional groups involved in the reactions, and on the kinetics of reactions. In the final analysis, the results of these interactions will dictate the kinds of changes occurring in the soil and its properties and behaviour.

8.3.1 Contaminants of Interest

8.3.1.1 Inorganic Contaminants

The inorganic contaminants of interest and concern in soil–water systems include the naturally occurring solutes obtained in soil formation processes, and elements originating from sources outside the region of the soil of interest. The common inorganic contaminants from outside, which are shown in the periodic table in Table 8.2, are sometimes referred as harmful heavy metals. Strictly speaking, heavy metals are those elements with atomic numbers higher than strontium (atomic number 38).

The more common heavy metals (*HMs*) associated with anthropogenic activities such as landfills, chemical waste leachates, tailings ponds, and sludges, include lead (*Pb*), cadmium (*Cd*), copper (*Cu*), chromium (*Cr*), nickel (*Ni*), iron (*Fe*), mercury (*Hg*), and zinc (*Zn*). Since metallic ions such as Cu^{2+}, Cr^{2+}, and so forth (M^{n+} ions) cannot exist in the porewater as individual metal ions, they will generally be coordinated, that is, chemically bound, to six water molecules. In their hydrated form, they will exist as $M(H_2O)_x^{n+}$, and since coordination with water is in the form of bonding with inorganic

TABLE 8.2

Periodic Table Showing Harmful Inorganic Elements of Interest and Concern as Contaminants

	1	2	3	4	5	6	7	8	9	10	11	12	13	14	15	16	17	18
	IA	IIA	IIIA	IVA	VA	VIA	VIIA	VIII	VIII	VIII	IB	IIB	IIIB	IVB	VB	VIB	VIIB	0
1																		
2		Be 4											B 5				F 9	
3																		
4				Ti 22	V 23	Cr 24	Mn 25	Fe 26	Co 27	Ni 28	Cu 29	Zn 30	Ga 31		As 33	Se 34	Br 35	
5		Sr 38		Zr 40		Mo 42					Ag 47	Cd 48	In 49	Sn 50	Sb 51	Te 52		
6	Cs 55											Hg 80	Tl 81	Pb 82	Bi 83			
7						U 92												

Note that uranium U is in the actinoid series.

anions, replacement of water as the ligand for M^{n+} can occur if the other ligand, which is generally an electron donor, can replace the water molecules bonded to the M^{n+}.

Ligands, which are defined as anions that can form coordinating compounds with metal ions, have free pairs of electrons as their characteristic feature. Water molecules forming a coordinating complex with a metallic ion are ligands, and the metallic ion M^{n+} would be identified as the central atom. The number of ligands attached to a central metallic ion is called the *coordination number*, and this coordination number for a metallic ion is generally the same, regardless of the type or nature of ligand involved. For example, the coordination number for Cu^{2+} is 4, as found in $Cu(H_2O)_4^{2+}$ and $CuCl^-$, and the coordination number for Fe^{3+} is 6, as found in $Fe(CN)_6^{3-}$ and $Fe(H_2O)_6^{3+}$. The common coordination numbers for heavy metals are 2, 4, and 6, with 6 being the most common. The arrangement of ligands around the central atom depends on the number of ligands involved in formation of the complex. Complexes with coordination number of 2 will have a linear arrangement of ligands, whereas complexes with a coordination number of 4 will generally have tetrahedral arrangement of ligands, and in some cases a square-planar arrangement of ligands. Ligands in complexes of coordination number 6 are arranged in an octahedral fashion. The more common inorganic ligands that will form complexes with metals include CO_3^{2-}, SO_4^{2-}, C^-, N_3^-, OH^-, SiO_3^-, CN^-, F^-, and PO_4^-.

Many of the transition metals have one oxidation state that is most stable. The most stable state for *Fe* is *Fe(III)* and the most stable states for *Co* and *Ni* are *Co(II)* and *Ni(II)*, respectively. The electronic configuration in the *d* orbitals is an important factor. Unpaired electrons that comprise one-half of the sets in *d* orbitals are very stable, which explains why *Fe(II)* can be easily oxidized to *Fe(III)* and why oxidation of *Co(II)* to *Co(III)* and *Ni(II)* to *Ni(III)* cannot be as easily accomplished. The loss of an additional electron to either *Co(II)* and *Ni(II)* still does not provide for one-half unpaired electron sets in the *d* orbitals.

Some heavy metals can exist in the porewater in more than one oxidation state, depending on the pH and redox potential of the porewater. Selenium *(Se)* can occur as SeO_3^{2-} with a valence of +4, and as SeO_4^{2-} with a valence of +6. There are two possible valence states for copper *(Cu)* in the porewater—valencies of +1 and +2 for *CuCl* and *CuS*, respectively. Chromium *(Cr)* and iron *(Fe)* present more than one ionic form for each of their two valence states—CrO_4^{2-} and $Cr_2O_7^{2-}$ for the valence state of +6 for *Cr*, and Cr^{3+} and $Cr(OH)_3$ for the +3 valence state for *Cr*, and Fe^{2+} and *FeS* for the +2 valence state for *Fe*, and Fe^{3+} and $Fe(OH)_3$ for the +3 valence state for *Fe*.

8.3.1.2 Organic Contaminants

Organic chemicals that find their way into the land environment originate from various chemical industrial processes, and as commercial substances for use in various forms. Commercial products that include organic solvents, paints, pesticides, oils, gasoline, creasotes, and greases, and so forth are some of the many sources for the chemicals found in contaminated sites. Of the million or so organic chemical compounds registered in the various Chemical Abstracts Services (CAS), many thousands of these are in commercial use. The more common organic chemicals found in contaminated sites fall into convenient groupings, which include

- *Hydrocarbons*—including the PHCs (petroleum hydrocarbons); the various alkanes and alkenes, and aromatic hydrocarbons such as benzene; MAHs (multicyclic aromatic hydrocarbons), for example, naphthalene; and PAHs (polycyclic aromatic hydrocarbons), for example, benzo-pyrene.
- *Organohalide compounds*—of which the chlorinated hydrocarbons are perhaps the best known. These include TCE (trichloroethylene), carbon tetrachloride, vinyl chloride, hexachlorobutadiene, PCBs (polychlorinated biphenyls), and PBBs (polybrominated biphenyls).
- *Oxygen-containing organic compounds*—such as phenol and methanol.
- *Nitrogen-containing organic compounds*—such as TNT (trinitrotoluene).

In the case of *nonaqueous phase liquids* (NAPLs), a useful classification scheme is to separate the light NAPLs (designated as LNAPLs) from the heavy NAPLs (DNAPLs) because of the manner in which these are transported in the subsurface regime. The LNAPLs are lighter than water, and will generally stay above the water table or float on top of it. On the other hand, since DNAPLs are heavier or denser than water, they are expected to sink through the water table and will come to rest at the impermeable bottom (bedrock). Examples of LNAPLs include gasoline, heating oil, kerosene, and aviation gas; and examples of DNAPLs include the organohalide and oxygen-containing organic compounds such as 1,1,1-trichloroethane, creasote, carbon tetrachloride, pentachlorophenols, dichlorobenzenes, and tetrachloroethylene.

8.3.2 Surface Properties and Interactions

The factors and properties of significance in interactions between contaminants and soils include the following:

- *Soil composition*—(a) type and distribution of the various soil fractions, and (b) physical and surface properties such as surface functional groups, cation exchange capacity (*CEC*), specific surface area (*SSA*), density, water content, and degree of saturation
- *Contaminants*—(a) types, distribution, and concentration of contaminants in the porewater, and (b) chemical and surface properties

- *Soil-contaminant systems*—Eh, pH of the system; microorganisms; and micro-environmental factors

8.3.2.1 Clay Minerals

The surface properties of both soil solids and contaminants are important factors in contaminant–soil interaction. In soils, the *SSA*, *CEC*, and functional groups of clay mineral particles, soil organics, and amorphous materials are the key attributes. The chemical properties of the functional groups will influence the surface acidity of clay mineral particles. Acid-base types of reactions characterize sorption interaction mechanisms involving short-range forces between clay particle surfaces and contaminant ions.

The clay mineral montmorillonite with its 2:1 layer-lattice structure has siloxane-type surfaces on both bounding surfaces, as compared to kaolinite mineral with its 1:1 structure that has a siloxane upper bounding surface and a gibbsite layer at the opposite bounding surface. Typically, a siloxane surface is defined by the basal plane of oxygen atoms that bounds the tetrahedral silica sheet. In view of the structural arrangement of the silica tetrahedral and the nature of the substitutions in the layers, siloxane-type surfaces are reactive surfaces (see Section 2.2 in Chapter 2). The regular structural arrangement of interlinked SiO_4 tetrahedra with silicon ions underlying the surface oxygens provides for development of cavities defined by six bounding oxygen ions in ditrigonal formation.

Surface silanol groups are weak acids, and when these groups characterize the surfaces of mineral particles, these surfaces will be hydrophilic. The amount of silanol groups on the siloxane bounding surface depends upon the crystallinity of the interlinked SiO_4 tetrahedra. Isomorphic substitution of ions in the tetrahedral and octahedral layers by lower valence ions will produce resultant charges on the siloxane surface, the result of which is a reactive particle surface. It follows that no isomorphous substitution in the tetrahedral layer will render the surface free of resultant charges. In silanol surfaces, the *OH* groups on the silica surface become the centres of adsorption of the water molecules; and if strong *H*-bonding is established between silanol groups and neighbouring siloxane groups, the acidity will be decreased. *H*-bonding can also be established with internal silanol groups if such groups are present. Broken edges of the layer-lattice minerals will result in hydrous oxide-types of edge surfaces.

There are two kinds of negative charges associated with the surfaces of soil particles: constant or permanent negative charges, and pH-dependent negative charges. The constant or permanent negative charges are due to isomorphous substitutions and/or site vacancies in the layer-lattice structures of the clay minerals and noncrystalline hydrous oxides. In the case of pH-dependent negative charges, we need to distinguish between inorganic and organic soil fractions as contributors to these charges. The inorganic fractions include both nonhydrolysable and hydrolysable types, with the nonhydrolysable pH-dependent negative charges associated with hydrous-oxide coatings on clay mineral particles. Hydrolysable pH-dependent negative charges are obtained from the structural OH_2 groups that result from protonation of hydroxyls at defects from removal of structural *Al*, *Fe*, and *Mg* (Huang and Jackson, 1966). Strictly speaking, we should refer to the surfaces of soil fractions associated with these charges as *variable-charge surfaces*, as opposed to pH-dependent surfaces.

8.3.2.2 Oxides and Soil Organic Matter

The nature and magnitude of charges for hydrous oxides and oxides are dependent upon their basic structure and the pH of the immediate environment characterizing the system. The term *oxides* is used to include the hydrous oxides, oxyhydroxides, and oxides. A characteristic feature of the various oxides such as iron, aluminium, manganese, titanium, and silicon is the fact that their surfaces essentially consist of broken bonds. The surfaces of the hydrous oxides of iron and aluminium will coordinate with hydroxyl groups, which will protonate or deprotonate, depending on the pH of the surrounding medium. Exposure of the Fe^{3+} and Al^{3+} on the surfaces promotes development of Lewis acid sites when single coordination occurs between the Fe^{3+} with the associated H_2O. This means that interactions of oxide surfaces with water will be between the unsatisfied bonds on the oxide surfaces and the hydroxyl groups of dissociated water molecules. Both negative and positive charges will exist on the oxide surfaces, the predominance of each being dependent on the pH of the system. When the sum of the negative charges equals the sum of the positive charges, the *point of zero charge* (*pzc*) is attained. This means that one could switch from a net positive charge to a net negative charge when one changes the pH of the system from pH ranges below the *pzc* to values above the *pzc* (see Section 2.5.2 in Chapter 2).

Carbon and nitrogen combine with oxygen and/or hydrogen to form the various types of surface functional groups associated with soil organic matter. Depending on the pH of the soil water, these functional groups can protonate or deprotonate, that is, they will develop positive or negative charges, depending on the pH of the soil–water environment and their respective acidity or basicity constants (i.e., pK_a or pK_b). The acidic properties associated with the soil organics are due to the carboxyl (*COOH*) group and the hydrogen in the oxygen-containing functional groups that can be dissociated. The carboxyl and phenolic *OH* groups are considered to be responsible for a significant portion of the source of negative charge. Hayes and Swift (1985) report a range of from 2 to 4 meq/g for soil organics, in comparison with the charge range of from 0.01 to 2 meq/g for clay minerals (see Section 2.4.2 in Chapter 2).

8.3.2.3 Surface Functional Groups of Organic Chemical Contaminants

The nature of the functional groups that characterize organic chemical compounds will essentially define how they bond with soil solids—primarily with clay minerals. The various factors of importance, vis-à-vis interactions with soil solids, that determine the nature of the functional groups in organic molecules include shape, size, configuration, polarity, polarisability, and water solubility. Organic molecules such as amine, alcohol, and carbonyl groups are positively charged by protonation. Figure 8.2 gives some examples of the more common types of functional groups of some common organic chemical compounds. It is interesting to note that most of these functional groups are similar to those that characterize soil organic matter.

The *hydroxyl (OH) functional group* consists of a hydrogen atom and an oxygen atom bonded together. The hydroxyl functional group is also the reactive surface functional group for most clay minerals, amorphous silicate minerals, metal oxides, oxyhydroxides, and hydroxides. In the case of organic chemical compounds, this functional group characterizes the *organic hydroxy compounds*. The two groups of hydroxy compounds include (a) *aliphatic hydroxy compounds* such as alcohols, and (b) *aromatic hydroxy compounds* such as phenols, which are compounds with a hydroxyl group attached directly to an aromatic ring. The alcohols are essentially hydroxyl alkyl compounds (R-OH) with a carbon atom bonded to the hydroxyl group. They can have different structures—primary, secondary, and tertiary—as shown in Figure 8.2, with examples of 1-butanol $CH_3CH_2CH_2CH_2OH$, 2-butanol

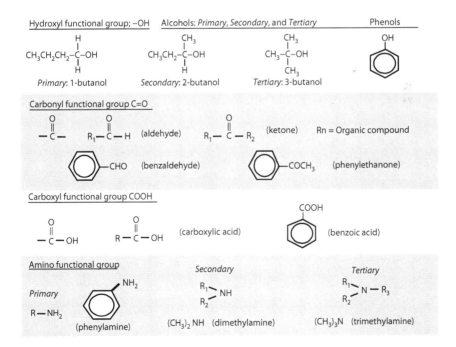

FIGURE 8.2
Some common surface functional groups for organic chemical compounds.

$CH_3CH_2CCH_3HOH$, and 3-butanol $(CH_3)_3COH$. The structural differences are important as they provide differences in chemical reactivity.

There is a subgroup of the *OH* functional group. This is the *phenolic functional group,* which consists of a hydroxyl attached directly to a carbon atom of an aromatic ring. Organic chemical compounds with phenolic functional groups can combine with other compounds such as pesticides, alcohol, and hydrocarbons to form new compounds, for example, anthranilic acid, cinnamic acid, ferulic acids, gallic acid, and *p*-hyroxy benzoic acid. The major types of phenolic compounds found in soil include pesticides, cyclic alcohols, napthols, and so forth.

In addition to the hydroxyl (*OH*) functional group, there are two other significant functional groups: (a) functional groups having a *C–O* bond, for example, carbonyl, carboxyl, methoxyl, and ester groups, and (b) nitrogen-bonding functional groups, for example, amine and nitrile groups. The *carbonyl (C = O) functional group* consists of a carbon atom bonded to an oxygen atom by two pairs of electrons (double bond). The compounds that contain the *carbonyl functional group,* called carbonyl compounds, include aldehydes and ketones (see Figure 8.3). Aldehydes are easily oxidized to the corresponding acids, whereas ketones are difficult to oxidize because of the absence of hydrogen attachment to the carbonyl group. Some typical aldehydes and ketones include formaldehyde (methanol), acetaldehyde (ethanol), and acetone (proponone). Most carbonyl compounds have dipole moments because the electrons in the double bond are unsymmetrically shared, and whilst they can accept protons, the stability of complexes formed between carbonyl groups and protons is very weak.

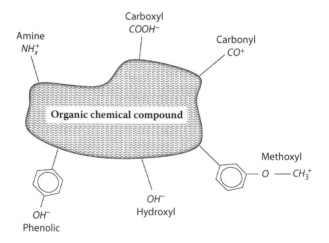

FIGURE 8.3
Schematic illustration of some of the more common surface functional groups associated with organic chemical compounds. These same functional groups are also found in most soil organic matter (see Figure 2.8 in Chapter 2).

The *carboxyl (COOH) functional group* is obtained through a combination of the carbonyl and hydroxyl groups into a single unit. The general formula for compounds containing this functional group is $R-CO_2H$, where R is an aliphatic residue. Carboxylic acids are typical of the carboxyl group. The different types of carboxylic acids include (a) lower acids, which are liquids with an unpleasant odour (e.g., formic and acetic acids) and are miscible with water, and (b) higher acids which are wax-like "solids" (e.g., oleic acid) and are almost insoluble. Whilst carboxyl compounds can become acidic by donating hydrogen ions to form basic substances, they are weak acids in comparison to inorganic acids.

The *amino NH_2 functional group* is found in primary amines. Some of the more common amines include dimethylamine $(CH_3)_2NH$, butylamine $CH_3CH_2CH_2CH_2NH_2$, and ammonia NH_3. These amines, together with others such as trimethylamine and amylamines, are used in the manufacture of ion-exchange resins, disinfectants, insecticides, dyes, soaps, cosmetics, and so forth. Depending on the nature of the functional group, the amines may be aliphatic, aromatic, or mixed. Aliphatic amines are stronger bases than ammonia, and aromatic amines are much weaker than aliphatic amines and ammonia.

8.3.2.4 Clay Particle Interaction with Organic Chemical Functional Groups

In respect of the *hydroxyl (OH) functional group* characterizing alcohols and phenols, the hydroxyl of alcohol can displace water molecules in the primary hydration shell of cations adsorbed onto clay minerals, depending on the polarizing power of the cation. The other mechanisms for adsorption of hydroxyl groups of alcohol are through hydrogen bonding and cation–dipole interactions. Most primary aliphatic alcohols form single layer complexes on clay minerals, with their alkyl chain lying parallel to the surfaces of the clay particles. Some short-chain alcohols such as ethanol can form double layer complexes with the clay minerals.

For the *carbonyl group* of ketones, adsorption onto clay minerals is a function of the nature of the exchangeable cation and hydration status of clay. Acetone is adsorbed both physically and by electrostatic interaction with montmorillonites. However, the electrostatic interaction between the cation and acetone is weakened when a hydration shell surrounds the cation. Acetone and nitrobenzene form double layer complexes with the clay particles, with linkage through a water bridge to the cation of the exchange complex.

For expanding layer-lattice minerals, interaction of the *carboxyl group* of the organic acids (e.g., benzoic and acetic acids) with the minerals is either directly with the interlayer cation or through the formation of a hydrogen bond with the water molecules coordinated to the exchangeable cation. Water bridging is an important mechanism in the adsorption process, together with the polarizing power of the cation. In addition to coordination and hydrogen bonding, organic acids can be adsorbed through the formation of salts with the exchangeable cations.

Amines can protonate in clays and can replace inorganic cations from the clay complex by ion exchange. They can be adsorbed with their hydrocarbon chain perpendicular or parallel to the surfaces of clay minerals, depending on their concentration. For example, ethylenediamine (EDA) is adsorbed onto montmorillonite by hydrogen bonding coordination and aniline is adsorbed onto clay particles by cation–water bridges.

Nonpolar organic chemical compounds such as those found in various petroleum fractions in petroleum hydrocarbons (*PHCs*) are primarily nonpolar organics with low dipole moments (generally less than one), and dielectric constants less than three. Interactions leading to adsorption of nonionic organic compounds by soil fractions is governed by the *CH* activity of the molecule—that is, activity arising from electrostatic activation of the methylene groups by neighbouring electron-withdrawing structures such as $C = O$ and $C = N$. Molecules possessing many $C = O$ or $C = N$ groups adjacent to methylene groups would be more polar and hence more strongly adsorbed than those compounds in which such groups are few or absent. The chemical structures of petroleum hydrocarbons such as monocyclic aromatic hydrocarbons (*MAHs*), and polycyclic aromatic hydrocarbons (*PAHs*) indicate that there are no electron-withdrawing units such as $C = O$ and $C = N$ associated with the molecules. Accordingly, the *PHC* molecules would be weakly adsorbed (mainly by van der Waals adsorption) by the soil functional groups, and do not involve any strong ionic interaction with the various soil fractions.

8.4 Contaminant Sorption Mechanisms

The mass transfer of contaminants such as ions, molecules, and compounds from the porewater of a soil–water system to the surfaces of soil particles, involves adsorption processes that include (a) Coulombic interactions, which are between nuclei and electrons and are essentially electrostatic in nature, and (b) interatomic bonds such as ionic, covalent, hydrogen, and van der Waals:

Coulombic—ion–ion interaction

Interatomic bond

   Ionic—electron transfer between atoms that are subsequently held together by the opposite charge attraction of the ions formed.

   Covalent—electrons shared between two or more atomic nuclei, that is, each atom provides one electron for the bond. When a bond is formed by the sharing of a pair of electrons provided by one atom, this is called *coordinate covalent bonding*.

Hydrogen—very strong intermolecular permanent dipole-permanent dipole attraction.

Van der Waals—dipole–dipole (Keesom), dipole-induced–dipole (Debye), and instantaneous dipole–dipole (London dispersion).

The result of the mass transfer of contaminants to the soil particles is referred to as partitioning. The term *sorption* is used to refer to the adsorption processes responsible for the partitioning of the dissolved contaminants in the porewater to the surfaces of the soil fractions. These adsorption processes fall into two groups as follows:

- *Physical adsorption* or *physisorption*—adsorption of contaminants as a result of ion-exchange reactions and van der Waals forces.
- *Chemical adsorption* or *chemisorption*—contaminant adsorption involving short-range chemical valence bonds.

The soil properties involved in sorption of contaminants are sometimes referred to as *assimilative properties*.

8.4.1 Physical Adsorption (Physisorption)

Physical adsorption (*physisorption*) of contaminants in the porewater by soil particles solids is classified as *nonspecifically* or *specifically* adsorbed, depending on whether the interactions occur in the diffuse ion layer or in the Stern layer. The counterions in the diffuse ion layer will reduce the potential ψ. These counterions are generally referred to as *indifferent ions*. They are held primarily by electrostatic forces and hence are *nonspecific*. This means to say that they are nonspecifically adsorbed, and whilst they will reduce the magnitude of ψ, they will not reverse the sign of ψ (Figure 8.4). The adsorption of most of alkali and alkaline earth cations by clay minerals is a good example of nonspecific adsorption. The process has been referred to as the outer-sphere surface complexation of ions by the functional groups associated with the clay particles (see Section 3.3.1 in Chapter 3).

Replacement of exchangeable cations involves cations interacting with the negative charge sites on clay minerals mainly through electrostatic forces. Ion exchange reactions occur with clay minerals and nonclay mineral soil fractions. The position of adsorbed cations on the surfaces of particles is called the *adsorption binding site* of soil particles for cations. In clay minerals, the adsorption binding sites refer to the position of exchangeable cations on clay minerals. The horizontal and vertical distribution of counterions for these minerals can be determined and demonstrated by molecular dynamics (MD) simulations using the interaction potential energy for a pair of atoms. According to Smith (1998), there are five adsorption binding sites on

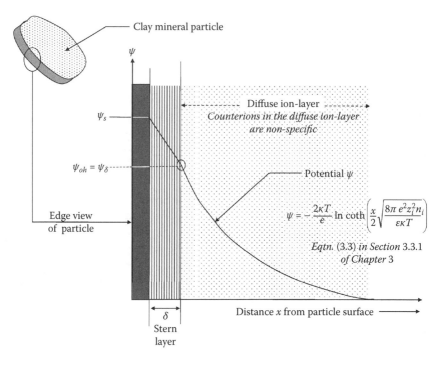

FIGURE 8.4
Generalized DDL model showing: (a) electrified interface with aqueous solution containing dissolved contaminants, (b) Stern layer, and (c) potential ψ. ψ_{oh} refers to the outer Helmholtz potential. This is taken to be equal to ψ_{δ} the Stern potential. ψ_s refers to the potential at the surface of the clay mineral particle (see Figure 3.5 in Chapter 3).

montmorillonite particle surfaces as follows: (a) a tetrahedral-layer Si atom, (b) a tetrahedral-layer Al atom, (c) the hexagonal cavity that is adjacent to a tetrahedral-layer Al atom, (d) the hexagonal cavity, and (e) the octahedral-layer Mg atom. These adsorption binding sites are the positions where the moving cations will frequently dwell for protracted periods on the clay mineral–water interface.

The simulated trajectories of Cs atoms in the interlayer water, obtained by Nakano and Kawamura (2006) using Equations (4.3) to (4.5) in Section 4.3.1, Chapter 4, and the interaction potential parameters listed in Table 8.3, are shown in Figure 8.5 (top left and bottom right pictures). Note that the identity of the various parameters shown in Table 8.3 can be found in Chapter 4. The black "lumps" shown in the pictures are the positions where the trajectories concentrate, strongly suggesting that these are adsorption binding sites. There appear to be three kinds of adsorption binding sites: (a) over the centre of a hexagonal cavity, (b) over a corner oxygen atom in a hexagonal cavity, and (c) over a side of a tetrahedron on the plane of the 2:1 sheet.

TABLE 8.3

Interaction Potential Parameters

Atom	Atomic Weight w 10^3kg. mol^{-1}	z e	a nm	b nm	c (kJ/ mol)$^{0.5}$nm^{-3}		
O_w	16.00	−0.92	0.1728	0.01275	0.05606		
H_w	1.01	0.46	0.0035	0.004400	0		
O_c	16.00	−1.125278	0.1868	0.01510	0.05524		
H_c	1.01	0.46	0.0074	0.00320	0		
Si	28.09	2.10	0.0987	0.00830	0		
Al	26.98	1.95	0.1089	0.00880	0		
Na	22.99	1.00	0.1314	0.0115	0.01637		
Ca	40.08	2.00	0.1494	0.0094	0.01228		
Cs	132.90	1.00	0.1884	0.01300	0.04501		

Atom–atom	D_1 kJ.mol^{-1}	β_1 nm^{-1}	D_2 kJ.mol^{-1}	β_2 nm^{-1}	D_3 kJ.mol^{-1}	β_3 nm^{-1}	r_3 nm
O_w-H_w	57394.9	74.0	−2189.3	31.3	34.74	128.0	0.1283
Si-O_c	205951.2	50.0	−13734.3	22.4	0	0	0
Al-O_c	151533.2	50.0	−8104.1	22.4	0	0	0
H_c-O_c	57394.9	74.0	−3277.6	31.3	34.74	128.0	0.1283

Atom– atom–atom	f_k 10^{-19}J	θ_o deg^o	r_m nm	g_r nm^{-1}
H_w-O_w-H_w	1.15	99.5	0.143	92.0
Si-O_c-Si	0.61	120.0	0.177	168

Source: Data from Nakano, M., and Kawamura, K., 2006, *Clay Sci.*, 12(Suppl. 2):76–81.
Note: The subscripts "w" and "c" associated with O (oxygen) and H (hydrogen) refer to "water" and "clay," respectively.

8.4.2 Chemical Adsorption (Chemisorption)

Chemical adsorption (chemisorption) refers to high affinity, specific adsorption of cations in the inner Helmholtz layer through covalent bonding. The two types of covalent bonding are (a) more or less equal sharing of electrons, generally identified as *covalent bonding*, and (b) coordinate-covalent bonding, where the shared electrons originate only from one partner. Cations penetrating the coordination shell of the structural atom are bonded by covalent bonds via O and OH groups to the structural cations, with valence forces that are of the type that bind atoms to form chemical compounds of definite shapes and energies. The chemisorbed ions are referred to as *potential-determining ions* (*pdis*) since they have the ability to influence the sign of ψ (see Figure 8.4). In contrast to electrostatic positive adsorption, much higher adsorption energies are obtained in chemical adsorption. Reactions, which can be either endo-thermic or exothermic, usually involve activation energies in the process of

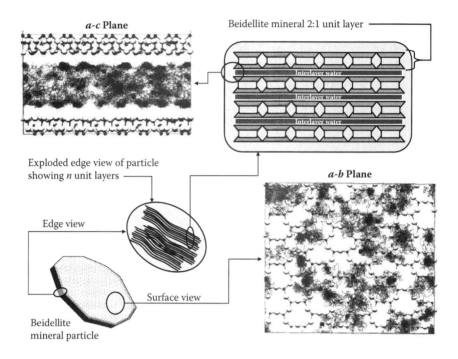

FIGURE 8.5

Trajectories (thread-like thin lines) of roaming *Cs* atoms: (a) in a-c plane, which is the vertical section of the 2:1 unit layer particle, and (b) in a-b plane (surface plane) for a hydrated *Cs*-beidellite with 7.5 H_2O that is equivalent to 2 or 3 layers of water. The tetrahedral-layer *O* atoms on the mineral surface are shaded in circles and the tetrahedral-layer *Si* atoms are small black circles. Trajectories concentrate in the positions that are (a) over a centre of hexagonal cavity, (b) over a corner oxygen atom in hexagonal cavity, and (c) over a side of tetrahedron on the plane of the 2:1 sheet.

adsorption, meaning that the energy barrier between the molecule being adsorbed and the surfaces of the clay particles must be surmounted if a reaction is to occur. Strong chemical bond formation is often associated with high exothermic heat of reaction, and the first layer is chemically bonded to the surface with additional layers held by van der Waals forces. Table 8.4 shows the sorption of heavy metals by several kinds of clay minerals where the various sorption mechanisms include chemisorption involving hydroxyl groups from broken bonds in the clay minerals, formation of metal-ion complexes, and precipitation as hydroxides or insoluble salts.

8.4.3 Specific Adsorption

Ions that are adsorbed onto the surfaces of soil particles (mainly clay minerals) by forces other than those associated with the electric potential within the Stern layer—for example, through covalent bonding—are said to be specifically

TABLE 8.4

Examples of Heavy Metal Retention by Some Clay Minerals

Clay Mineral	Chemisorption	Chemisorption at Edges	Complex Adsorption	Lattice Penetration[a]
Montmorillonite	Co, Cu, Zn		Co, Cu, Zn	Co, Zn
Kaolinite			Cu, Zn	Zn
Hectorite			Zn	Zn
Brucite			Zn	Zn
Vermiculite			Co, Zn	Zn
Illite	Zn	Zn, Cd, Cu, Pb		
Phlogopite			Co	
Nontronite			Co	

Source: Adapted from Bolt, G.H., 1979, *Soil Chemistry, Part B: Physico-Chemical Models*, Elsevier, Amsterdam, 479 pp.

[a] Lattice penetration = lattice penetration and imbedding in hexagonal cavities.

adsorbed. The ions involved in *specific adsorption* are potential-determining ions (*pdis*) since they have the ability to influence the sign of ψ. Whilst the cations that are specifically adsorbed in the inner part of the Stern layer will lower the point of zero charge (*pzc*), the anions specifically adsorbed will, on the other hand, shift the *pzc* to a higher value (Arnold, 1978). Sposito (1984) refers to specific adsorption as the effects of inner-sphere surface complexation of the ions in solution by the surface functional groups associated with the clay fractions.

8.4.4 Rate-Limiting Processes—Heavy Metals

As discussed in the previous sections, the types of sorption processes are controlled by the sorption properties of the soil and the heavy metal contaminants themselves. What has not been discussed is the point that the rate of sorption is also dependent on the type and distribution of the soil fraction in the test soil. Depending on the distribution of the various soil fractions, and depending on the nature of the soil fractions, sorption rates can be rapid or slow. Metal sorption kinetics related to the various oxides and soil organic matter are relatively rapid (Sparks, 1995), whereas sorption rates by clay minerals will be influenced by the nature of the interlayer characteristics. Unconfined montmorillonites can sorb metals more rapidly than vermiculites because the absence of physical restriction on the montmorillonites permits expansion of the interlayer space and allows for entry of the metals into this space. In contrast, interlayer spaces in vermiculites are restricted, and hence will impede movement of the metals in sorption processes. However, if montmorillonites are confined—that is, if montmorillonite interlayer expansion is severely constrained—sorption of the metals will become less rapid.

Interdiffusion of counterions can be considered a rate-determining step in ion exchange. This means to say that when a counterion *A* diffuses from its location in the DDL region (i.e., the region within the ion exchanger) into the solution, a counterion *B* from the solution must move into the space formerly occupied by counterion *A*. The *ion exchanger* is generally identified as the region where the ions are controlled by DDL-type forces. The process of diffusion of counterions *A* and *B* is the interdiffusion of counterions between an ion exchanger and its equilibrium solution. There are at least two rate-determining steps:

- Particle-type diffusion—interdiffusion of counterions within the ion exchanger (DDL region) itself
- Film-associated diffusion—interdiffusion of counterions in the Stern layer

The many factors and processes, such as diffusion-induced electric forces, selectivity, specific interactions, and nonlinear boundary conditions, make it difficult to develop and specify rate laws that apply diffusion equations to ion-exchange systems. The fluxes of different ionic species are both different and coupled with one another, making it difficult to specify one characteristic constant diffusion coefficient that will describe the flux rate of the different ionic species. Stochiometry of ion exchanges requires conservation of electroneutrality between the counterions and the charged clay particle surfaces. For electroneutrality to be preserved, the different electric phenomena established must be considered in the determination of the various diffusion processes.

8.5 Laboratory Determination of Partitioning

In practice, it is not always feasible or necessary to fully delineate between the various mechanisms and forces involved in the sorption of contaminants by the soil fractions in a particular piece of soil mass. In most instances, it is only necessary to have knowledge of the sorption potential or capability of the soil for a particular set of contaminants—that is, quantitative capability of the soil to sorb the spate of contaminants introduced to the soil. In essence, what is desired is the sorption characteristics of a soil for a particular set of contaminants. The term *partitioning*, when used in the context of soil contamination, refers to the mass transfer of contaminants in the porewater onto the surfaces of soil particles by sorption forces and mechanisms discussed in the previous sections.

The two popular types of laboratory tests for determination of the sorption characteristics of a soil are (a) batch equilibrium adsorption isotherm tests, and (b) contaminant leaching tests using soil columns or cells. The results obtained from these two types of tests differ in respect to "what is measured" (see Figure 8.6). Batch equilibrium adsorption isotherm tests are tests that use individual and separate batches of soil suspensions or soil solutions consisting of soil particles totally dispersed in a solution containing the contaminants of interest. This means that the contaminants in the aqueous solution can interact directly with all the soil particles and their surfaces. Adsorption isotherms, obtained as information concerning the adsorption potential of a candidate soil for a target contaminant, provide one with a basis for comparison between different soils and/or different target contaminants. When multispecies contaminants are used as the target species, the proportions of each of the species comprising the multispecies mixture become very important. One needs to have knowledge of how selective or preferential sorption and speciation will influence sorption of each species.

Soil column or soil cell leaching tests utilize compact or intact natural soil samples. With introduction of leachate containing contaminants of interest to the samples, flow of the leachate through the soil allows the contaminants to interact with exposed particle surfaces. The presence of microstructural units in the soil means that not all available particle surfaces will interact with

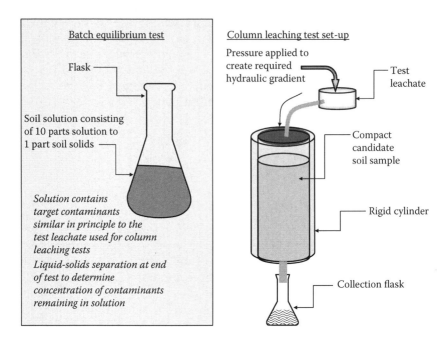

FIGURE 8.6
Batch equilibrium test for determination of adsorption isotherms and column leaching test set-up—showing the differences in test protocols and test sample. Note that many jurisdictions suggest a choice of different soil solution ratios, ranging from 10:1 to 50:1 for batch equilibrium testing—depending on soil composition whether the contaminants are inorganic or organic chemicals. The preference for inorganic contaminant testing is for soil solutions of 10 parts solution to 1 part soil; that is, 10:1.

the contaminants. The proportion of soil particle surfaces exposed to interaction with the contaminants in the leachate can vary anywhere from 15% to 70%, depending on the type of soil, density, and soil structure. Information obtained using test leachates permits one to obtain an appreciation of "what might happen" in the field. The rationale for this type of test is founded on the assumption that the soil sample in the test column is representative of "what is in the field," and that leaching of a test leachate through the column will essentially replicate the field condition.

It is often not possible to apply hydraulic gradients that replicate field conditions in laboratory leaching columns—especially in cases of percolation of surface water and fluid flow in landfill and engineered waste containment barriers. In most cases, hydraulic gradients used for leaching column tests or conductivity cells will far exceed field hydraulic gradients, leaving one to ask the question as to how excessive gradients can possibly provide one with simulated field situations. A further complication in extrapolation of laboratory test results to field behaviour is the variability of soil structure and properties. That being said, it is nevertheless useful for one to conduct leaching column tests, if for no other reason, to test various kinds of soils under a variety of conditions, which will provide one with a base for comparison, and thus allow one to compare results on a qualitative basis.

8.5.1 Batch Equilibrium Tests and Adsorption Isotherms

Batch tests on individual samples of a candidate soil are conducted on soil solutions prepared with previously air-dried soil ground to uniform powdery texture for mixing with individual contaminant solutions with various concentrations of the contaminants. Whilst there are similarities in the basic elements of batch equilibrium testing between inorganic contaminants and organic chemicals, there are marked differences between the two in the manner in which test contaminant solutions are prepared and introduced to the soil sample. The objectives of the tests for both kinds of contaminants are nevertheless the same—that is, determination of the sorption characteristics of the soil for the target contaminant.

8.5.1.1 Inorganic Contaminants

The results obtained in characterization of sorption are usually expressed as *adsorption isotherms.* The contaminant sorbed or removed from the soil solution is identified as the *adsorbate,* and the soil particles involved in the sorption of the contaminants from the soil solution are called the *adsorbent. Adsorption isotherms* are generally defined as characteristic relationships between adsorbate and absorbent. In the context of soils, adsorption isotherms are characteristic relationships that describe adsorption of target contaminants by candidate soils to contaminant concentrations remaining in solution. Figure 8.7 shows the elements of the test protocols for tests involving inorganic contaminants. The graphical representation of the relationship between the equilibrium concentration of contaminants (adsorbate) remaining in solution and the equilibrium concentration of contaminants sorbed by the soil particles is shown in the bottom right-hand corner of the figure.

The typical set of characteristic equilibrium isotherms shown in Figure 8.8 includes the relationship for the *constant adsorption* curve given as: $c^* = k_1 c$, where c refers to contaminant concentration in solution and c^* refers to sorbed contaminant concentration. The sorbed concentration of contaminants c^* is related to the equilibrium concentration of contaminants remaining in solution via a constant k_1. The obvious limitation of the constant adsorption relationship is the fact that this relationship predicts limitless adsorption of contaminants—an unlikely proposition inasmuch as there are a limited number of sorption sites available in the soil sample. Experience has shown the equilibrium Freundlich and Langmuir adsorption isotherm relationships, shown in the figure, to be more realistic isotherms in comparison to the constant adsorption relationship. The emphasis on the word *equilibrium* in association with the isotherm relationship discussed here is deliberate.

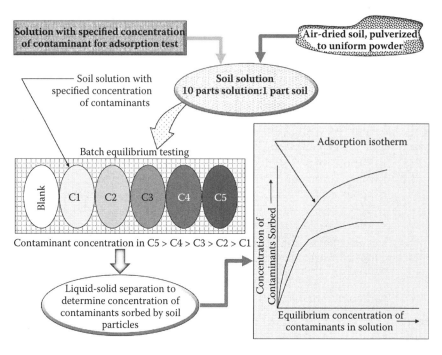

FIGURE 8.7
Batch equilibrium procedure for determination of adsorption isotherms for inorganic contaminants. The five batches of soil solution shown on the left, starting with batch C1, have progressively increasing concentrations of contaminants in the solution.

Nonequilibrium sorption of contaminants will be discussed in a later chapter when transport and fate of contaminants are considered. Note that whilst the Freundlich and Langmuir relationships are improvements over the constant adsorption relationship, these relationships could, theoretically, also permit unlimited sorption of contaminants by the soil being tested.

The simple procedure to determine the constants in the often-used Freundlich relationship shown in the figure ($c^* = k_2 c^m$) is to express the relationship in a logarithmic form, $log\ c^* = log\ k_2 + m\ log\ c$, as shown graphically in the bottom-right hand portion of Figure 8.8. The illustration shows the Freundlich constants k_2 and m to be the intercept on the ordinate and the slope of the line, respectively. It is important to note that the concentration of contaminants or solutes in the solution (i.e., the parameter used to describe the variation along the abscissa) has been expressed either as the equilibrium concentration of solutes (contaminants) or the initial concentration of solutes. If the abscissa is plotted in terms of the concentration of solutes in the leachate solution used, as opposed to the equilibrium concentration (i.e., concentration remaining in solution), we will obtain two different adsorption isotherm curves, as demonstrated in the graphical plot shown in Figure 8.9.

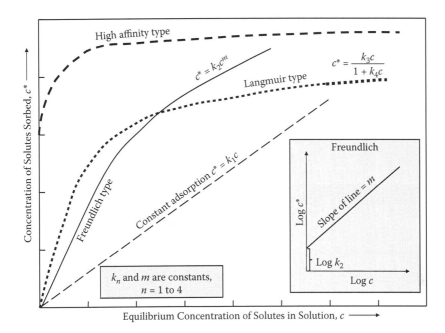

FIGURE 8.8
Batch equilibrium procedure for determination of adsorption isotherms for inorganic contaminants. The five batches of soil solution shown on the left, starting with batch C1, have progressively increasing concentrations of contaminants in the solution.

Also shown in the graph is the pH of the clay-solution in relation to the concentration of solutes in the solution. The pH of the solution decreases as the amount of Pb (in nitrate form) is added to the solution. The initially high pH at low Pb input concentrations is the result of the high pH of the illitic clay. This is particularly significant, not only because of the marked differences in the characteristic shapes and slopes of the isotherms, but also because slope k_d of the chosen curve is used as the distribution coefficient in the "diffusion-dispersion" contaminant transport relationship. Regardless of which form is used for the abscissa, the units used for sorbed concentration (ordinate) and the concentration of solutes in the solution should be consistent with each other. Coles and Yong (2006) show that there is no need to exclude any data when full contaminant uptake occurs if one uses the *initial concentration* isotherm relationship.

With soils containing such fractions as oxides of aluminium and other types of amorphous materials, their higher specific surfaces and types of functional groups will have considerable influence on the character of the adsorption isotherms. It is important to bear in mind that most standard procedures assume that when interaction of the contaminants and soil particles occur, chemical equilibrium conditions are achieved within a 24-hour

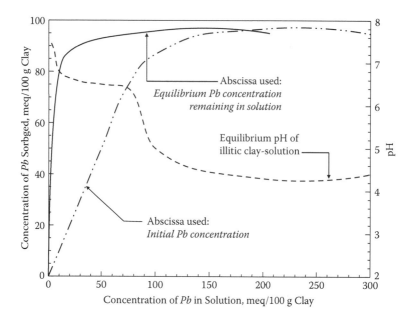

FIGURE 8.9
Characteristic adsorption isotherms obtained from batch equilibrium tests of different types of soils. Right-hand inset shows the log-plot format for the often-used Freundlich relationship for determination of the constants k_2 and m.

period. For soils that do not contain soil organic matter (*SOM*), the 24-hour period used is generally acceptable. For soils without *SOM*, equilibrium is generally reached within a 3- to 4-hour period. However, when *SOM* is present, equilibrium may not be reached for a few days, depending on the nature and proportion of *SOM* present in the soil. It is generally wise to obtain proper characterization of the soil to be used for the batch equilibrium tests to avoid problems with analysis of test results obtained.

Figure 8.10 shows the results of batch equilibrium test with initial *Pb* concentration of 200 ppm, introduced into the soil solution as a $PbCl_2$ salt, in relation to *Pb* removed from the aqueous phase and to pH. The *Pb* precipitation boundary shows that when the pH of the system reaches about 4, precipitation of *Pb* becomes an important component in the removal of *Pb* from the aqueous phase. A significant portion of the soluble ions will be sorbed by the soil solids as shown in the diagram, and the ions remaining in solution would either be hydrated or would form complexes, giving one Pb^{2+}, $PbOH^+$, and $PbCl^+$.

8.5.1.2 *Organic Chemical Contaminants*

The tests for determination of adsorption isotherms for organic chemical contaminants are essentially similar to the procedures and techniques used to

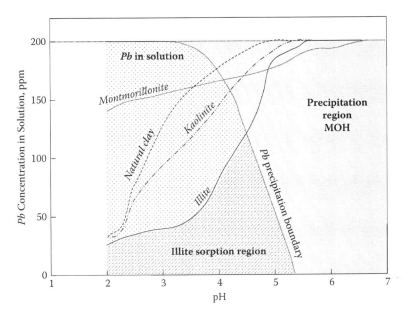

FIGURE 8.10
Influence of type of soil and pH on *Pb* removed from the soil solution—interpreted as partitioning of *Pb*—in soil suspension tests using different types of soils. *Pb* introduced into soil solution as $PbCl_2$. *Pb* precipitation boundary was determined from separate tests on $PbCl_2$.

determine the adsorption isotherms for inorganic chemical contaminants. There are some differences, however, in the preparation of the test soil solution. In most jurisdictions—for example, USEPA and OECD—the choice of soil-solution ratio is perhaps one of the most critical issues. Much of the concern lies in the amount of particle surfaces exposed to the test solution. This is expected, and is in fact not dissimilar to the concerns raised in inorganic contaminant sorption testing of soils. OECD (2000) notes that "the distribution of a chemical between soil and aqueous phases is a complex process depending on a number of different factors: the chemical nature of the substance, the characteristics of the soil, and climatic factors such as rainfall, temperature, sunlight and wind."

Preparation of the contaminant solution generally requires dissolution of the test substance in a 0.01 M solution of calcium chloride ($CaCl_2$) in distilled or deionised water. The rationale for this dissolution procedure is to obtain a solution that would minimize cation exchange and also to improve liquid–solids separation at a later stage.

The distribution coefficient, identified as k_d, refers to the ratio of the concentration of organic chemical contaminants retained by the soil fractions to the concentration of chemical contaminants in the porewater (aqueous phase). The linear relationship for partitioning of organic contaminants gives us $C_s = k_d C_w$, where C_s refers to concentration of the organic chemical contaminants retained by the soil particles, and C_w refers to the concentration remaining in the aqueous phase (porewater), respectively. As we have pointed out previously in respect to partitioning of inorganic contaminants, the linear relationship is not really tenable since it means that one can have infinite partitioning. Soil suspension tests with target organic chemical contaminants can yield three types of adsorption isotherms that describe the partitioning of organic chemicals.

The Freundlich isotherm for organic chemical contaminants is given as: $C_s = k_1 C_w^n$. As before, k_1 and n are the Freundlich parameters. When $n = 1$, a linear relationship between C_s and C_w is obtained. When $n < 1$, the retained organic chemical decreases proportionately as the available organic chemical increases. This would indicate that all available retention processes are being exhausted, and when $n \ll 1$, one would expect that retention capacity is fully taxed. The reverse is true when $n > 1$, that is, the retention capacity of the clay keeps increasing as more organic chemicals are retained. This can happen in situations where initial retention of the chemicals by the soil will disrupt the microstructure, thus releasing more reactive particle surfaces for interaction with the organic chemicals. This concern has also been articulated in the guidelines issued by certain regulatory bodies—that is, concern that such disruption of soil structure would render analyses of adsorption isotherms problematical.

Studies using paddy soils in Japan (Kibe et al., 2000) with organic contents ranging from 1.2 to 6.8%, *SSA* from 6.2 to 34 m²/g, and *CEC* from 12 to 45 meq/100 dry soil, were conducted in interaction with different herbicides such as esprocarb ($C_{15}H_{23}NOS$), pretilachlor ($C_{17}H_{26}CINO_2$), simetryn ($C_8H_{15}N_5S$), and thiobencarb ($C_{12}H_{16}CINOS$). The Freundlich-type relationships obtained had n values ranging from 1.0 to 1.6 (Kibe et al., 2000). Experiments conducted using batch equilibium tests with aniline (C_6H_7N) and trichloroethylene (*TCE, C_2HCl_3*) in interaction with 6% bentonite produced Freundlich-type adsorption isotherm results after 7 and 10 days for both chemicals. Other types of isotherms have been obtained with different types of organic chemicals and soils—not all of them easily "fitted" by a Freundlich relationship. For example, in a different set of studies, Langmuir-type adsorption isotherms were found for tetracycline ($C_{22}H_{24}N_2O_8$). The sorption of an herbicide, atrazine, in Wisconsinan oxidized till and other soils and sediments reported by Moorman et al. (2001) is shown in Figure 8.11. The k_d adsorption values were determined as 0.43 ± 0.25 for the Wisconsin oxidized till, 0.51 ± 0.02 for the Okoboji loess, and 0.55 ± 0.24 for the alluvium.

Organic molecules are generally varied in nature, that is, they have varied sizes, shapes, molecular weight, and so forth. By and large, these molecules are less polar than water, and their water solubility will influence or control their partitioning. In addition, their water solubility will also influence the various processes such as oxidation/reduction, hydrolysis, and biodegradation that result in their transformation. Figure 8.12 shows the influence of water solubilities (*ws*) on the adsorption isotherms for three organic chemicals: naphthalene ($C_{10}H_8$), 2-methyl naphthalene ($C_{11}H_{10}$), and 2-naphthol

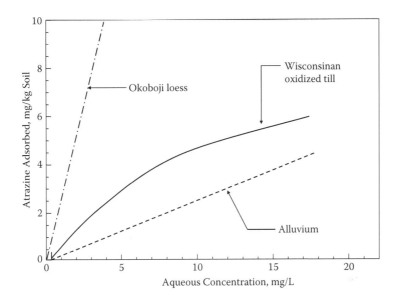

FIGURE 8.11
Adsorption of atrazine on various soils. (Interpreted from data reported by Moorman, T.B., et al., 2001, *Soil Sci.*, 166:921–929.)

($C_{10}H_8O$). Because the water solubilities of naphthalene and 2-methyl naphthalene are closely similar, it is not surprising to see that their isotherms also demonstrate close similarity. The 2-naphthol ($C_{10}H_8O$), which is an intermediate along the pathway of reaction sequence of naphthalene by *Cunninghamella elegans*, with a water solubility of about 30 times greater than the other two organic chemicals, shows a significantly different isotherm. This demonstrates that an organic chemical with higher water solubility will have a greater amount of that chemical remaining in the aqueous phase— in contrast to another organic chemical with a much lower water solubility. Adsorption studies involving petroleum hydrocarbons (PHCs) and surfaces of clay particles show that adsorption occurs only when the water solubility of the PHCs is exceeded, and that the hydrocarbons are accommodated in the micellar form.

Weakly polar (resin) and nonpolar compounds (saturates and aromatic hydrocarbons) of PHCs develop different reactions and bonding relationships with the surfaces of soil particles. Most hydrocarbon molecules are hydrophobic and have low aqueous solubilities. Partitioning of PHCs onto soil surfaces occurs to a greater extent than in the aqueous phase. Weakly polar compounds are more readily adsorbed onto soil particle surfaces in contrast to nonpolar compounds. The adsorption of nonpolar compounds onto soil particle surfaces is dominated by weak bonding (van der Waals attraction), and is generally restricted to external soil particle surfaces

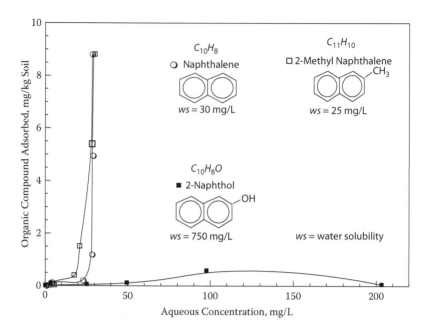

FIGURE 8.12

Adsorption isotherms for naphthlene, 2-methyl naphthlene, and 2-naphthol with kaolinite. (Data from Hibbeln, K.S., 1996, Effect of kaolinite and cadmium on the biodegradation of naphthalene and substituted naphthalenes, M.Sc. thesis, McGill University, Montreal, Canada.)

primarily because of their low dipole moments (less than 1) and their low dielectric constants (less than 3) (Yong and Rao, 1991).

8.5.2 Column Leaching Tests

Column leaching tests are generally conducted to determine the attenuation of contaminants as they leach through the test samples. Whilst there are no explicit specifications on the kinds of equipment such as sizes and materials, and exact test protocols, there are guidelines issued by various jurisdictions regarding procedures, methodology, and general types of columns to be used. Both the USEPA (method 835-340) and OECD (test method 312) recommend that the columns be made of *suitably inert material* such as glass, stainless steel, PVC, and so forth. They also recommend that the minimum dimensions for the test columns be 4 cm in diameter and 35 cm in height. The various pieces of information collected from the column leaching tests, such as *contaminant sorption profiles* (Figure 8.13) and *contaminant breakthrough curves*, are designed to inform one of (a) the sorption and contaminant attenuation characteristics of the soil as leachates are introduced to the soil, and (b) the ability of the test soil to function as a contaminant barrier material.

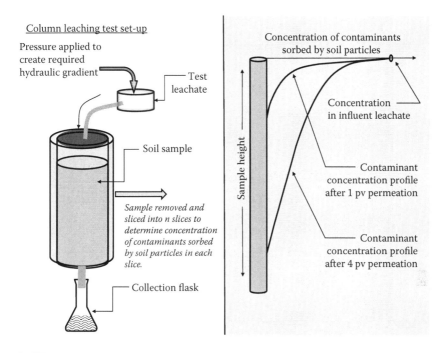

FIGURE 8.13

General procedure for column leaching tests to determine contaminant sorption profiles. Replicate samples are permeated with leachate and samples tested after permeation of 1 to 4 to 6 pore volumes (pv) of test leachate. Samples are removed after the pre-determined permeation and sectioned to determine the concentration of contaminants sorbed by the particles in each section. The sorption profiles shown in the right-hand side of the figure show typical profiles after 1 and 4 pore volume permeations of test leachate. One pore volume refers to the whole pore volume included in a test sample.

Unlike batch equilibrium testing, column leaching tests use compact soil samples obtained as either laboratory-prepared samples or as samples retrieved from the field. To determine contaminant sorption profiles, replicate samples are used to allow sacrifice of samples after permeation of different quantities of leachate. Sectioning of individual samples at the end of each test for analysis of contaminants in the porewater tells one about the partitioning of contaminants. Figure 8.14 is a schematic of the general procedure and intent for column leaching tests for heavy metal contaminants. For the soil particles, laboratory analyses of the ions associated with the clay minerals allows one to determine the concentration of exchangeable and extractable ions.

In the case that the test samples are aggregate soil and compact clay, the total surface area of soil particles and microstructural units interacting with the contaminants in the leachate will be less than those in batch equilibrium tests. This is because the presence of microstructural units will mean that

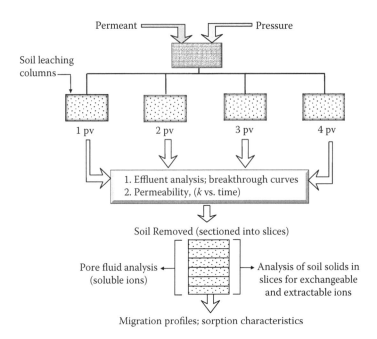

FIGURE 8.14
General test procedure for leaching column tests for determination of partitioning of heavy metals. Replicate samples are subject to increasing amounts of influent leachate apportioned in quantities expressed in terms of pore volumes (pv).

the resultant total exposed particle surface areas would be considerably less in comparison to the sum total of individual particle surface areas, as would be the case of batch equilibrium testing. Figure 8.15 compares the adsorption isotherm for a kaolinite with the sorption curve obtained for the same kaolinite in column leaching tests.

A point of importance to note in column leaching tests is the hydraulic gradients used. If the intent of the leaching tests is to understand "what happens in the field," it would be useful to ensure that the hydraulic gradients used for the test are not greater than field hydraulic gradients. However, experience shows that, in many instances, these field hydraulic gradients are frequently too small for laboratory tests where the time scales for test completion are relatively short. If such is the case, the imposed laboratory hydraulic gradients should not exceed those gradients that result in non-laminar steady state porewater flow.

Sampling and analysis of the leachate output from the leaching columns after various quantities of discharge measured in terms of pore volumes (pv) will provide one with information on the concentration of contaminants in the output leachate. One can construct a relationship between the concentration of contaminants in the collected discharge to the number of

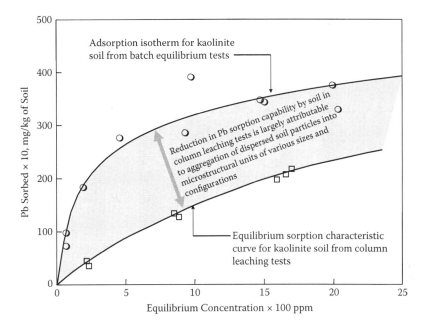

FIGURE 8.15
Comparison of *Pb* sorption curves obtained from batch equilibrium and column leaching tests for kaolinite soil. The difference in sorption capacity between the adsorption isotherm and the equilibrium sorption curve obtained from column leaching tests can be largely ascribed to the compact soil sample microstructure effects on available particle surfaces for interaction with the influent contaminant leachate. (Data from Yong, R.N., 2001, *Geoenvironmental Engineering: Contaminated Soils, Pollutant Fate, and Mitigation,* CRC Press, Boca Raton, FL, 307 pp.).

throughput leachate pore volumes (pv). Since many variables such as length of column, concentration, amount of leachate used, and leaching time have to be considered in the analysis, a useful procedure is to use a normalization procedure for data representation. By expressing the discharge concentration of contaminants in relation to the initial concentration used as input, and relating this to the number of throughput pore volumes as shown in Figure 8.16, we can determine sorption capability soil. The *breakthrough point* is defined as the pore volume throughput at the 50% relative concentration c/c_i, where c is the throughput concentration of contaminants at a particular time or a particular pore volume discharge, and c_i is the concentration of contaminants in the influent leachate. The curves in the diagram are called *breakthrough curves*. Three idealized types of breakthrough curves are shown in the figure: poor, good, and high sorption capabilities. For comparison, the breakthrough curves for a *Pb* contaminant introduced as $Pb(NO_3)_2$ at pH values below *Pb* precipitation, for a kaolinite, an illitic clay, and a natural clay from Lachenaie, Québec, Canada, are shown. Since all three soils do not possess soil organics and amorphous materials, the dominant

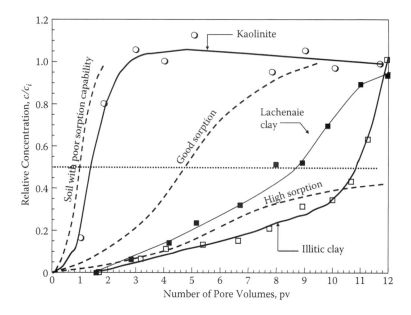

FIGURE 8.16

Ideal characteristic breakthrough curves for soils with poor, good, and high sorption capabilities. For comparison, actual *Pb* breakthrough curves from column testing with a length:diameter ratio of 2 for a kaolinite, an illitic clay, and a natural clay from Lachenaie, Québec, Canada, are shown. *c* represents the concentration of contaminants (in the collection beaker) after a number of pore volumes discharged from the sample, and c_i represents the initial contaminant concentration of the influent leachate.

sorption process is most likely to be the exchange mechanisms similar to those operative in characterizing the *CEC* of the soils. Using this as a guide, it would appear that the low *CEC* for kaolinite would allow breakthrough to occur rapidly (at about 1.4 pv). Since the *CEC* values for the Lachenaie clay (60 meq/100 g soil) and the illite clay (50 meq/100 g soil) are relatively close to each other, it is not surprising to see that the breakthrough points for the two soils are close to each other—8.6 pv for the Lachenaie clay and about 10.5 pv for the illitic clay.

8.5.2.1 Breakthrough Curves

Breakthrough curves describe the relative movement of contaminants and their carrier in test samples. Figure 8.17 shows a schematic of the contaminant profiles of the sorbed contaminants with soil composition and contaminants in their aqueous carrier (leachate) after a specific time *t*. There are three "fronts" shown in the diagram. The *contaminant concentration front moving in concert with the aqueous front* shown after time *t* is obtained when both aqueous carrier and contaminants move at the same rate. This combined

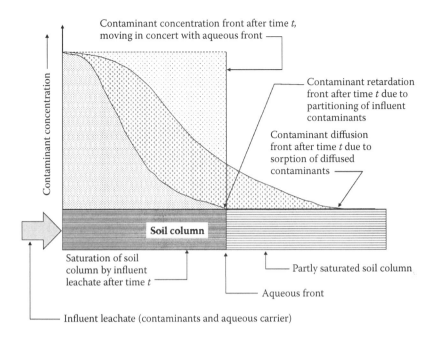

FIGURE 8.17
Movement of contaminants and aqueous carrier (leachate) after time *t*. The *contaminant concentration front moving in concert with the aqueous front* is obtained when both aqueous carrier and contaminants move at the same rate. In the case of the *contaminant diffusion front*, the contaminants diffuse along the porewater boundary layers of the partly saturated portion of the column. The *contaminant retardation front* is the result of partitioning of contaminants.

contaminant and aqueous front is obtained when the contaminants show no diffusive capability, and when there is no physicochemical interaction between the contaminants and the reactive surfaces of the soil particles.

In the case of the contaminant diffusion front, the contaminants diffuse along the porewater boundary layers in the partly saturated portion of the soil column, where influent leachate has yet to penetrate, and because of their diffusivity capability, the front can lead the aqueous front in the same time frame. The contaminant retardation front is observed at aqueous front and lags behind the contaminant diffusion front because of the effects of partitioning and interactions between solutes and soil particles. One should note that there is an adsorption region in the partly saturated soil column in front of the aqueous front of influent leachate after time *t* due to sorption of the diffused contaminants. By definition, breakthrough of contaminants starts at a time when porewater, together with diffused contaminants in the partly saturated soil column, reach a relative contaminant concentration of 0.5 (see Figure 8.16). This defines the *breakthrough point* and should not be confused with complete breakthrough of contaminants. *Complete breakthrough* of contaminants occurs when discharge of contaminants begins to occur at the outlet end of the soil column. Breakthrough curves primarily express the sorption and attenuation characteristics of soils—between influent contaminants and soil fractions—and are determined for situations of leachate penetration into soils.

8.6 Partitioning and Soil Composition

8.6.1 Inorganic Contaminants

The various soil fractions in a soil have different roles to play in respect to partitioning of contaminants. The inorganic contaminants of concern, discussed in this section, are the heavy metals, and the primary question addressed is, How are the partitioned heavy metals distributed in the affected soil? The term *distributed in a soil* refers to the proportion of heavy metals retained or sorbed by each type of soil fraction. A general procedure for determination of the distribution of partitioned heavy metal contaminants in a soil is the use of a selective sequential extraction technique commonly known as the *selective sequential extraction (SSE)* technique.

8.6.1.1 Selective Sequential Extraction (SSE) Technique

Application of the selective sequential extraction (*SSE*) technique for removal of sorbed heavy metals from individual soil fractions requires the use of chemical reagents chosen for their capability in selectively destroying the bonds established between heavy metal contaminants and specific individual soil fractions. A proper choice of chemical reagents for selective bond destruction is key to the success of the technique. Two particular points need to be noted:

- The use of aggressive chemical reagents will not only destroy the bonds, but could also erode or damage the structure of the various particles constituting the soil fractions. Extractants (reagents) that have a history of use in routine soil analyses are the best types of reagents. These are available and are classified as concentrated inert electrolytes, weak acids, reducing agents, complexing agents, oxidizing agents, and strong acids.

- Since a chemical reagent chosen to destroy the bond between heavy metals and a target soil fraction may have a small unintended collateral destructive effect on the structure of another soil fraction, one should start with the least aggressive reagent, to avoid or to minimize collateral damage.

Figure 8.18 shows a schematic view of the *SSE* procedure. The top portion of the schematic shows the sequence of steps taken to determine the amount of heavy metals associated with the various soil fractions. The "boxes" at the bottom portion of the drawing show the kinds of reagents that are used to achieve removal of the heavy metals "bonded" to each type of soil fraction. The numbers in the boxes indicate the order in which the reagents are used. Thus, for example, the #1 *Exchangeable metals* box indicates the types of reagents that can be used for the first heavy metals extraction sequence in step number 1. The *Soil solution* box shows the soil solution composed of reagent (e.g., KNO_3 or $MgCl_2$) and soil particles.

The atomic adsorption (*AA*) analyses of the supernatants at the various sequences of extraction are generally portrayed in terms of metal–soil fraction association—that is, heavy metals associated with a particular soil fraction. It is important to realize that the results obtained from *SSE* testing of soils pertaining to the distribution of contaminants amongst the various soil fractions must be considered as *operationally defined.* This is because the results obtained are dependent on the choice of laboratory techniques and reagents used in the *SSE* test. The measurements should be considered to be

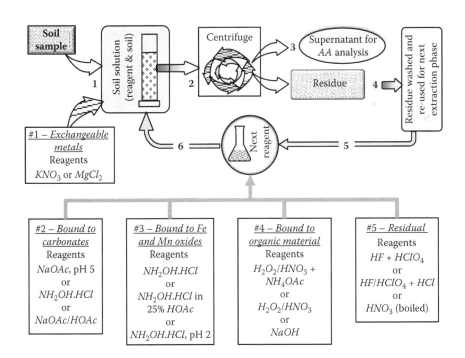

FIGURE 8.18
General laboratory procedure for implementation of *SSE* procedure in determination of partitioning of heavy metals onto different clay fractions. The reagents listed in the boxes at the bottom of the figure are a sampling of the kinds of reagents that can be used.

more qualitative than quantitative. They are nevertheless useful in that they provide an insight into the distribution of the partitioned heavy metals.

There is some concern that redistribution of the extracted metals will occur during the sequential extraction stage, meaning that metals being extracted by the first reagent, for example, will be sorbed (resorbed) by the other soil fractions. Should this occur, one would obviously record lower values for extracted metals by the first reagent, which will affect the results of metals extracted in the later stages of reagent application. One could argue that a portion of the resorbed metals can and will be released upon application of the second or third reagent. This will mean higher amounts of released metals will be obtained in association with the second reagent or third. The resorption study of sequentially extracted metals by Ho and Evans (2000) that included *Pb, Cu,* and *Zn* showed that 20% to 30% of the *Cu* and *Zn* released from the acid-soluble faction of the soil was resorbed by the reducible mineral oxide fraction. It should be noted that it is not always easy to distinguish between sorption-retention (physisorption and chemisorption) of heavy metals and precipitation-retention (i.e., retention by precipitation mechanisms). The *Pb, Cu, Zn,* or *Cd* precipitated or coprecipitated as natural carbonates can be released if the immediate environment is acidified. The European production standard in respect to *SSE* protocols issued by the Bureau Communautaire de Référence (Community Bureau of Reference [*BCR*]) differentiates between (a) *acid-extractable* (0.11 M acetic acid), (b) *reducible* (0.1 M hydroxylamine hydrochloride acidified to pH2 with nitric acid), and (c) *oxidizable* (1.0 M ammonium acetate extraction after oxidation by 8.8 M hydrogen peroxide). Ure et al. (1993) gives details of *BCR* harmonization of extraction techniques in *SSE* testing.

Application of the reagents in the steps indicated in Figure 8.18 with the procedure shown at the upper portion of the figure provides information regarding metals associated or bonded to the various soil fractions based on the following rationale:

- *Exchangeable metals* (#1 in Figure 8.18): The metals extracted through the use of neutral salts as ion-displacing extractants such as $MgCl_2$, $CaCl_2$, KNO_3, and $NaNO_3$ are considered to be exchangeable metals. It is not uncommon for the literature to refer to these released metals as *"in the exchangeable phase,"* that is, meaning that they are nonspecifically adsorbed and ion exchangeable and can be replaced by competing cations. The soil fractions associated with these exchangeable metals are clay minerals, organics, and amorphous materials. If the salt solutions are applied at neutral pH, one would expect that, at most, only minimal dissolution of carbonates would result. In the procedure used, for example, by Yong et al. (2001b) for extraction of the metals in the exchangeable phase from estuarine alluvium, 1 M KNO_3 was used, with pH adjusted to the natural pH of the soil. There is little evidence to suggest that dissolution of the soil solids occurs because of the neutral electrolytes. Pickering (1986) showed that $MgCl_2$ sediment leachate contained only low levels of *Al, Si,* and organic carbon, confirming the weakness of the neutral salts in interaction with the clay surfaces, sulphides, and organic matter. If the neutral salt solutions are applied at neutral pH, dissolution of *Fe* or *Mn* oxides is not expected, and we would only expect minimal dissolution of carbonates. Other types of salts such as NH_4Cl and NH_4OAc may dissolve considerable amounts of compounds such as $CaCO_3, MgCO_3, BaCO_3, MgSO_4$. Reagents such as $CaSO_4$ and NH_4OAc can cause some dissolution of *Mn*-oxyhydrates and metal oxide coatings.

- *Bound to carbonates* (#2 in Figure 8.18): The metals precipitated or co-precipitated as natural carbonates can be released by application of acidified acetate as the extractant. A solution of 1M HOAc-NaOAc (pH 5) is generally sufficient to dissolve calcite and dolomite to release the metals bound to them without dissolving organic matter, oxides, or clay mineral particle surfaces.

- *Bound to Fe and Mn oxides* (#3 in Figure 8.18): The metals released in this sequence of extractant treatment include both metal contaminants attached to amorphous materials and metals from poorly crystallized metal oxides such as *Fe, Al,* and *Mn* oxides. The metal oxides include ferromanganese nodules, ranging from completely crystalline to completely amorphous. These occur as coatings on detrital particles and as pure concretions. Their varying degree of crystallization results in several types of association with the heavy metals: exchangeable forms via surface complexation with functional groups (e.g., hydroxyls, carbonyls, carboxyls, amines, etc.) and interface solutes (electrolytes), moderately fixed via precipitation and co-precipitation (amorphous), and relatively strongly bound. The reagents selected for oxyhydrates should not attack either the silicate minerals or the organic matter. Chester and Hughes (1967) used a combination of an acid reducing agent (1 *M* hydroxylamine hydrochloride) with 25% (v/v) acetic acid for the extraction of ferromanganese oxides. Similarly, Yong et al. (2001b) used 0.04 *M* $NH_2OH.HCl$ in 25% (v/v) *HOAc* with occasional agitation for 6 hours at 96°C.

- *Bound to organic matter* (#4 in Figure 8.18): The general technique used to release metals bound to organic matter as a result of oxidation of the organic matter is to use oxidants at levels well below their (organic matter) solubilities. Since the binding mechanisms for metals in association with organic matter include complexation, adsorption, and chelation, it is expected that some overlapping effects will be obtained with those methods designed to release exchangeable cations. Yong et al. (2001b) considered this as a three-phase procedure. In Phase (i), 3 mL of 0.02 M HNO_3 together with 5 mL of 30% H_2O_2 was adjusted to pH 2 with HNO_3 and agitated occasionally for 2 hours at room temperature. For Phase (ii), 3 mL of 30% H_2O_2 at pH 2 was used. This was intermittently agitated for 3 hours at a temperature of 85°C. For Phase (iii), 3 mL of NH_4OAc in 20% (v/v) HNO_3, diluted to 20 was used. This was continuously agitated at room temperature for 30 minutes.

- *Residual* (#5 in Figure 8.18): The amount of metal contained in the residual fraction is generally not considered to be significantly large. The metals are contained within the lattice of silicate minerals, and can become available only after digestion with strong acids at elevated temperatures. Determination of the metal associated with this fraction is important in completing mass balance calculations.

Most sequential extraction techniques have been developed for the study of sorption of cationic species by soil fractions. In the case of arsenic, a naturally occurring contaminant that is mainly found in the anionic form, one should use *SSE* procedures similar to those adopted for studies of phosphorus retention. In the three *SSE* procedures studied and reported by Van Herreweghe et al. (2003), schemes *I* and *II*, which were based on the phosphorus-like protocol, removed more arsenic than the *BCR* method. The schemes were (a) *BCR* technique, (b) scheme *I*, with the following steps: extractants of NH_4Cl, NH_4F; $NaOH$; diothionite-citrate-bicarbonate (*DCB*) and H_2SO_4; and finally, concentrated $HCl/HNO_3/HF$, and (c) scheme *II*, with the following steps: water-anion-exchange membranes; NH_4F; $NaOH$; diothionite-citrate-bicarbonate (*DCB*) and H_2O_2/HNO_3; and finally, concentrated $HCl/HNO_3/HF$). This was due mainly to the use of *DCB* for amorphous and crystalline oxide extraction, in comparison to the use of hydroxylamine hydrochloride for amorphous oxide only for extraction of arsenic. The use of *NaOH* to remove easily reducible *Fe* bound to arsenic resulted also in the extraction of most of the arsenic in highly contaminated samples. The anionic exchange membranes used in the first step of scheme *II* were not beneficial, and it is recommended that they should not be used.

8.6.2 Organic Chemical Contaminants

Partitioning of organic chemical contaminants in a soil–water system is determined in terms of the relative fugacity of the organic chemical compound in an organic solvent and in water. The fugacity of the organic chemical in the organic solvent or in water is related to its chemical potential, and at equilibrium, the chemical potentials in the organic solvent and water must be equal. With this understanding, we can define the equilibrium partition coefficient

k_{os} as the ratio of c_{os}/c_w, where the concentration of an organic chemical in an organic solvent is defined as c_{os}, and the concentration of that same organic chemical in water defined as c_w. The organic solvent that is generally used in determination of equilibrium partition coefficients is *n*-octanol because of its very low solubility in water. Because *n*-octanol is amphiphilic, that is, part lipophilic and part hydrophilic, it can accommodate organic chemicals with the various kinds of functional groups. The dissolution of *n*-octanol in water is about 8 octanol molecules to 100,000 water molecules in an aqueous phase, that is, a ratio of about 1 to 12,000 (Schwarzenbach et al., 1993). Since water-saturated *n*-octanol has a molar volume of 0.121 L/mol as compared to 0.16 L/mol for pure *n*-octanol, this close similarity allows us to ignore the effect of the water volume on the molar volume of the organic phase. The partition coefficient determined with this solvent is called the *octanol-water partition coefficient k_{ow}*.

A key element in the determination of k_{ow} octanol–water partition coefficient is the efficiency achieved in mixing of the chemicals to achieve equilibrium concentrations, as noted from thermodynamic requirements for equilibrium partitioning. There are different techniques to achieve this thorough mixing, as shown in published literature. The EPA Office of Prevention, Pesticides and Toxic Substances guideline OPPTS 830.7550, for determination of the partition coefficient, uses the shake flask method. The similar shake flask method is given in OECD test method 107. The slow stirring method adopted by OECD as a guideline is given as OECD test method 123. Another procedure is the use of high performance liquid chromatography (HPLC) for determination of the partition coefficient (OECD test number 117).

The k_{ow} octanol–water partition coefficient has been widely adopted in studies of the environmental fate of organic chemicals. Good correlations have been obtained between solubilities of organic compounds and their *n*-octanol–water partition coefficient k_{ow}. In addition, it is sufficiently correlated not only to water solubility, but also to soil sorption coefficients. The relationship for the *n*-octanol–water partition coefficient k_{ow} in terms of the solubility S has been reported by Chiou et al. (1977, 1982) as

$$\log k_{ow} = 4.5 - 0.75 \log S \text{ (ppm)} \tag{8.1}$$

Organic chemicals with k_{ow} values less than 10 generally have high water solubilities and small sorption coefficients, and are considered to be relatively hydrophilic. On the other hand, organic chemicals with a k_{ow} values greater than 10^4 generally have low water solubilities, and are very hydrophobic and not very water-soluble. Solvent systems that are almost completely immiscible are fairly well behaved, and if the departures from ideal behaviour exhibited by the more polar solvent systems are not too large, a thermodynamic treatment of partitioning can be applied to determine the distribution of the organic chemical without serious loss of accuracy.

Aqueous concentrations of hydrophobic organics such as polyaromatic hydrocarbons (PAHs), and compounds such as nitrogen and sulphur heterocyclic PAHs, and some substituted aromatic compounds indicate that the accumulation of the hydrophobic chemical compounds is directly correlated to the organic content (soil organic matter, SOM) of a soil. A large proportion (by weight) of SOM is carbon, and as noted in previous chapters, the SOM functional groups are similar to most of the organic chemicals. They occupy a position in-between water and hydrocarbons insofar as polarity is concerned, and because of their composition and structure, they are well suited for hydrophobic bonding with organic chemical pollutants.

Studies have shown that whereas the variability in sorption coefficients between different soils may be due to characteristics of soil fractions (surface area, cation exchange capacity, pH, etc.), and the amount and nature of the organic matter present, a good correlation of sorption can be obtained with the proportion of organic carbon in the soil. The partition coefficient k_{ow} can be related to the organic content coefficient k_{oc}, which is, in essence, a measure of the hydrophobicity of the chemical contaminant. This coefficient can be determined as follows (Olsen and Davis, 1990):

$$k_{oc} = \frac{k_d}{f_{oc}} = 1.724 k_{om} \tag{8.2}$$

where f_{oc} is the organic carbon content in the organic matter in the soil–water system, k_d is the distribution coefficient discussed in the previous subsection, and k_{om} is the distribution coefficient for the organic matter (see Section 8.5.1). Values for k_{om} and k_{oc} for a number of organic chemicals can be found in the various handbooks dealing with environmental data for such chemicals. Because of competing sorption sites from other soil fractions, soils with organic matter content of less than one percent by weight can give high values for k_{oc}. It has been suggested that the relationship given as Equation (8.2) should not be used when $f_{oc} < 1$. McCarty et al. (1981) gives a critical minimum level for the organic carbon content $f_{oc\text{-}cr}$ as

$$f_{oc\text{-}cr} = \frac{SSA}{200(k_{ow})^{0.84}} \tag{8.3}$$

where SSA denotes the specific surface area of the soil.

The graphical relationship shown in Figure 8.19 uses some representative values reported in the various handbooks (e.g., Verscheuren, 1983; Montgomery and Welkom, 1991) for log k_{ow} and log k_{oc}. The values used for log k_{ow} are considered to be the mid-range results reported from many studies. Not all log k_{oc} values are obtained as measured values. Many of these have been obtained through application of the various log k_{oc}-log k_{ow} relationships reported in the literature, for example, Kenaga and Goring (1980) and

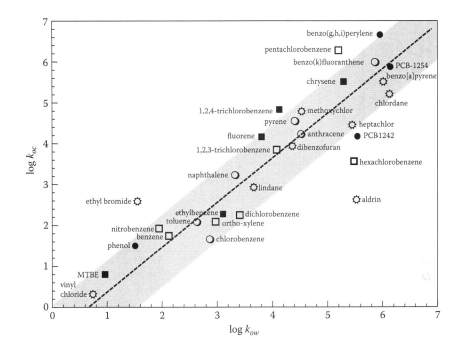

FIGURE 8.19

Relationship of log k_{oc} and log k_{ow} for several organic chemical compounds. A number of the coefficients are estimated values based on observations reported by several researchers. The shaded band encompasses a large majority of the chemical compounds. The sources of information for the graph include the various handbooks on groundwater chemicals and organic chemicals.

Karickhoff et al. (1979). The approximate relationship shown by the dashed line in Figure 8.20 is given as

$$\log k_{oc} = 1.06 \log k_{ow} - 0.68 \tag{8.4}$$

Several relationships between f_{oc} and k_{ow} have been proposed in the literature, for example,

$$\log k_{oc} = \log k_{ow} - 0.21$$

$$\log k_{oc} = 1.029 \log k_{ow} - 0.18 \tag{8.5}$$

$$\log k_{oc} = 0.72 \log k_{ow} + 0.49$$

The first relationship shown in Equation (8.5) was reported by Karickhoff et al. (1979) in respect to 10 PAHs, whilst the second one referring to pesticides was reported by Rao and Davidson (1980). The relationship describing the group containing chlorinated benzenes which also includes methylated benzenes has been reported by Schwarzenbach and Westall (1981).

Instead of using the octanol–water partition coefficient k_{ow} and the organic content coefficient k_{oc}, the accommodation concentration of hydrocarbons in water is sometimes used to inform one of the partitioning tendency of organic substances between the aqueous phase and soil fractions. Hydrocarbon molecules with lower accommodation concentrations in water (i.e., higher k_{oc} values) would be partitioned to a greater extent onto the soil fractions than in the aqueous phase. The results of Meyers and his coworkers (1973, 1978) show that there exists a general inverse relationship between the accommodation concentration of the hydrocarbons and the proportion (percent) adsorbed. This means to say that the lower the accommodation concentration of the hydrocarbon in water, the greater the tendency of the organic compound to be associated with the reactive surfaces of the soil fractions. This means that the aromatic fraction of petroleum products (which are the most toxic) would have the least affinity for the reactive surfaces associated with the soil fractions. A study of adsorption data of hydrocarbons shows that anthracene is substantially adsorbed, as can be confirmed by the high k_{oc} value and the very low solubility of the organic compound in water. The higher accommodation concentrations of the aromatic hydrocarbons inhibit their association with the soil fractions.

8.7 Concluding Remarks

One of the most significant concerns facing soil scientists and soil engineers is the threat to the health of the soil–water system posed by the various environment stressors described in Section 5.2 of Chapter 5. Of the many different types of stressors described, perhaps the most insidious ones are those that pollute the soil–water system. The sorption properties of soils, together with the actions of microorganisms in soils, provide the soil–water system with the tools to ameliorate the health of threatened soil–water systems. These tools constitute the most important environmental soil properties. The main focus of this chapter has been directed towards the sorption potential of soil.

As a means of understanding how soil sorption properties are important in protecting the health of the soil–water system, we have been primarily concerned with the interactions between soil fractions and both inorganic contaminants, represented primarily by heavy metals, and organic chemical compounds, particularly with the processes that lead to partitioning. Some of the key points are contained in the following notes:

1. Processes involved in heavy metal sorption and retention include ion exchange, precipitation as a solid phase (oxide, hydroxides, carbonates), and complexation reactions.

2. With respect to clays, we note that the two kinds of surface charge reactions, that is, reactions associated with the net negative charge and pH-dependent reactions due to hydration of broken bonds, are responsible for the major sets of reactions.

3. Metallic ions are generally coordinated (chemically bound) to water molecules, and the water molecules that form the coordinating complex are the ligands; when the ionic activity of the metal solutes exceeds their respective solubility products, precipitation of these metals as hydroxides and carbonates is likely. In most instances, alkaline conditions are favourable for precipitation of heavy metals. These will form new substances in the pore water or will be precipitated onto soil particles.

4. Precipitation is a two-stage process consisting of (1) nucleation, and (2) particle growth. Some of the key factors involved in precipitation of heavy metals include (a) pH of the soil-water system, (b) concentration and species of heavy metals, (c) nature of inorganic and organic ligands, and (d) precipitation pH of the heavy metal pollutants.

5. Partitioning occurs as a result of mass transfer of the contaminants from the pore water of a soil–water system to the surfaces of the soil fractions. The two types of partitioning tests that appear to be the most popular are (a) batch equilibrium adsorption isotherm tests and (b) column or cell leaching tests.

6. Batch equilibrium adsorption isotherm tests work with soil solutions, and column or cell leaching tests work with compact or intact soil samples. Adsorption isotherms refer to characteristic curves relating adsorption of individual target contaminants by candidate soils to the available concentration of the target contaminant.

7. The results obtained in SSE procedures are operationally defined.

8. Partitioning of PHCs onto soil surfaces occurs to a greater extent than in the aqueous phase, resulting in lower environmental mobility and higher retention of the PHCs. Weakly polar compounds are more readily adsorbed onto soil surfaces in contrast to nonpolar compounds.

9. The water solubility of organic molecules will influence or control the partitioning of the organic pollutant, and the transformations occurring as a result of various processes associated with oxidation/reduction, hydrolysis. and biodegradation.

10. The k_{ow} octanol–water partition coefficient has been widely adopted as a significant parameter in studies of the environmental fate of organic chemicals.

References

Arnold, P.W., 1978, Surface-electrolyte interactions, in Greenland, D.J., and Hayes, M.H.B. (Eds.), *The Chemistry of Soil Constituents*, John Wiley & Sons, New York, pp. 355–401.

Bolt, G.H., 1979, *Soil Chemistry, Part B: Physico-Chemical Models*, Elsevier, Amsterdam, 479 pp.

Chester, R., and Hughes, R.M., 1967. A Chemical technique for the separation of manganese minerals, carbonate minerals and absorbed trace elements from pelagic sediments, *Chem. Geol.*, 2:249–262.

Chiou, G.T., Freed, V.H., Schmedding, D.W., and Kohnert, R.L., 1977, Partition coefficient and bioaccumulation of selected organic chemicals, *Environ. Sci. Technol.*, 11:5.

Chiou, G.T., Smedding, D.W., and Manes, M., 1982, Partition of organic compounds on octanol-water systems, *Environ. Sci. Technol.*, 16:4–10.

Coles, C.A., and Yong, R.N., 2006, Use of equilibrium and initial metal concentrations in determining Freundlich isotherms for soils and sediments, *Eng. Geol.*, 85:19–25.

Contaminant hydrogeology, Macmillan Publishing Co. New york, 458 p.

Fetter, C. W., 1993. *Contaminant hydrogeology,* Macmillan Publishing Co., New York, 458 p.

Hibbeln, K.S., 1996, Effect of kaolinite and cadmium on the biodegradation of naphthalene and substituted naphthalenes, M.Sc. thesis, McGill University, Montreal, Canada.

Ho, M.D., and Evans, G.J., 2000, Sequential extraction of metal contaminated soils with radiochemical assessment of readsorption effects, *Environ. Sci. Technol.*, 34:1030–1035.

Huang, P.M. and Jackson, M.L., 1996, Communication in *Nature*, 211:779.

Karickhoff, S.W., 1984, Organic pollutants sorption in aquatic system, *J. Hydraulic Engineer*, 110:707–735.

Karickhoff, S.W., Brown, D.S., and Scott, T.A., 1979, Sorption of hydrophobic pollutants on natural Sediments, *Water Res.*, 13:241–248.

Kibe, K., Takahashi, M., Kameya, T., and Urano, K., 2000, Adsorption equilibriums of principal herbicides on paddy soils in Japan, *Sci. Total Environ.*, 263:115–125.

Knox, R.C., Sabatini, D.A., and Canter, L.W., 1993, *Subsurface transport and fate processes,* Lewis Publishers, CRC Press, Boca Raton, FL, 430 pp.

MacDonald, E., 1994, Aspects of competitive adsorption and precipitation of heavy metals by a clay soil, M. Eng. thesis, McGill University, Montreal, Canada.

McCarty, P.L., Reinhard, M., and Rittman, B.E., 1981, Trace organics in groundwater, *Environ. Sci. Technol.*, 15(1):40–51.

Meyers, P.A., and Quinn, J.G., 1973, Association of hydrocarbons and mineral particles in saline solution, *Nature*, 244:23–24.

Meyers, P.A., and Oas, T.G., 1978, Comparison of associations of different hydrocarbons with clay particles in simulated seawater, *Environ. Sci. Technol.*, 132:934–937.

Montgomery, J.H., and Welkom, L.M., 1991, *Groundwater chemicals desk reference*, Lewis Publishers, Michigan, 640 pp.

Moorman, T.B., Jayachandran, K., and Reungsang, A., 2001, Adsorption and desorption of atrazine in soils and subsurface sediments, *Soil Sci.*, 166:921–929.

Nakano, M., and Kawamura, K., 2006, Adsorption sites of Cs on smectite by EXAFS analyses and molecular dynamics simulations, *Clay Sci.*, 12(Suppl. 2):76–81.

OECD, 1983, guidelines for the testing of chemicals, 117, partition coefficients by reversed phase chromatography, www.oecd-ilibrary.org (June 2010).

OECD, 2000, OECD guidelines for the testing of chemicals, 106: adsorption-desorption using a batch equilibrium method, www.oecd-ilibrary.org (June 2010).

OECD, 2004, OECD Guidelines for the testing of chemicals, 312: leaching in soil columns, www.oecd-ilibrary.org (June, 2010).

OECD, 2006, Guidelines for the testing of chemicals, 123, (1-octanol.water): slow-stirring method, www.oecd-ilibrary.org (June, 2010).

Pickering, W.F., 1986, Metal-ion speciation—soils and sediments (a review), *Ore Geol. Rev.* 1:83–146.

Rao, P.S.C., and Davidson, J.M., 1980, Estimation of pesticide retention and transformation parameters required in nonpoint source pollution models, in M.R. Overcash and J.M. Davidson (Eds.), *Environmental Impact of Nonpoint Source Pollution*, Ann Arbor Science, Ann Arbor, MI, pp. 23–27.

Smith, D.E., 1998, Molecular computer simulations of the swelling properties and interlayer structure of cesium montmorillonite, *Langmuir*, 14:5959–5968.

Sparks, D.L., 1995, Kinetics of metal sorption reactions, in *Metal Speciation and Contamination of Soil*. H.E. Allen, C.P. Huang, G.W. Bailey, and A.R. Bowles (Eds.), Lewis Publishers, Boca Raton, FL, pp. 35–58.

Sposito, G., 1984, *The Surface Chemistry of Soils*, Oxford University Press, New York, 233 pp.

Schwarzenbach, R.P., Gschwend, P.M., and Imboden, D.M., 1993, *Environmental Organic Chemistry*, John Wiley & Sons, New York, 681 pp.

Schwarzenbach, R.P., and Westall, J., 1981, Transport of non-polar organic compounds from surface water to ground water: laboratory sorption studies, *Environ. Sci. Technol.*, IS (II): 1360–1367.

Ure, A.M., Quevauviller, P.H., Muntau, H., Griepink, B., 1993, Speciation of heavy metal in soils and sediments. An account of the improvement and harmonisation of extraction techniques undertaken under the auspices of the BCR of the Commission of the European Communities, *Int. J. Environ. Anal. Chem.*, 51:135–151.

USEPA, 1991, Batch-type procedures for estimating soil adsorption of chemicals, EPA/530-SW-87-008-F.

USEPA, 1996, Product property test guidelines, Partition coefficient (n-octanol/water) shake-flask method, OPPTS 830.7550, EPA 712–C–96–038.

USEPA, 2008, Fate, transport and transformation test guidelines, OPPTS 835.1240, Leaching studies, EPA/712-C-08-010.

Verscheuren, K., 1983, *Handbook of Environmental Data on Organic Chemicals*, 2nd ed., van Nostrand Reinhold, New York, 1310 pp.

Yong, R.N., Elmonayeri, D.S., and Chong, T.S., 1985, The effect of leaching on the integrity of a natural clay, *Eng. Geol.*, 21:279–299.

Yong, R.N., and Rao, S.M., 1991, Mechanistic evaluation of mitigation of petroleum hydrocarbon contamination by soil medium, *Can. Geotech. J.*, 28:84–91.

Yong, R.N., Mohamed, A.M.O., and Warkentin, B.P., 1992, *Principles of Contaminant Transport in Soils*, Elsevier, Amsterdam, 327 pp.

Yong, R.N., 2001, *Geoenvironmental Engineering: Contaminated Soils, Pollutant Fate, and Mitigation*, CRC Press, Boca Raton, FL, 307 pp.

Yong, R.N., Yaacob, W.Z.W., Bentley, S.P., Harris, C., and Tan, B.K., 2001, Partitioning of heavy metals on soil samples from column tests, *Eng. Geol.*, 60:307–322.

Yong, R.N., and Mulligan, C.N., 2004, *Natural Attenuation of Contaminants in Soils*, Lewis Publishers, CRC Press, Boca Raton, FL, 319 pp.

9

Mobility and Attenuation of Contaminants

9.1 Introduction

The mobility of contaminants in the ground refers to the transport and attenuation of contaminants in the subsurface soils. In this chapter, the transport mechanisms and fate or conversion processes of contaminants constitute the main subjects of interest from the viewpoint of the controls or influence of contaminants on the soil–water system. Figure 9.1 gives a simple example of some of the basic issues relating to contaminant mobility. The left-hand sketch in the figure shows the source of contaminants, that is, discharge of leachate from a landfill. The axis in the contaminant plume identifies the $x = 0$ location used in the right-hand schematic graph, which portrays contaminant concentration distribution in relation to the distance x from the source at a particular time t. The *dispersion* distribution of contaminants identifies the situation where the agents involved in the transport of contaminants from the source to a point x are (a) the dilution effect of the soil water and groundwater, and (b) the physical obstruction to the mobility of contaminants, resulting in retarding the flow of contaminants.

The *dispersion and retention* distribution of contaminants essentially add another agent of influence to the mobility of contaminants. Specifically, this other agent is the contaminant retention capability of the soils derived from the sorption properties of the soil and acting in concert with the surface properties of the contaminants, as discussed in the previous chapter. Retention of the contaminants in the leachate stream occurs at a rate commensurate with the reaction mechanisms associated with the environmental properties of the soil, until the carrying capacity of the soil is reached—at which point no further retention occurs. In addition, abiotic and biotic attenuation or conversion processes of contaminants occur in the leachate steam.

Contaminant–soil interactions and reactions essentially govern the transport processes that include (a) diffusive or dispersive flow or movement of contaminants, (b) advective transport, and (c) abiotic reactions and partitioning. Other processes such as biotic reactions are generally considered as being influential in the longer term. Whilst these biotic reactions play a minor role in governing transport of contaminants—to a great extent because of the

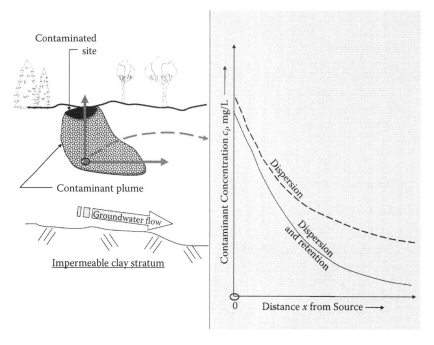

FIGURE 9.1

Schematic diagram showing distribution of contaminant concentration in subsurface soil due to dispersion and dispersion combined with sorption. The left-hand drawing shows the source of the contaminants and location of the axis near the bottom of the contaminant plume. The contaminant concentration spatial $c_i - x$ distribution shown in the right-hand sketch portrays the information from the axis source and at a distance x from the source at a particular time t.

long-term time scale of their reactions—their biggest role is in their determination of the fate of organic chemical contaminants, and in the alteration and transformation of clay minerals and natural organic matter in the soil.

9.2 Interactions and Mobility

9.2.1 Inorganic Contaminants

9.2.1.1 Soil Fractions and Sorption Sites

As one would intuitively expect, the amount of contaminants retained by the soil, as the contaminants in the leachate stream move through a soil, is related to the amount of sorption sites available. This is demonstrated by the results from soil suspension experiments reported by Yong and MacDonald (1998) shown in Figure 9.2. In this instance, the illitic clay consisting primarily of illite clay minerals with about 5.7% carbonates and 1.2% amorphous oxides

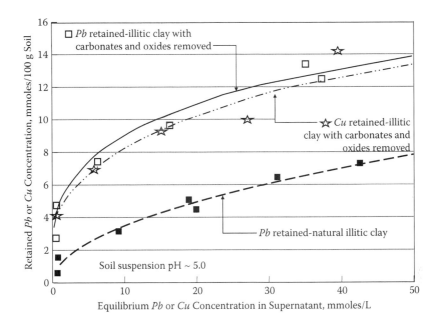

FIGURE 9.2
Lead (*Pb*) or copper (*Cu*) retention by natural illitic clay and illitic clay with carbonates and oxides removed in soil suspension tests. Lead nitrate [*Pb*(*NO*$_3$)$_2$] and copper nitrate [*Cu*(*NO*$_3$)$_2$] solutions were used as the source of *Pb* and *Cu,* respectively. (Data from Yong, R.N., and MacDonald, E.M., 1998, E.A. Jenne (Ed.), *Adsorption of Metals by Geomedia*, Academic Press, San Diego, chap.10, pp.230–254.)

was used in its natural state for the soil suspension. The effect of increase/decrease in available sorption sites is obtained by removing carbonates and oxides from the illitic clay.

The heavy metal contaminant species used were in the form of either *Pb*(*NO*$_3$)$_2$ or *Cu*(*NO*$_3$)$_2$. The adsorption test results for the natural illitic clay are portrayed in terms of the concentration of *Pb* adsorbed or retained by the soil solids in relation to the concentration of *Pb* remaining in the aqueous phase of the soil solution (supernatant). For the illitic clay with carbonates and oxides removed, either *Pb*(*NO*$_3$)$_2$ or *Cu*(*NO*$_3$)$_2$ solutions were used, and contaminant retention by the clay was either *Pb* or *Cu*. Removal of the carbonate and amorphous oxides shows that more *Pb* is adsorbed by the carbonate-free and oxide-free soil. Removal of amorphous coatings around particles that possessed positive surface charges and restructuring of microstructural units constituting the microstructure of the soil (see Figure 2.6 in Chapter 2) made available more sorption sites. Removal of the carbonates from the soil decreases the buffering capability of the soil, as shown in the next section. Retention of *Cu* for the carbonate-free and oxide-free illitic clay shows some influence from differences in contaminant species. This will be discussed in more detail in a later section in respect to contaminant selectivity and complexation.

9.2.1.2 pH and Chemical Buffering

The chemical state of the soil–water system plays an important role in the interactions between contaminants and soil particles, particularly with heavy metal contaminants. Precipitation of the metals can occur at pH values in excess of the precipitation pH of the metals. Because the porewater in the soil is in contact with the charged surfaces of clay minerals, increases of H^+ in the porewater may not contribute to the decrease of pH of the soil–water system because of ion exchanges with the adsorbed cations. Correspondingly, increases of OH^- in the porewater may not increase the pH of the soil–water system because of dissociation of H^+. These phenomena in soil porewater are called the *chemical buffering action* of the soil and are a manifestation of the chemical buffering capacity of the soil. This means to say that the chemical buffering capacity of a soil refers to its capability to accept inputs of acids or bases without significant changes in its pH status.

The *soil buffering capacity* is defined as the number of moles of H^+ or OH^- that must be added to the clay to raise or lower the pH of the soil by one pH unit. To determine the buffering capacity of a soil, a corresponding soil suspension is titrated with a strong acid or base to provide one with a titration curve for analysis. The titration curves for a montmorillonite and an illite, and a blank solution are shown in Figure 9.3. By plotting the negative inverse slope of a titration curve against the relevant pH, one obtains the buffering capacity or buffering potential of the soil. This means that the *soil buffering capacity*, defined as β, is determined in terms of the changes in the amount of hydrogen ions H^+ or hydroxyl ions OH^- added to the system as follows:

$$\beta = dOH^- / dpH = dH^+ / dpH.$$

The soil buffering capacity curves in Figure 9.3 show that when the pH of illitic soil–water system is greater than 4, its capacity for chemical buffering is higher than the montmorillonite. This is very interesting and informative since the illite soil has a smaller *CEC* than the montmorillonite soil. What this tells us is that the high resistance in pH change in illite is due not only to adsorption of H^+ onto the exchange sites on the clay particles, but also to the neutralization of H^+ by the carbonates in the illite soil. However, it should be noted that over the pH range of soil buffering effectiveness—about pH 4.2 to 7.2 for illite and about pH 4.2 to 8.6 for montmorillonite—the total amount of soil buffering, as determined by the area under the curve, is roughly identical between the two clays.

9.2.1.3 pH and Precipitation

The influence of pH and chemical buffering capacity has several implications in studying the interactions between the heavy metal contaminants and soil fractions. Take, for instance, the situation where a contaminant in the form of a *Pb* solution is used as the input leachate interacting with a clay, such as

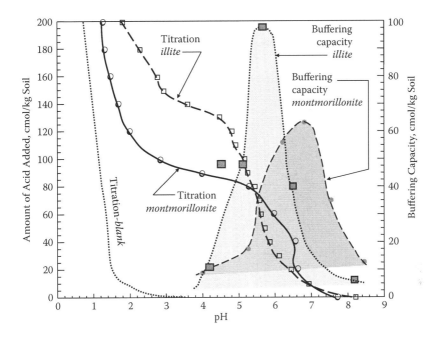

FIGURE 9.3
Titration and buffering capacity results for two clays and a *blank*.

one of those shown in Figure 9.3. When the acidity of the clay is raised due to the increasing amounts of contaminants applied to the clay, high amounts of *Pb* can be retained if the clay still possesses a sufficiently high buffering potential, presuming that the pH of the clay–water system is below the precipitation pH of *Pb*.

How significant is the role of pH in evaluation of test results of heavy metal contaminant-soil interaction? Figure 9.4 shows the partitioning data obtained by Phadungchewit (1990) in respect to an illitic soil and four different heavy metals: *Pb, Cd, Cu,* and *Zn,* where the concentration of each heavy metal species used for the tests was maintained at 1 cmol/kg soil. The precipitation boundaries for *Zn* and *Pb* shown in the figure come from the study conducted by MacDonald (1994). The influence of pH on retention and precipitation of *Pb, Cu, Zn,* and *Cd* by the soil is well demonstrated, and as seen in the figure, a major portion of the partitioned heavy metals can be attributed to precipitation of the metals. Failure to account for the different mechanisms involved in partitioning of the heavy metal contaminants in tests such as batch equilibrium tests conducted to determine adsorption isotherms can lead one to overanticipate the retention capability of candidate soils.

Precipitation of the heavy metals as hydroxides and carbonates will occur when the ionic activity of heavy metal solutes exceeds their respective solubility products—generally resulting in the formation of new substances. Alkaline conditions favour precipitation of heavy metals. Precipitates can

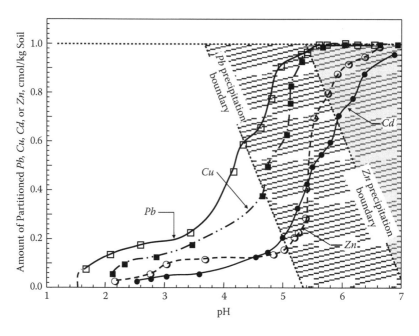

FIGURE 9.4

Partitioning of heavy metals by illitic clay in soil solution tests using single species metals. Note actual amount of heavy metal retained and amount removed from aqueous solution through precipitation as mechanisms involved in the partitioning of the heavy metals (Partitioning results reported by Phadungchewit, 1990, and precipitation boundaries results reported by MacDonald, 1994).

also be attached to clay particles from a two-stage process involving nucleation and particle growth. The critical factors involved in precipitation of heavy metals include (a) the pH of the soil–water system, (b) the nature of heavy metal contaminants such as concentrations and species, and whether one has single and/or multiple species of heavy metals, (c) the nature of inorganic and organic ligands in the porewater, and (d) the precipitation pH of each and every heavy metal species in the porewater. Because of the two-stage process for nucleation and growth of precipitates, it is more accurate to talk about the *onset of precipitation*. Precipitation behaviour of individual species of metals is affected by other ions in solution, and in a soil–water system where other ions and reactive soil fractions exist, one would expect that the precipitation phenomenon will become more complicated.

9.2.1.4 Selectivity, Affinity, and Preferential Sorption

The results shown in Figure 9.4 indicate higher retention of *Pb* at all the pH values below *Pb* precipitation pH. Above the Pb precipitation pH, partitioning of *Pb* is still higher than all the other metals. A vertical line drawn through pH 4.2 (roughly equal to the precipitation pH of *Pb*) shows preferential retention of the order of *Pb* > *Cu* > *Zn* ≈ *Cd* by the illitic clay. This preferential retention order by the soil is called *selectivity*. It is also sometime referred to as *affinity* or *preferential sorption*, that is, affinity for (or preferential sorption of) *Pb* over *Cu* and *Zn*. The greater the sorption preference for a particular metal, the lesser its mobility. The selectivity order shows preference for *Pb* over *Cu, Zn,* and *Cd*, meaning that *Pb* will be less mobile in comparison to the other metals. We also note that *Zn* will be more mobile because it is less preferentially sorbed, and thus is more available for movement in the pore water.

Various studies by researchers have shown that sorption selectivity for heavy metals does exist, and that this is in part determined by the type of soil fractions. The various test results reported by researchers in respect to retention of heavy metals in soil solution studies show that total retention (that is, 100% retention) of heavy metals at higher pH values are attributable to precipitation of the metals when the pH values exceed the precipitation pH of the metals. The Coles et al. (2000) study shows precipitation of *Pb* occurs at about 2 pH units lower than that of *Cd*, forming $Pb(OH)_2$ and $Cd(OH)_2$, respectively. Tests involving montmorillonites show that the selectivity order for montmorillonite for pH values below about 4 appears to be *Pb* > *Cu* > *Zn* > *Cd*, similar to illite. Results obtained from reactions at pH values below 3 are not quantitatively reliable because of dissolution processes.

In general, selectivity is influenced by ionic size/activity, soil type, and pH of the system. The selectivity order reported in the literature confirms that selectivity order depends on the soil type, pH environment of soil-contaminant interaction, and polarizing power of the metal cations. According to Evans (1989), the order of selectivity is primarily due to the increase in hydrated radii of the cations. This means to say that metal cations with small ionic radii have larger hydrated radii because of greater attraction of the O–H dipoles in water to the charged ions, reflecting the polarizing capability of the cations. In the case of divalent metal ions, when the concentrations applied to a soil are similar, a correlation between effective ionic size and selectivity order may be expected (Elliott et al., 1986). The ease of exchange, or the strength with which metallic ions of equal charge are held within the clay matrix, is generally inversely proportional to the hydrated radii, or proportional to the unhydrated radii (Bohn, 1979). If one predicts a selectivity order on the basis of unhydrated radii, one should obtain

$$Pb^{2+}(0.120 \text{ nm}) > Cd^{2+}(0.097 \text{ nm}) > Zn^{2+}(0.0.074m) > Cu^{2+}(0.072 \text{ nm}),$$

where the measurements in brackets refer to the unhydrated radii. The Yong and Phadungchewit (1993) study shows a general selectivity order to be *Pb* > *Cu* > *Zn* > *Cd*, somewhat different from the ranking based on unhydrated radii. However, it agrees well with the ranking based on the *pk* of the first hydrolysis product of the metals, where *pk* is defined as the logarithm of the reciprocal of the dissociation constant for electrolytes.

Applying *K* as the equilibrium constant when *n* = 1 to the reaction in the relationship

$$M^{2+} (aq) + nH_2O \leftrightarrow M(OH)_n^{2-n} + nH^+,$$

one obtains a selectivity order with the ranking scheme as follows:

$$Pb(6.2) > Cu(8.0) > Zn(9.0) > Cd(10.1)$$

where the numbers in the brackets refer to the *pk* values.

9.2.1.5 Ligands in Porewater

Discussions in the previous chapter and at the outset of this chapter have indicated that environmental factors are important in the development of soil properties and behaviour. The chemistry of the soil–water system, which is always changing, has a significant role in the interactions between contaminants and soil particles. Continuing the discussion on heavy metal contaminants, it is seen that the types of inorganic and organic ligands in the porewater are factors that need consideration in determining retention of heavy metal contaminants. The *Cd* retention results shown in Figure 9.5 for a kaolinitic clay exposed to two different chloride solutions, *NaClO₄* and *NaCl*, indicate that speciation or formation of complexes have changed the amount of *Cd* sorbed by the kaolinite soil. The results shown in the figure (from Yong

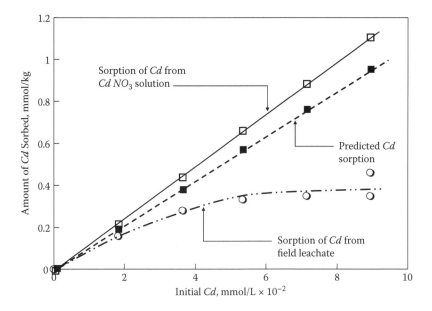

FIGURE 9.5
Sorption of *Cd* from solution of *CdNO₃*, and leachate from an operating landfill, by a kaolinitic clay. Predicted sorption refers to the results calculated using equilibrium constants. (Data from Yong, R.N., and Sheremata, T.W., 1991, *Can. Geotech. J.*, 28:378–387.)

and Sheremata, 1991) refer to sorption of *Cd* from solutions comprised of $CdNO_3$ salt with different concentrations of *Cd*, together with sorption from an actual landfill leachate containing not only *Cd*, but also *Mg*, *K*, *Na*, SO_4^{2-}, *Zn*, *Cu*, *Pb*, *Cl*-, amongst other constituents in the leachate.

Since mononuclear complexes formed between a central metal ion and a number of anions or ligands may be positive, negative, or neutral, one would expect *Cd* to combine with chloride ions to form complexes such as: $CdCl^+$ $CdCl_2^0$, $CdCl_3^-$, and $CdCl_4^{2-}$. The *predicted sorption* curve shown in the figure refers to the theoretical amount of *Cd* sorbed if the mononuclear complexes $CdCl^+$, $CdCl_2^0$, and $CdCl_3^-$, together with $Cd(OH)^+$ and $Cd(OH)_2$, are formed. In the following relationships, given as Equation (9.1), that describe the formation of these mononuclear complexes, the last two relationships are related to the hydrolysis of *Cd*.

$$Cd^{2+} + Cl^- \Leftrightarrow CdCl^+$$

$$CdCl^+ + Cl^- \Leftrightarrow CdCl_2^0$$

$$CdCl_2^0 + Cl^- \Leftrightarrow CdCl_3^- \qquad (9.1)$$

$$Cd^{2+} + OH^- \Leftrightarrow CdOH^+$$

$$CdOH^+ + OH^- \Leftrightarrow Cd(OH)_2$$

The equilibrium constants for Equation (9.1) are given as:

$$K_1 = \frac{(CdCl^+)}{(Cd^{2+})(Cl^-)}$$

$$K_2 = \frac{(CdCl_2^0)}{(CdCl^+)(Cl^-)}$$

$$K_3 = \frac{(CdCl_3^-)}{(CdCl_2^0)(Cl^-)} \qquad (9.2)$$

$$K_4 = \frac{(CdOH^+)}{(Cd^{2+})(OH^-)}$$

$$K_5 = \frac{(Cd(OH)_2)}{(CdOH^+)(OH^-)}$$

Note that the parentheses in the relationships given in Equation (9.2) denote the activities of the ions in solution and that the degree of complexation in respect to the first three relationships in Equation (9.1) will depend on the values of the equilibrium constants K_1, K_2, and K_3, respectively. One can conclude that the difference between the *predicted sorption* curve and the actual *Cd*-leachate curve is due to the fact that other constituents in the landfill leachate have successfully competed for the available sorption sites presented by the soil fractions.

When the formation of complexes between heavy metals and ligands happens in the aqueous phase as *speciation*, we obtain a competition between the complexes and heavy metals for sorption of the heavy metals on the soil fractions. In leachates escaping from waste containment landfills, it is not unusual for these leachates to contain all kinds of inorganic and organic ligands, which in turn will interfere with heavy metal sorption by the soil fractions because of formation of complexes that are soluble but unable to be sorbed onto the surfaces of the soil fractions. In particular, Cl^- ions, sulphates, and organics may form complexes with heavy metals, thereby creating interference with their adsorption by soil fractions (Doner, 1978; Benjamin and Leckie, 1982; Garcia-Miragaya and Page, 1976). The leaching cell experiments of Doner (1978) show that the transport of Ni, Cu, and Cd in soils can be as much as four times greater in the presence of Cl^- than in the presence of ClO_4^-. This shows evidence that Cl^- ions form complexes that are unable to be adsorbed, and heavy metals are more easily transported in the form of complexes. On the other hand, since ClO_4^- does not form complexes with these metals, heavy metals are adsorbed by the soil fractions, resulting in a decrease in their transport through the soil (Hester and Plane, 1964; Klanberg et al., 1963). The experimental results from Yong and Sheremata (1991) in respect to Cl^- and ClO_4^- influence are shown in Figures 9.6 and 9.7.

The results shown in Figure 9.6 demonstrate the effect of the mononuclear complexes formed. The information presented in the figures indicate that Cd^{2+} and $CdCl^+$ were able to be sorbed by the kaolinitic soil, whereas $CdCl_2^0$

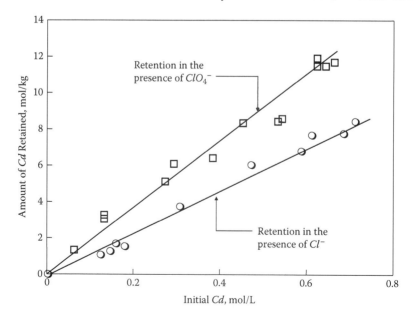

FIGURE 9.6
Cd retention by kaolinitic clay using two chloride solutions related to initial *Cd* concentration.

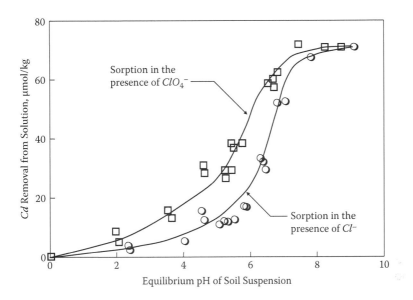

FIGURE 9.7
Sorption of *Cd* by kaolinite soil from two chloride solutions in relation to pH.

and $CdCl_3^-$ were not able to be sorbed. Whilst some research evidence claims that Cd^{2+} is sorbed more readily than $CdCl^+$ (Garcia-Miragaya and Page, 1976; Benjamin and Leckie, 1982), other studies indicate that ion pairs such as $CdCl^+$ are sorbed onto clay mineral surfaces (Sposito and Mattigold, 1979), and that neutral and negative ion pairs, such as $CdCl_2^0$ and $CdCl_3^-$, are not measurably sorbed by kaolinite. In regard to the speciation for the kaolinite clay, the amount of *Cd* not sorbed by the kaolinite soil particles in the presence of Cl^- is due to (a) the decrease in activity due to $NaCl$, (b) competition from Na^+ for adsorption sites, and (c) complexation of Cd^{2+} as negative and neutral chloride complexes.

In regard to the mobility of *Cd* and other heavy metals, it is important to note that as the pH increases, there is a tendency for the heavy metals to be removed from solution as hydroxides, or to precipitate onto the clay surface. The results presented in Figure 9.7 show that the greater tendency for *Cd* to remain in solution in the presence of Cl^- than in the presence of ClO_4^- at pH values higher than the isoelectric point (*iep*) is because of the competition of Cl^- with OH^- for formation of complexes with Cd^{2+} that are unable to be sorbed by the kaolinite. By implication, one can assume that the same tendencies exist for other heavy metal contaminants.

9.2.2 Organic Chemical Compounds

The aqueous concentrations of hydrophobic organics, such as polyaromatic hydrocarbons (*PAHs*), in soil–water systems, are highly dependent on adsorption–desorption equilibria with sorbents in the systems (see Chapter 8). It is important to note that in adopting a linear equilibrium approach to sorption of organic molecules, one needs to consider that (a) adsorption–desorption of many hydrophobic organic molecules occurs over a time scale of many months, and (b) ionization of organic compounds decreases adsorption, leading to increases in mobility of the compounds. An example of the latter phenomenon can be seen in acidic compounds such as phenols and organic acids where the loss of a proton in solution forms anions that tend to become very water soluble because of their charge (Zachara, 1986). Chlorophenols undergo substantial ionization at neutral pH, and acidic compounds tend to ionize more as the pH increases.

9.2.2.1 NAPLs Mobility

The mobility or migration of *NAPLs* depends, to a large extent, on the product source quantity that is released, presumably on the ground surface, and the physical and surface properties of the soil through which the *NAPLs* would move. Most *NAPLs* are partially miscible in water. The locations of the water table, capillary fringe, and bedrock are of considerable importance in the flow patterns of the *NAPLs* in the subsoil. It is recalled from the previous chapter that the light *NAPLs* (*LNAPLs*) include hydrocarbon fuels such as gasoline, heating oil, kerosene, jet fuel, and aviation gas, and that dense *NAPLS* (*DNAPLs*) include chlorinated solvents such as carbon tetrachloride, 1,1,1-trichloroethane, chlorophenols, chlorobenzenes, trichloroethylene, tetrachloroethene, and coal tar, creosote, and *PCBs* (see Section 8.3.1 in Chapter 8).

The density differences that distinguish between *LNAPLs* and *DNAPLs* create distinct patterns in their mobility in soil, as seen in the schematic illustration given in Figure 9.8. *LNAPLs* will essentially stay afloat on the water surface that establishes the water table, with some dissolved chemical plume escaping into the water at the contact interface. *DNAPLs* will plunge through the vadose zone and come to rest on an impermeable surface such as that established by the rock mass shown in the figure. The characteristics of movement of the *NAPLs* through the vadose zone are influenced by the quantity of released product (that is, released *NAPLs*) and both the porosity and permeability of the soil in this zone. Small quantities of released product tend to be held locally and will migrate downward through the aid or percolating rainwater. Large quantities of released product on the other hand will move downward more or less rapidly dependent on the mass effect produced by the released product, especially in the vadose zone with a big crack and fissure (Section 7.4.4, in Chapter 7). The natural water content of the soil in the vadose zone plays a significant role in whether or not bonding between the product and soil fractions occur. For partly wet soils in the vadose zone, water hydration layers surrounding soil particles will hinder bonding between the product and soil particles. Completely dry soils

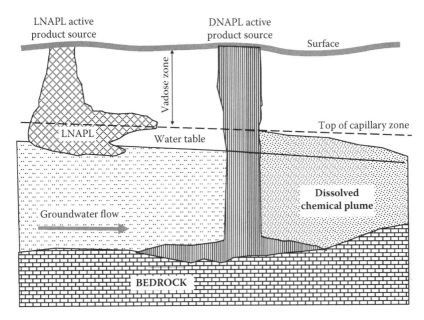

FIGURE 9.8
Schematic illustration of *LNAPL* and *DNAPL* penetration plumes from active product sources.

on the other hand provide particle surfaces that could invite bonding with the product.

The movement and distribution of *LNAPLs* in the capillary zone are mainly lateral because of the spreading effect obtained as the *LNAPLs* come to rest on top of the water table. Note that the literature uses the term *capillary zone* interchangeably with *capillary fringe*. A lens effect is produced when more and more product arrives, at which time some penetration of the *LNAPLs* into the water table may occur. Since *LNAPLs* are lighter than water, buoyant forces are important. They serve to maintain the water level by keeping the product on top of the surface. The characteristic shape of the *LNAPL* movement in the capillary zone is often called a *LNAPL pancake*, the configuration and extent of which are dependent on the permeability of the soil, the percolation rate, and the degree of water saturation. In general, the greater the permeability of the soil, the greater will be the lateral spread of the product and the thinner will be the pancake.

In the saturated zone, that is, the zone under the water table, the movement of *NAPLs* is generally viewed in terms of two different classes of substances: (a) miscible or dissolved substances, and (b) immiscible substances. The mobility of miscible or dissolved substances will be discussed in a later section dealing with the transport and fate of contaminants. In respect to immiscible substances, the low solubility, high density, and low viscosity properties of *DNAPLs* will cause them to move downward through the saturated zone, to come to rest on an impermeable surface, as shown schematically in Figure 9.8.

9.2.2.2 Vapour Phase and Volatilization

The higher the vapour pressure of the organic chemical compound, the more likely it is to evaporate in soil pores. Mobility of the vapour phase of these chemicals depends on the permeability of the soil in the unsaturated zone. Lower gas permeability strata will likely mean that the vapour phase will move by diffusion mechanisms, whereas higher gas permeability strata will mean that movement of the vapour phase will be more in line with advective forces. At equilibrium between the organic compound and the vapour phase, the equilibrium partial pressure of a component is directly related to the mole fraction and the pure constituent vapour pressure. Raoult's law may be written as $P_i = x_i P_i^o$, where P_i is the partial pressure of the constituent, x_i is the mole fraction of the constituent, and P_i^o is the vapour pressure of the pure constituent. Because Raoult's law is only applicable when equilibrium conditions exist, and when the mole fraction of a constituent is greater than 0.9, this in effect limits its use to concentrated NAPLs.

For a vapour in equilibrium with its solution in some solvent, the equilibrium partial pressure of a constituent is directly related to the mole fraction of the constituent in the aqueous phase. Henry's law may be written as $P_i = H_i X_i$, where P_i is the partial pressure of the constituent, X_i is the mole fraction of the constituent in the aqueous phase, and H_i is Henry's constant for the constituent. In general, the concentration of any single molecular species in two phases that are in equilibrium with each other will bear a constant ratio to each other so long as their activity coefficients remain relatively constant, assuming that there are no significant solute-solute interactions and no strong specific solute-solvent interactions.

A factor that bears consideration in the assessment of the mobility of organic chemical compounds is effusion or emission of the vapour phase to the atmosphere—that is, *volatilization*. Estimations of volatilization of a chemical from soil to air are by and large made on the basis of mathematical predictions based on models of the physical processes relying on Raoult's and Henry's laws. The rate at which an organic chemical volatilizes is a function of several factors such as soil properties, chemical properties of both soil and chemical involved, and immediate environmental conditions. To be more precise, the rate of volatilization of an organic molecule from an adsorption site associated with a soil particle to a vapour phase in soil–air and then to the atmosphere is a function of many physical and chemical properties of both the chemical and the soil, and also in the process involved in moving from one phase to another, as for example (a) organic compound in soil → compound in solution, (b) compound in solution → compound in vapour phase in soil–air, and (c) compound in vapour phase in soil–air → compound in atmosphere.

The water content of a soil affects volatilization losses through competition of adsorption sites associated with the soil fractions. For nonpolar and weakly polar compounds, preferential sorption of water by soil particles will displace previously sorbed chemicals. However, once the hydration layer around a soil particle is formed, the vapour density of a weakly polar compound in the soil–air is greatly increased, meaning that more intake of water by the soil will not materially contribute to desorption of previously sorbed chemicals. What we learn from this is that whilst a compound can be sorbed onto dry soil fractions (particularly soil organic matter), thereby reducing its volatilization rate, wetting of the soil will result in displacement of the compound and hence allow volatilization to occur.

Volatilization of organic chemicals from the soil surface can occur through a process that is enhanced by the *wick effect* or *wick evaporation,* meaning that the evaporation of the chemical transported to the soil surface is in concert with the evaporation of water that is brought to the soil surface by capillary forces acting on the water in the soil. The volatilized rate of the chemical from the soil surface remains approximately constant and rapid so long as capillary water continues to flow upward to keep the soil moist. The amount of chemical volatilized is related to the time needed to dry the soil sufficiently to reduce the vapour density of the chemical. It is recalled that dry soil allows for enhanced bonding with most chemicals.

9.3 Mobility and Attenuation

The overriding concern in the management of soil quality, to protect the health of biotic species and the environment, is to be able to anticipate where and when specific kinds of threats will impact on the soil–water system of interest. Most of the analytical/computer models developed to provide information on specific threats and the nature of their impact on soils fall into the class of models dealing with the *transport and fate of contaminants*. For brevity in communication, the term *models* is used to include all the various kinds of analytical, simulation, and computer-generated models. It goes without saying that the viability, validity, accuracy, and applicability of models are entirely dependent on how well they "understand and represent what is happening in the total system." This requires an understanding of all the independent and interdependent processes acting on the total contaminant–soil–water system and the outcome of these processes in respect to the transport and fate of contaminants in the ground. The purpose of the discussion in this section is to examine the attenuation of contaminants in the soil–water system in the context of the various processes that seek to interfere with their mobility through partitioning and other processes that result in attenuation of the contaminant concentrations.

9.3.1 Physicochemical Forces and Transport of Contaminants

The movement of contaminants in a soil is conditioned by various stressors originating from external and internal sources. Stressors that originate from internal sources are linked to the physical and chemical properties of the soil fractions, and their actions generally result in physical and chemical reactions with the contaminants entering the soil–water system. Figure 9.9 is a schematic that illustrates the actions of thermodynamic forces on contaminants in a soil–water system. In this particular instance, the example shows the actions relative to the reactive surfaces of two clay particles. These reactions are depicted as F forces in the top portion of the figure, which together with a chemico-osmotic flow component portraying the results of physicochemical reactions at the bottom portion of the figure, provide the basis for development of a relevant analytical model. The influent contaminant concentration of solutes c_i is greater than the concentration c at the outflow end, and the chemico-osmotic flow shown at the right-hand of the diagram occurs as a result of this difference in concentration of solutes. The chemico-osmotic influence on counter flow is more significant for soils containing greater

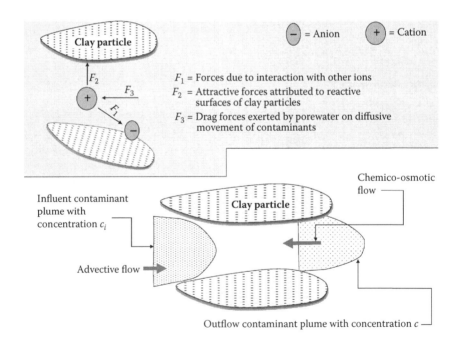

FIGURE 9.9
Schematic diagram portraying physicochemical forces and interacting ions in transport of contaminants through a clay. The top diagram shows the thermodynamic F forces acting a contaminant ion. The bottom diagram shows the counter chemico-osmotic flow and influence of drag forces on contaminant flow profile.

surface-active clay particles, especially when these soils have low porosities and low fluid permeability.

The three sets of thermodynamic F forces acting on the cation shown in Figure 9.9 are

- F_1—*Forces due to interaction with other ions.* Because of the differing sizes and activities of the various ions in the contaminant plume, the mobilities of these ions will be different. The result of the interactions between the ions can be seen in terms of their effect on the activity coefficients of the ions and their diffusion coefficients. This means that these interactions will accelerate diffusion of ions with lower mobilities and a reduction in diffusion of other ions.

- F_2—*Attractive forces attributable to the reactive surfaces of clay particles and the drag forces of the pore fluid.* Increasing the concentration of cations in the porewater increases cations in the diffuse ion-layer, and as ionic concentration increases, overall activity decreases, resulting in decreased diffusivity of the ions.

- F_3—*Drag forces exerted by the porewater on the diffusive movement of the contaminant solutes.* This is given as $F_3 = 6\pi\eta r v_c$, where η is the viscosity of the porewater and is a function of the concentration of ions in the porewater, r is the radius of the moving contaminant solute, and v_c is the velocity of the moving contaminant solute. In the relationship reported by Jones and Dole (1929), the viscosity is expressed as: $\eta = \eta_o (1 + a\sqrt{c} + bc)$, where η_o represents the viscosity of the pure porewater (solvent), c is the concentration of ions, and a and b are coefficients to be determined from experiments. The term $a\sqrt{c}$ accounts for the retardation forces due to electrostatic interaction of the contaminant ions, and the term bc accounts for the interactions between the porewater and the contaminant ions. The drag forces increase with increasing porewater viscosity.

In the simplest form of analysis, we can consider the thermodynamic forces F_1 and F_2 acting on the contaminant ions to be primarily described as functions of the concentration of contaminants in the porewater. Furthermore, we can also consider that the diffusion coefficient of each solute component is a function of its own concentration in the porewater. Chemical and biological interactions and other factors, such as multispecies ions, ligands, speciation, and complexation, are considerations that constitute more sophisticated and complex sets of analyses. The phenomenological relationships between the rates of flow (fluxes J) and the thermodynamic forces F responsible for the fluxes for the situation shown in Figure 9.9 can be described by a power series from the second postulate of irreversible thermodynamics (Onsager, 1931). In a near equilibrium state, it is sufficient to use the first power in the series, that is, $J_i = \Sigma L_{ij} F_j$, where L_{ij} denotes the phenomenological coefficients. The relationships in one dimension are given as follows:

$$J_w = L_{pp} \frac{\partial \psi_p}{\partial x} + L_{pc} \frac{\partial \psi_c}{\partial x}$$

$$J_c = L_{cp} \frac{\partial \psi_p}{\partial x} + L_{cc} \frac{\partial \psi_c}{\partial x} \tag{9.3}$$

where J_w is the fluid flux, J_c is the contaminant solute flux relative to water, L_{pp}, L_{pc}, L_{cp}, and L_{cc} are the phenomenological coefficients, and $\partial \psi_p / \partial x$, $\partial \psi_c / \partial x$ are the thermodynamic forces due to porewater pressure u and concentration gradients, respectively, and are described as follows:

$$\frac{\partial \psi_p}{\partial x} = V_w \frac{\partial(-u)}{\partial x}$$

$$\frac{\partial \psi_c}{\partial x} = \frac{RT}{c} \frac{\partial(-c)}{\partial x} \tag{9.4}$$

where c is the concentration of contaminant solutes in the porewater, V_w is the water molar volume, R is the gas constant, and T is the absolute temperature. From Equations (9.3) and (9.4) we obtain:

$$J_w = L_{pp} V_w \frac{\partial(-u)}{\partial x} + L_{pc} \frac{RT}{c} \frac{\partial(-c)}{\partial x}$$

$$J_c = L_{cp} V_w \frac{\partial(-u)}{\partial x} + L_{cc} \frac{RT}{c} \frac{\partial(-c)}{\partial x} \tag{9.5}$$

L_{pp} can be determined from the first relationship in Equation (9.5) using Darcy's relationship for the situation when $\partial(-C)/\partial x = 0$. This gives us $L_{pp} = k/p_w V_w n$, where k is the Darcy coefficient, ρ_w is the specific weight of water, and n is the porosity. Similarly, L_{cc} can be determined from the second relationship in Equation (9.5) when $\partial(-u)/\partial x = 0$ using Fick's first law. This give us $L_{cc} = (c/RT) D$ where D is the diffusion coefficient.

To take into account the physicochemical processes, we can define the coefficient of osmosis k_c as $k = (V_w RT / c) L_{pc}$ This coefficient can be determined through parameter estimation from experimental results. The coefficient of ionic restriction k_{ir} is given as follows: $k_{ir} = (n V_w / c) L_{cp}$ From the third postulate of irreversible thermodynamics, we can evaluate k_{ir} in terms of k_c, and expressing the porewater pressure u as $u = \rho_w h$, where h is the hydraulic head; Equation (9.5) can be written as follows:

$$J_w = \frac{p_w k}{V_w n} \frac{\partial(-h)}{\partial x} + \frac{k\pi}{V_w} \frac{\partial(-c)}{\partial x}$$

$$J_c = \frac{p_w c k_{ir}}{n} \frac{\partial(-h)}{\partial x} + D \frac{\partial(-c)}{\partial x} \tag{9.6}$$

The second term on the right-hand side (right hand side) of the first relationship in Equation (9.6) is the chemico-osmotic effect responsible for deviations from Darcy's law, and the first term on the right hand side of the second relationship in Equation (9.6) represents the restrictions of movement of the ions relative to that of the porewater as a function of the effect of negatively charged surfaces of the clay particles.

The mass conservation for contaminant ions or solutes in an infinitesimal volume of fluid in soil pore states that:

$$\frac{\partial c}{\partial t} + \frac{\partial J_c^*}{\partial x} + \frac{\rho_s}{n}\frac{\partial c^*}{\partial t} \tag{9.7}$$

where ρ_s is the dry density of the clay, c^* is the concentration of contaminants sorbed or desorbed, and J_c^* is the contaminant flux relative to fixed coordinates. Yong and Samani (1987) have obtained the flux of contaminants relative to a fixed coordinate system from the preceding equations and from the application of mass conservation of contaminants have reported the following relationship:

$$\frac{\partial c}{\partial t} + \left(\frac{k_{tr}}{k} + 1\right)V.\frac{\partial c}{\partial x} + \frac{k_\pi k_{ir}}{2k}\frac{\partial^2 c}{\partial x^2} = \frac{\partial}{\partial x}\left(D\frac{\partial c}{\partial}\right) \pm \frac{\rho_s}{n}\frac{\partial c^*}{\partial t} \tag{9.8}$$

where $V_x = J_w V_w$. Derivation of this relationship assumes that the system is homogeneous and uniform, meaning that point values for advective velocity, sorbed contaminant concentration, and porosity are considered to be representative of the values of the system.

9.3.2 Modelling for Macroscopic Transport of Contaminants

There are several approaches that can be adopted in developing models for analysis and prediction of the mobility and attenuation of contaminants in a soil–water system. As mentioned previously, these models fall into the class known as transport and fate models, and most of them are *equilibrium models*— meaning that they rely on equilibrium reactions between contaminants and soil fractions being achieved instantaneously during the transport process.

How well do transport and fate models simulate "what is happening out there"? Much obviously depends on how well the basis functions represent the various processes and mechanisms of (a) adsorption–desorption, (b) kinetics of sorption and desorption, and (c) other partitioning processes and mechanisms such as speciation, complexation, precipitation, and others involved in the transport of the contaminants being studied.

9.3.2.1 *Diffusion and Advection*

Generally speaking, contaminant transport in soils can be by diffusion or dispersion and also by advective flow, depending on the magnitude of the hydraulic gradient. In unsaturated soils, volumetric water content is an important factor in contaminant transport. Contaminant flux q_c through an unit area of soil mass can be expressed by an equation of Fickian type as follows:

$$q_c = -D_c \theta \frac{\partial c}{\partial x} + q_w c$$

(9.9)

where c is the concentration of contaminants in the porewater, D_c is the contaminant diffusivity or dispersive coefficient, θ denotes volumetric water content, q_w refers to water flux through a unit area of soil mass, and $q_w c$ denotes advective contaminant transfer through the unit area of soil mass. As it is difficult to directly measure the velocity v in the soil pores (that is, porewater velocity), the volumetric flux q is used to determine the amount of contaminant and water transferred per unit time per unit area of soil, that is, $q = v \theta$.

The continuity equation of contaminant transport in a small soil mass that includes soil solid and pore can be written as follows:

$$\frac{\partial \theta \cdot c}{\partial t} = -\frac{\partial q_c}{\partial x} - \frac{\rho_d}{\rho_w} \frac{\partial c^*}{\partial t} - S - \lambda \cdot \theta \cdot c$$

(9.10)

where c^* is the concentration of contaminants sorbed by the soil, ρ_d is the dry density of the soil, ρ_w is the density of water, and λ is the decay coefficient with dimension $1/t$. The term $\lambda \cdot (\theta \cdot c)$ is the decay rate of radioactive contaminants or organic compounds that are transformed in proportion to their concentrations. For contaminants that are not radioactive or compounds that are not transformed, this term can be ignored. The contaminant–soil interactions and reactions resulting in the net sorption of contaminants c^* have been described in the previous chapter as partitioning—that is, the mass transfer of contaminants (solutes in the porewater) onto the soil particles.

S is the storage term that accounts for *sources* and *sinks* in the soil–contaminant–water system. In clays, especially, microstructural units are features that dominate the macroscopic structure of the clay. The storage term S and the concentration of contaminants sorbed by the soil fractions c^* are impacted by the properties of these microstructural units—that is, these microstructural units serve as sources s_i and sinks s_j for solutes in general and metal contaminants in particular. Taking into account all the sources from $i=1$ to $i=a$ and sinks from $j=1$ to $j=b$, and the solutes and/or metal contaminants involved in the sources/sinks, the storage term S can be written as

$$S = \sum_{i=1}^{a} S_i - \sum_{j=1}^{b} S_j$$

(9.11)

Combining Equations (9.9) and (9.10), the general equation for contaminant transfer in unsaturated soils is obtained as follows:

$$\frac{\partial(\theta \cdot c)}{\partial t} = \frac{\partial}{\partial x}\left(D_c\theta\frac{\partial c}{\partial x}\right) - \frac{\partial\left(q_w \cdot c\right)}{\partial x} - \frac{\rho_d}{\rho_w}\frac{\partial c^*}{\partial t} - S - \lambda \cdot \theta \cdot c \qquad (9.12)$$

The second term on the right-hand side of the equation that refers to the contribution of advective transport can be ignored when advective transport is vanishingly small—as can be determined from the Peclet number. The Peclet number, which can be used to determine whether advective transport is important, is defined as the ratio of the contribution of advective transport to the contribution of molecular diffusion, that is, $P_e = v_L d / D_0$, where D_0 is the molecular diffusion coefficient of a specified contaminant in an infinite dilute solution, d is the average soil particle diameter, and v_L is the longitudinal solution flow velocity (advective flow). For $P_e <$ 0.1, advective influence on transport of contaminants is considered to be diminished—as shown in Figure 9.10. For $P_e < 10^{-2}$, the effects of advective velocities on the transport of contaminants in the pore water may be discounted or ignored. The transition zone between dominantly diffusive transport and predominantly advective transport of the dissolved species occurs in the range of Peclet numbers between 10^{-2} and 10, and when $P_e >$ 10, advection becomes the dominant mechanism of contaminant transport.

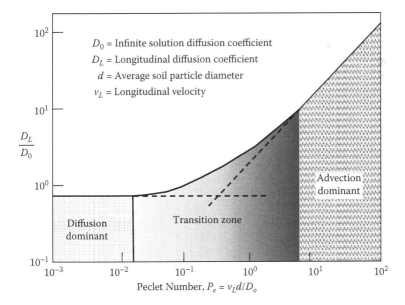

FIGURE 9.10
Diffusion and advection dominant flow regions for contaminants in relation to Peclet number. (Adapted from Perkins and Johnston, 1963.)

In the fully saturated state, the commonly cited one-dimensional relationship used to describe contaminant transport in soil–water systems, for contaminants that are not radioactive or organic compounds that are transformed in proportion to their concentrations, is generally written as follows:

$$\frac{\partial c}{\partial t} = D_c \frac{\partial^2 c}{\partial x^2} - v \frac{\partial c}{\partial x} - \frac{\rho_d}{n\rho_w} \frac{\partial c^*}{\partial t} + S \qquad (9.13)$$

where n refers to the porosity and S refers to the source term. Equation (9.13) can be obtained by applying relations of $q_w = v\theta$ and $\theta = n$ into Equation (9.12).

The partitioning c^* term in Equation (9.13) that refers to the concentration of contaminants removed from the pore fluid (and presumably sorbed by the soil fractions) is most often correlated to their respective equilibrium adsorption isotherms such as those discussed in Section 8.4 of the previous chapter. If one chooses a constant adsorption isotherm with k_d as the distribution coefficient, that is, $c^* = k_d c$, one will obtain the following:

$$\left(1 + \frac{\rho_d}{n\rho_w} k_d\right) \frac{\partial c}{\partial t} = D_L \frac{\partial^2 c}{\partial x^2} - v \frac{\partial c}{\partial x} + S \qquad (9.14)$$

where the first bracket term in the equation refers to the *retardation factor R*, that is, $R = [1 + (\rho_d/n\rho_w) \cdot k_d]$, and finally, one obtains the commonly quoted relationship:

$$R \frac{\partial c}{\partial t} = D_L \frac{\partial^2 c}{\partial x^2} - v \frac{\partial c}{\partial x} + S \qquad (9.15)$$

A criticism of the use of a constant adsorption isotherm is that it allows for the unrealistic situation of infinite adsorption of contaminants. More popular choices for characteristic adsorption isotherms include the Freundlich isotherm and variations thereof.

9.3.2.2 Irreversible and Reversible Sorption/Partition Models

Precipitation, dissolution, hydrolysis, acid-base reactions, complexation, and sorption—to name a few—are some of the interactions/reactions that will impact on the sorption/partition processes. In such situations, one should use kinetic models that describe the rate of sorption of contaminants by the soil fractions. Kinetic sorption models can be either *irreversible* or *reversible* first, second, third order, and so forth kinetic sorption models.

Irreversible kinetic sorption models are used in situations where contaminants sorbed by the soil fractions are tightly held by the soil fractions and cannot be displaced or remobilized by further incoming leachate contaminants. For partitioned inorganic contaminants to be immobile, one requires retention mechanisms such as specific adsorption, chemisorptions involving hydroxyl groups from broken bonds in clay minerals, formation of metal–ion complexes, and precipitation as hydroxides or insoluble salts. Studies on Ni^{2+} adsorption by montmorillonite reported by Xu et al. (2008) showed that equilibrium adsorption was achieved in a matter of hours. Desorption experiments conducted on the samples failed to detach previously sorbed metals, meaning that chemical adsorption and surface complexation were responsible for retention of metal. The standard value of enthalpy registered in the adsorption process indicated that adsorption was endothermic, a similar conclusion reached previously in the study on the sorption kinetics of cesium and strontium ions on a zeolite reported by El-Rahman et al. (2006).

Reversible kinetic sorption models are used for situations where detachment and mobility of previously sorbed contaminants by incoming leachate contaminants occurs. These are kinetic models that account for rates of adsorption of contaminants and rates of desorption of previously sorbed contaminants. The question of the degree of reversibility of the sorbed contaminants is a recurring one, since this impacts directly on the subsequent rate and magnitude of contaminant sorption–desorption phenomena. Figure 9.11 illustrates the problem of retention capability in relation to the kinetics of contaminant sorption–desorption. The rate of desorption of sorbed contaminants is a function of (a) the sorption characteristics of the particular contaminant with the soil fractions, (b) the overall chemical composition of the contaminating

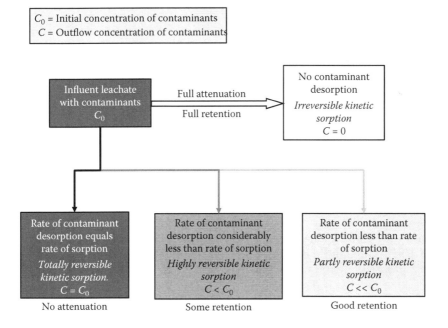

FIGURE 9.11
Schematic illustration of contaminant leachate outflow results due to irreversible, partly reversible, and totally reversible kinetic sorption of contaminants.

leachate, (c) the hydrodynamic stressors responsible for leachate penetration into the soil–water system, and (d) the assimilative capacity of the soil.

In the schematic shown as Figure 9.11, the *totally reversible kinetic sorption* situation shown in the bottom left corner of the figure represents the worst case scenario, probably obtained by the influent leachate passing through a granular soil where no retention of contaminant occurs, that is, concentration of contaminants $C = C_0$. The *highly reversible kinetic sorption* situation shown in the middle bottom portion of the figure assumes that the rate of desorption of previously retained contaminants is less than the rate of sorption. The discharge leachate flow (outflow) will show a concentration of contaminants C less than the entry concentration C_0, that is, $C < C_0$. In real-life situations, continued inflow of the contaminant leachate into the soil under consideration will eventually fully exhaust the carrying capacity of the soil, at which time the discharge or outflow concentration C will be equal to the inflow concentration C_0.

The *partly reversible kinetic sorption* situation is similar in many respects to the highly reversible kinetic sorption situation with the exception that a greater concentration of contaminants is retained in the sorption–desorption process, resulting in outflow concentrations of contaminants considerably less than the inflow concentrations, that is, $C \ll C_0$. Again, in real-life situations, if inflow of contaminant leachates continues, the full carrying capacity of the soil involved will be reached—albeit at a considerably longer time period than the previous situation when $C < C_0$. The ideal situation is one where no desorption of initially sorbed contaminants occurs. Whilst this might happen in the initial stages of inflow of a contaminant leachate, the reality is that if the inflow continues for a long period of time, sooner or later, the full carrying capacity of the soil will be reached. What is required of kinetic adsorption–desorption models is the answer to "When will full carrying capacity be reached?"

9.3.2.3 Coupled Flow of Water and Contaminants

Analyses or modelling of contaminant transport in soils have been conducted macroscopically on the basis of coupled flow of porewater and contaminants through unit areas of soil mass (including soil solids and pores). The contribution of individual flows arising from all driving forces to the total flow is studied using phenomenological relations written in terms of the general flux and the driving force as $J_i = \sum_{k=1}^{n} L_{ik} X_k$, where $I = 1, 2, \ldots, n$, J_i is the flux or flow of i, X_k is the driving force of k ($k = 1, 2, \ldots, n$), and L_{ik} are the phenomenological coefficients.

From Onsager's fundamental theorem, the reciprocal relations $L_{ik} = L_{ki}$ ($i, k = 1, 2, \ldots, n$) are obtained when a proper choice is made for fluxes and forces. When flux occurs for a nonequilibrium state, the deviation ΔS of the entropy from its equilibrium value will result in changes in the system. The local entropy production per unit time $\Delta S / \Delta t = \sigma$ in a soil volume is given as the sum of the products of fluxes and their conjugate forces in irreversible processes:

$$\frac{\Delta S}{\Delta t} = \sigma = \sum_{i}^{n} J_i X_i \tag{9.16}$$

A proper choice of the set of flows and conjugate forces can be seen in Equation (7.1) in Chapter 7.

Defining a dissipation function Φ as $\Phi = T\Sigma$, which is a measure of the rate of local dissipation of free energy by irreversible processes, and choosing the entropy flow as one of the conjugate flows, we obtain (cf. Equation [7.2] in Chapter 7):

$$\Phi = J_s \cdot grad(-T) + \sum_{i}^{n} J_i \cdot grad(-\mu_i) \tag{9.17}$$

where J_s is the entropy flow, J_i signifies material flow, and μ_i is the chemical potential of the *i*th matter. The dissipation function is also expressed by the sum of products of flows (including entropy flow) and conjugate forces. For an isothermal condition, since $grad(-T) = 0$, the dissipation function will be obtained as $\Phi = \Sigma_i^n J_i \cdot grad(-\mu_i)$, and the phenomenological relationship $J_i = \Sigma_{k=1}^{n} L_{ik} X_k$ can be determined between the flows and forces on water and contaminants—provided that the chemical potential of the soil water (porewater) can be separated into the chemical potential of water and contaminant. These phenomenological relations can be specified in the partly saturated state after rearrangement as Equation (9.18), taking into account (a) the chemical potential of water expressed in terms of the matric potential ψ_m, and (b) the chemical potential of contaminant expressed in relation to the concentration of contaminants:

$$J_\theta^* = -L_{\theta\theta}\frac{\partial\psi_m}{\partial x} - L_{\theta c}\frac{\partial c}{\partial x}$$

$$J_c^* = -L_{c\theta}\frac{\partial\psi_m}{\partial x} - L_{cc}\frac{\partial c}{\partial x} \tag{9.18}$$

where J_θ^* is the flux of water that does not include flow due to the gravitational force, that is, $J_\theta^* = J_\theta - q_g$, where J_θ is the total water flux through a unit area of soil under the gravitational force, and q_g is the water flux due to the gravitational force that is given by $-k(\theta)$ (see Equation [3.24] in Chapter 3). J_c^* is the contaminant flux without inclusion of the advective transport of contaminants, that is, $J_c^* = J_c - q_w c$, where J_c is the total transport of contaminants through a unit area of soils (see Equation [9.9]); ψ_m is the matric potential of water; and c is the concentration of contaminant; $\partial\psi_m/\partial x$ and $\partial c/\partial x$ are the driving forces of water and contaminant, respectively. $L_{\theta\theta}$, $L_{\theta c}$, $L_{c\theta}$, and L_{cc} are the phenomenological coefficients. It requires $L_{\theta c} = L_{c\theta}$ to be made available.

Combining with the continuity law in a small soil element one can obtain the expression for water and contaminant transport in macroscopic soil mass as follows:

$$\frac{\partial \theta}{\partial t} = \frac{\partial}{\partial x}\left\{\left(D_{\theta\theta} + D_{\theta c}\right)\frac{\partial \theta}{\partial x}\right\} + \frac{\partial k(\theta)}{\partial x},$$

$$\frac{\partial(\theta \cdot c)}{\partial t} = \frac{\partial}{\partial x}\left\{\left(D_{c\theta} + D_{cc}\right)\frac{\partial c}{\partial x}\right\} - \frac{\partial\left(q_w c\right)}{\partial x} - \frac{\rho_d}{\rho_w}\frac{\partial c^*}{\partial t} - S - \lambda \cdot \theta \cdot c$$

(9.19)

where $D_{\theta\theta} = L_{\theta\theta}\left(\partial\psi_m/\partial\theta\right)$, $D_{c\theta} = L_{c\theta}\left(\partial\psi_m/\partial c\right)$, $D_{cc} = D_c\theta$, and $D_{\theta c} = L_{\theta c}\left(\partial c/\partial\theta\right)$ are moisture, contaminant–moisture, contaminant, and moisture–contaminant diffusivity coefficients, respectively; S is the storage term that accounts for *sources* and *sinks*; and θ is the volumetric water content. The first equation refers to water transport in a partly saturated state. For clays, the second term in the right-hand side of the first equation, that is, the gravitational effect for water flow, can be ignored. The equations in Equation (9.19) take into account (a) the effect of contaminant concentration on the water diffusivity coefficient governing water flow, and (b) the effect of the matric potential on the contaminant diffusivity coefficient governing contaminant transport. The mathematical forms of the equations are the same as Equation (3.29) shown in Chapter 3 and also as Equation (9.12).

The choice of the phenomenological driving forces and coefficients can be performed using several techniques. Elzahabi and Yong (1997) chose the osmotic potential ψ_π of the porewater as the driving force of contaminant and defined the phenomenological coefficients in relation to contaminant transport using van't Hoff's law $\psi_\pi = RTc$. The functional form for the phenomenological coefficients is determined on the basis of experimental information on the distribution of contaminants along columns of test samples with the aid of an appropriate mathematical technique—for example, *identification technique* used by Yong and Xu (1988) for evaluation of the phenomenological coefficients.

9.4 Microstructure and Diffusive Transport

The macroscopic view of movement of contaminants in a soil, in a deterministic sense, assumes that flow of contaminants is a somewhat homogeneous process. The advective-diffusion type of analysis described in the previous section, and even the sorption–desorption type of analysis, do not always pay the required attention to the many impedances to movement, many of which are due to the nonuniform physical features and reactive surfaces of the soil fractions that characterize the soil. A way in which some of these features and factors can be incorporated in the analyses is to account for the effect of the structural components of the soil, meaning that the microstructural units are combined to form the macrostructrure of the soil.

9.4.1 Soil Structure Control

Soil structure plays a very important role in the control of contaminant transport in a soil, through the sizes and arrangement of pores, and also through the interactions between contaminants and soil fractions. To account for soil structure effect, we need to go back and look at why pore spaces in soils are not uniform, and what this means in the context of movement of contaminants in soils. Pore spaces in soils are not uniform because of many factors, such as (a) the different types of soil fractions constituting the soil, (b) soil origin and formation, (c) irregular shapes and sizes of soil solids, and (d) distribution of microstructural units of varied sizes. The microstructural units are generally composed of clay mineral particles, meaning that pore spaces will consist of interlayer spaces. Some pore spaces could be nonconducting, and continuity between the pore spaces may not exist. Many micropores, that is, pore spaces between soil particles in microstructural units, are too small and will not permit easy movement of contaminants because of prohibitive energy requirements. The large differences in sizes between the macro and micropores render them significantly different in how they allow water and contaminants to move through the pore spaces. Because of the presence of reactive surfaces of soil particles in the microstructural units, these units will become sources and sinks for contaminants, as shown in the mechanistic model in Figure 9.12.

The fluid flow example portrayed in the figure shows advective flow of contaminants through the channels defined by the macropores channels and diffusive contaminant transport in the micropores of the microstructural units acting as sinks and sources. The movement of contaminants in the interlayer spaces shown in the top right-hand corner of the figure will be determined by the interactions between contaminants and the surface charges of the mineral particle, as described in the previous chapters. The sink/source phenomena represented by the microstructures has sometimes been referred to as the *stagnant region* in reference to transport of contaminants into or out of the microstructural units by many researchers. The term *stagnant* may be interpreted as being too restrictive if one interprets it as being an immobile region for contaminants. In actuality, this should be called the *partly stagnant* region since some movement of contaminants or solutes in and out of the region occurs. One could define a contaminant diffusivity coefficient D_s in a manner suggested previously by Paissioura (1971), Rao et al. (1980), and Wagenet (1983):

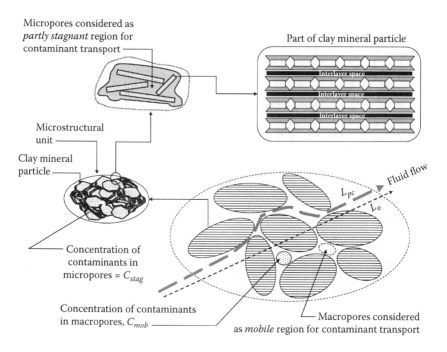

Micropores considered as *partly stagnant* region for contaminant transport

Part of clay mineral particle

Interlayer space
Interlayer space
Interlayer space

Microstructural unit

Clay mineral particle

Concentration of contaminants in micropores = C_{stag}

Concentration of contaminants in macropores, C_{mob}

L_{pc}

L_o

Fluid flow

Macropores considered as *mobile* region for contaminant transport

FIGURE 9.12
Fluid flow through a representative soil volume. Microstructural units composed of clay mineral particles function as sinks and sources. In terms of transport of contaminants, the micropores in the microstructural units are considered as *partly stagnant* regions, and the macropores are considered as *mobile* regions. In the partly stagnant regions, the interlayer spaces in the clay mineral particle shown in the upper-right drawing may be highly stagnant regions.

$$D_s = (D_m + D_h + D_{stag}), \qquad (9.20)$$

where D_s is the longitudinal diffusion–dispersion coefficient, D_m is the molecular diffusive coefficient of the contaminant under consideration, D_h is the advective–dispersive coefficient, and D_{stag} is a dispersion coefficient associated with contaminants diffusing from the micropores (partly stagnant region) to the mobile regions in the macropores.

The *longitudinal diffusion–dispersion coefficient* D_s, which represents flow in the direction of the advective velocity, is often called the *longitudinal diffusion coefficient*. The molecular diffusive coefficient D_m is a function of the volumetric water content of the soil, the infinite dilute solution diffusion coefficient D_0, and a tortuosity factor τ, and is expressed as $D_m = D_0\tau$. The tortuosity τ is generally taken as the ratio of the path length of connected pore channels conducting the contaminants L_{pc} to the straight-line path length L_o. In the schematic diagram shown in Figure 9.12, L_{pc} is the length of the broad dashed-line, and L_o is the straight-line length of that same dash-line. Given the tortuous nature of the channels made by connecting macropores, the path length of connected pore channels will always be greater than the straight-line path, that is, $L_{pc} > L_o$, and this ratio will be greater than one. A commonly used value for this factor is $\sqrt{2}$.

The advective dispersive coefficient of contaminants D_h is expressed as $D_h = \alpha v$, where α is a dispersivity parameter, and v is the advective velocity. The dispersion coefficient D_{stag}, which is associated with diffusion of contaminants from stagnant to the mobile regions, can be determined using the expression given by Paissioura (1971) and Rao et al. (1980) as $D_{stag} = v^2 r^2 (1-\phi)/15D_{es}$, where r is the average equivalent diameter of the soil solids, ϕ = pore water fraction in the conducting region (equivalent to volumetric water content θ), and D_{es} is the effective diffusion coefficient in the stagnant region.

9.4.2 Molecular Diffusion Coefficient

The molecular diffusion of contaminants in microstructural units will be influenced by the nature of micropores and macropores and their distribution. In partly saturated clays, contaminant diffusion is the predominant mechanism of transport of porewater contaminants via transport in film boundaries and in water-filled micropores of the microstructural units. Since clays comprise the bulk of these microstructural units, the properties of interlayer water feature prominently in water and contaminant transport, as illustrated in Figure 9.12. At very low water contents, contaminants in the micropores of the microstructural units will move as film boundary transport. Continuity of water films formed by these hydration layers will establish contact between adjoining particles in the microstructural units, and Brownian activity of contaminants is a significant factor to contaminant transport in the film boundaries and in the interlayer water of partly saturated clays.

The technique of Manheim and Waterman (1974) for estimating the molecular diffusion of contaminants with the electrical resistance-conductance uses a formation factor f, which describes the ratio of the electrical resistance of a brine-filled porous medium R_p to that of the brine R_w occupying the same volume as the bulk porous medium. The relationship for f is given as $f = R_p / R_w = 1/\phi^n$, where $\phi = A_{pc}/A_o$ and where the power n ranges from 2.5 to 5.4, A_{pc} is the fractional cross sectional area available for conductance of electric current, and A_o is the gross cross sectional area that corresponds to the water content. The molecular diffusive coefficient obtained with this technique is given as $D_m = D_0\phi^2$.

The examples of approximations for D_m are listed as follows:

$$D_m = D_0\tau \text{ Simplified tortuosity model} \tag{9.21}$$

$$D_m = D_0\phi^2 \text{ Mannheim and Waterman (1974) Lerman (1979)} \tag{9.22}$$

$$D_m = D_0\left(1-\frac{r}{r_p}\right)^2\left[1-2.10\frac{r}{r_p}+2.09\left(\frac{r}{r_p}\right)^3-0.95\left(\frac{r}{r_p}\right)^5\right] \text{ Renkin (1954)} \tag{9.23}$$

For macropores and for micropores in the range of nanometres, the relationship proposed by Renkin (1954) includes consideration of tortuosity of pores and the boundary drag offered by the walls of the pore spaces. In Equation (9.22), r is the radius of the dissolved species of contaminants, and r_p represents the radius of a typical pore space, and when r_p approaches r in size, $D_m \cong 0$.

A common parameter used in most approaches to determine molecular diffusive coefficients is the infinite dilute solution diffusion coefficient D_0. The studies of both Nernst (1888) and Einstein (1905) on the movement of suspended particles controlled by the osmotic forces in a solution provide us with the following relationships for D_0:

$$\text{Nernst-Einstein: } D_o = \frac{uRT}{N} = u\kappa T \tag{9.24}$$

$$\text{Einstein-Stokes: } D_o = \frac{RT}{6\pi N\eta r} = 7.166 \times 10^{-21} \frac{T}{\eta r} \tag{9.25}$$

$$\text{Nernst: } D_o = \frac{RT\lambda}{F^2 |z|^2} = 8.928 \times 10^{-10} \frac{T\lambda}{|z|^2}, \tag{9.26}$$

where u is the absolute mobility of a contaminant, R is the universal gas constant, T is the absolute temperature, N represents Avogadro's number, κ is Boltzmann's constant, λ is the conductivity of the target contaminant, r is the radius of a hydrated contaminant, η is the absolute viscosity of the fluid, z is the valence of the ion, and F is Faraday's constant. The relationships reported as Equations (9.24) to (9.26) show that the infinite solution diffusion coefficient D_0 is a product that includes ionic radius, absolute mobility of the ion, temperature, viscosity of the fluid medium, valence of the ion, equivalent limiting conductivity of the ion, and so forth. A compilation of values on various species of contaminants and their respective D_0 under various conditions can be found in Li and Gregory (1974), Jost (1960), and Lerman (1979), amongst others, and experimental values for λ for many major ions at various temperatures can be found in Robinson and Stokes (1959).

9.5 Attenuation of Organic Chemicals

Modelling the mobility of organic chemicals in the ground requires knowledge of the sorption–desorption characteristics of the chemical in question as in a case of inorganic chemicals. This is not a simple task since one needs to consider factors such as nonlinearities and hysteresis as the age of the organic chemical in the ground increases. Redistribution of the previously sorbed chemical can occur, as in the case of polyaromatic hydrocarbons

(*PAHs*), where distribution of the previously sorbed *PAHs* were covalently bound to humic acids or coupled to metal-oxide surfaces (Bollag, 1992). Transformations of the original organic chemical can occur, resulting in conversion of the chemical into one or more resultant products by processes that can be abiotic, biotic, or a combination of both processes. These will have different properties—especially in respect to their interactions with soil fractions—thus making predictions of how organic chemicals will move through a soil–water system and how fast and where very difficult. We define the term *persistence* as "the continued presence of a pollutant in the substrate." To provide the answers to these two important questions, one needs to consider the basic elements involved in the many kinds of reactions and processes that act on the organic chemical compounds in the soil, since the outcome of these activities will produce products that will impact directly on the mobility characteristics of the chemicals.

Transformed organic chemical compounds resulting from biotic processes, defined as intermediate products on the path towards complete mineralization, are generally classified as degraded products. On the other hand, transformed products derived from abiotic processes in general do not classify as being intermediate products along the path to mineralization. Strictly speaking, products resulting from abiotic processes should be called *conversion products* instead of transformed products, since complete mineralization of the original organic chemical will not be achieved with abiotic processes. It is not always easy to distinguish between the two different types of products because some of the converted products may become amenable to biotic transformations, resulting in transformations that will include a combination of both types of processes.

Those organic chemical compounds that are resistant to transformation by abiotic and biotic processes are define as *persistent organic chemical compounds* that are resistant to conversion by abiotic and/or biotic transformation processes. *Recalcitrant organic chemical compounds* are those persistent organic chemical compounds that are totally resistant to conversion by abiotic and/or biotic transformation processes. The persistence of organic chemical pollutants in soils depends on at least three factors: (a) the physicochemical properties of the pollutant itself, (b) the physicochemical properties of the soil (that is, soil fractions comprising the soil), and (c) the microbial forms present in the soil that can degrade or assimilate the organic chemical pollutants. The abiotic reactions and transformations resulting are therefore sensitive to factors (a) and (b). All of the factors are important participants in the dynamic processes associated with the activities of the microorganisms in the biologically mediated chemical reactions and transformation processes.

9.5.1 Abiotic and Biotic Processes

Abiotic conversion processes occur without the mediation of microorganisms. These processes include chemical reactions such as hydrolysis, oxidation-reduction, and photochemical reactions. The subject of photo-chemical processes will not be addressed in these discussions since these constitute a very minor part of the processes involved. *Biotic* transformation processes are biologically mediated transformation reactions, and include associated chemical reactions arising from microbial activities. The major difference between the products obtained from these two processes (abiotic and biotic) is the fact that abiotic conversion products are generally other kinds of organic compounds, which explains why the products are called *conversion products*. This is quite distinct from the transformation products that come from the mineralization of organic chemical compounds resulting from biotic processes.

Complete mineralization of an organic chemical compound can occur if the compound is a primary substrate, as opposed to transformation result-ing from partial degradation of the compound due to biological processes. Biologically mediated transformation processes are the only types of pro-cesses that can lead to mineralization (conversion to CO_2 and H_2O) of the organic chemical compounds. When complete mineralization does not occur in the transformation processes, the products obtained are called intermediate products. Conversion products obtained from abiotic processes and transformed products obtained from biotic processes can themselves become recalcitrant, as shown, for example, with tetrachloroethylene, that is, perchloroethylene (*PCE*, C_2Cl_4).

Transformed products can be obtained from biotic processes under either aerobic or anaerobic conditions. The products obtained under anaero-bic conditions will be different from the products obtained under aerobic conditions, since transformation processes under aerobic conditions are oxidative, whereas transformational processes under anaerobic conditions a more likely to be reductive processes. The various processes under aero-bic conditions include hydroxylation, epoxidation, and substitution of *OH* groups on molecules, and the processes under anaerobic conditions include hydrogenolysis, H^+ substitution for Cl^- on molecules, and dihaloelimination (McCarty and Semprini, 1994).

9.5.1.1 Hydrolysis

In the chemical reaction between an organic chemical and water, defined as a *hydrolysis reaction*, the water molecule or OH^- ion replaces groups of atoms (or another atom) in the organic chemical, and a new covalent bond with the OH^- ion is formed, with cleavage of the bond and the *leaving group X* in the reacting organic molecule. No change in the oxidation state of the organic molecule is involved in the conversion. The term *neutral hydrolysis* is used to refer to nucleophilic attack by H_2O, to distinguish it from acid-catalyzed and base-catalyzed hydrolysis where catalytic activity is accomplished by the H^+ and OH^- ions, respectively, a distinction made necessary since both acid-catalysis and base-catalysis impact both the pathway and rate of hydrolysis kinetics directly. A *nucleophile* is an electron-rich reagent (nucleus-liking species) containing an unshared pair of electrons, whilst an *electrophile* has an electron-deficient (electron-liking species) reaction site and forms a bond by accepting an electron pair from a nucleophile. Some common inorganic nucleophiles include HCO_3^-, ClO_4^-, NO_3^-, SO_4^{2-}, Cl^-, HS^-, OH^-, and H_2O. The products of hydrolysis reactions are generally compounds that are more polar in comparison to the original chemical compound, and will therefore have different properties.

The surface acidity of clay minerals in soils can catalyze hydrolysis reactions. The surface acidity of kaolinite minerals, for example, is derived from the surface hydroxyls on the octahedral layer of the mineral particles, whereas the surface acidity of montmorillonites is due to isomorphous substitution and to interlayer cations. Measurements on surface acidity of many clay minerals have shown that these can be at least anywhere from two to four units lower than that of bulk water (Mortland, 1970; Frenkel, 1974). Soil-catalyzed hydrolysis reactions can be significant because they can affect the hydrolysis half-lives of the reacting organic chemicals, meaning that they can affect the kinetics of hydrolysis. Figure 9.13 shows the effect of moisture content on the acidity of a kaolinite, using data reported by Solomon and Murray (1972). Since the surface acidity is significantly reduced as the moisture content of the soil is increased, there will be a corresponding decrease in

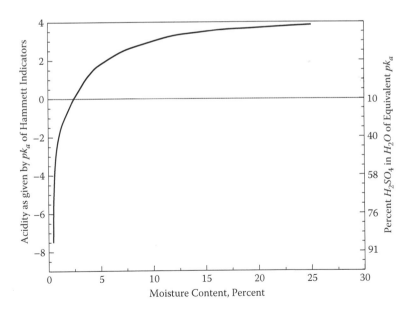

FIGURE 9.13
Effect of moisture content on surface acidity of kaolinite. pk_a is the dissociation constant for the conjugate acid. (Adapted from Solomon, D.H., and Murray, H.H., 1972, *Clay Clay Miner*, 20:135–141.)

the catalytic activity. In the case of montmorillonites, since the layer of water molecules next to the charged layer-lattice sheet is strongly polarized, loss of protons occurs. The charge and nature of the cations affect the degree of catalytic activity since these cations influence the polarizing power and the degree of dissociation of the water in the inner Helmholtz plane (adsorbed water). The surface acidity of montmorillonites increases in conjunction with increases in the valency of the exchangeable cations.

In respect to the kinetics of hydrolysis, Jeffers et al. (1989) report that hydrolysis reactions for perchloroethylene (*PCE*, also known as tetrachloroethylene) and trichloroethylene (*TCE*) do not appear to be significant, most likely due to the fact that their hydrolysis half-lives are 9.9 and 1.3 years in water of 20°C, respectively. However, halogenated alkanes such as chloroethane and 1,2 dibromoethane are known to undergo hydrolysis when there is insignificant biodegradation. For pH values between 2 and 11, methyl fluoride, methyl chloride, and allyl chloride have hydrolysis half-lives of 10 years, 100 days and 80 h, respectively (USEPA, 1998). Clays can also increase the rates of hydrolysis by catalysis (Vogel et al., 1987), and dissolved organic matter and metal ions can also influence hydrolysis rates.

9.5.1.2 Oxidation-Reduction Reactions

Oxidation-reduction reactions occur in interactions between organic chemical compounds and soil fractions under abiotic and biotic conditions. In contrast to transformations occurring through nucleophilic replacement reactions where no net transfer of electrons occurs, electron transfer occurs in oxidation-reduction (redox) reactions. One needs to consider the nature and result of electron transfer between the interacting participants (chemical compounds, microorganisms, and soil fractions). The chemical reaction process defined as *oxidation* refers to a removal of electrons from the subject of interest, and *reduction* refers to the process where the "subject (electron acceptor or *oxidant*)" gains electrons from an electron donor (*reductant*). One cannot readily distinguish between abiotic and biotic redox reactions since redox conditions are more likely than not to be the product of factors that include microbiological processes. The number of functional groups of organic chemical pollutants that can be oxidized or reduced under abiotic conditions is considerably smaller than those under biotic conditions (Schwarzenbach et al., 1993). Quantification of reaction rates is difficult because the interactions between the organic chemicals occur with both microorganisms and the different soil fractions. This makes determination of reaction pathways almost impossible.

Abiotic redox reactions of organic chemical compounds in soil systems occur when electron acceptors are present. Soils containing clay minerals can function as electron acceptors (oxidizing agents) since the structural elements of clay minerals such as *Al, Fe, Zn*, and *Cu* can transfer electrons to the surface-adsorbed oxygen of the clay minerals. The release of these electrons

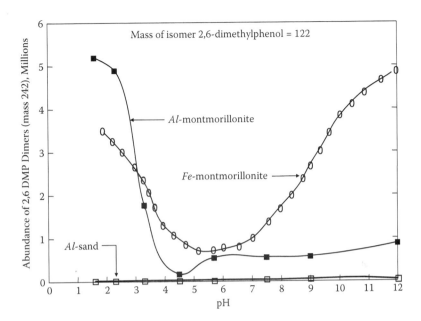

FIGURE 9.14
Oxidation of 2,6-dimethylphenol (DMP) by *Al*-montmorillonite, *Fe*-montmorillonite, and *Al*-sand. (Data from Yong, R.N., et al., 1997, *Appl. Clay Sci.*, 12:93–110.)

as hydroperoxyl radicals ($-OOH$) allows them to function as electron acceptors, which means to say that these radicals can abstract electrons from the organic chemical compounds. The results reported by Yong et al. (1997) of the oxidation of a phenol by *Fe*-montmorillonite and *Al*-montmorillonite, shown in Figure 9.14, indicate that in addition to the electron-abstracting capability of the hydroperoxyl radicals, (a) the partially coordinated aluminium on the edges of the clay particles accepted electrons from the phenol, and (b) the oxidation-reduction properties of the exchangeable cations also contributed to phenol polymerization through coupling of radical cations with the phenol. The results shown in the figure indicate that more effective oxidation of the 2,6-dimethylphenol (2,6-xylenol, $C_8H_{10}O$) is obtained by the *Fe*-montmorillonite, presumably because of the greater oxidizing capability of the *Fe(III)*. The intermediate product formed is a 2,6-dimethylphenol dimer of mass 242, as shown by the degree of "abundance" on the ordinate of the graph in the figure.

9.5.2 Organic Halogen Compounds

Organic halogen compounds are transformed by dehydrohalogenation and dehalogenation with the mediation of microorganisms. The term *dehydrohalogenation* refers to the removal of a halogen atom from a halogenated alkane or from a carbon atom, followed by removal of a hydrogen atom from an adjacent carbon, thus yielding an alkene. Although chloroethane can be dehydrohalogenated (Jeffers et al., 1989), monohalogenated aliphatics, in general, do not undergo this reaction. Bromines are more reactive than chlorines. Dehydrohalogenation reactions reported by Jeffers et al. (1989) include *1,1,1-trichloroethane* (1,1,1-TCA, CH_3CCl_3) and *1,1,2-trichloroethane* (1,1,2-TCA, Cl_2CHCH_2Cl), that go to *1,1-dichloroethane* (DCA, CH_3CHCl_2) with bacteria of *Clostridium* species under anaerobic conditions and *Pseudomonas flourescens* under aerobic conditions, in addition to 1,1,1,2- or 1,1,2,2-tetrachloroethane ($ClCH_2CCl_3$ or $Cl_2CHCHCl_2$) and pentachloroethane (Cl_2CHCCl_3), which go to trichloroethylene (TCE, C_2HCl_3) and perchloroethylene (PCE, C_2Cl_4).

Anaerobic dehalogenation of tetrachloroethylene or perchloroethylene (PCE, C_2Cl_4) goes to trichloroethylene (TCE, C_2HCl_3) with bacteria of *Methanosarcina* species, to 1,2-dichloroethylene (DCE, $C_2H_2Cl_2$) and to vinyl chloride (VC, C_2H_3Cl). The structural changes and the changes in the properties of the intermediate products are shown in Figure 9.15. Beginning with PCE, where the log k_{oc} value indicates good partitioning to the soil fractions, degradation of the PCE to TCE and onward to VC shows that the log k_{oc} values diminish considerably to a very low value for the vinyl chloride. As the PCE continues to degrade, more of the chemical substance is released into the

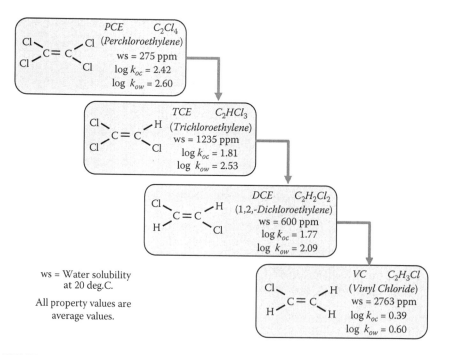

FIGURE 9.15
Schematic showing pathway of degradation of PCE to VC, with corresponding values for water solubility (ws) and partition coefficients. *Cd* retention by kaolinitic clay using two chloride solutions related to initial *Cd* concentration.

porewater, meaning that more of it becomes environmentally mobile. This is particularly true for *VC*, which has low values of log k_{oc} and high water solubility values (see Figure 8.19 in Section 8.6.2 in Chapter 8). Under aerobic conditions, *TCE* goes to CO_2 with methane-oxidizing bacteria, nitrification bacteria (*Nitrosomonas europaea*), propane-oxidizing bacteria (*Mycobacterium vaccae*), *Burkholderia cepacia* G4, and *Pseudomonas putida* F1.

In general, which organic halogenated compound proceeds in anaerobic or aerobic reactions depends on the structures of the halogenated compounds and the conditions or state of the soil environment. The compounds that are highly halogenated undergo anaerobic reactions because they are acidic. The metals that catalyze anaerobic reactions are heavy metals such as *Ni, Fe, Cr,* and *Co.* The compounds that are less halogenated undergo aerobic reactions. The rate of transformation under aerobic reactions is higher than in anaerobic processes. The transformations of halogenated compounds proceed with a mixed stock of various kinds of microorganisms that act in different transformation modes in soils.

9.6 Concluding Remarks

The following are some critical issues and points that bear serious consideration:

1. It would be a mistake to make simple generalizations on heavy metals retention based on soil suspension tests. One needs to consider a number of factors in assessment of metal–soil interaction, such as (a) mechanisms contributing to sorption of the heavy metals, (b) types of soil fractions involved in interaction with the metals, (c) types and concentrations of the metals, and (d) pH and redox environments. Preferential sorption or selectivity is generally not the same for any two soils, since this is very closely related to the nature and distribution of the reactive surfaces available in the soil. The amounts of heavy metal contaminants retained, and the pH influence on retention for any metal, appear to be markedly affected by the presence of other metallic ions in the aqueous phase.

2. In metal-contaminated soils, hydrolysis reactions can be catalyzed by heavy metals sorbed by the soil fractions. The possible mechanisms involved include (Larson and Weber, 1994; Stone, 1989) (a) the sorbed heavy metals which function as Lewis acids, and these can coordinate the hydrolyzable functional groups of the organic chemical in the soil, thus making them more electrophilic; and (b) nucleophilic attack by the metal hydroxo groups associated with the clay mineral surfaces.

3. In metal-ion-catalyzed hydrolysis reactions, coordination of the lone pair electrons (of the oxygen) results in polarization of the carbonyl

functional group, which in turn will make it more susceptible to nucleophilic attack by H_2O or OH^- (Larson and Weber, 1994). Direct polarization processes can accelerate hydrolysis rates by four orders of magnitude (Buckingham, 1977).

4. Formation of a metal-coordinated nucleophile that is more reactive than a corresponding free nucleophile is also possible with metal-ion-catalyzed hydrolysis (Plastourgou and Hoffmann, 1984). The increased acidity of water molecules results in production of OH^-.

5. In a *biologically mediated redox* reaction, the metabolic process is generally *catabolic* (that is, energy releasing) and the result is a transfer of electrons from the organic carbon, which results in the oxidation of the organic chemical compound. Common electron acceptors in the soil system are oxygen, nitrates, sulphates, Fe^{3+}, Mn^{4+}, and other trace metals.

6. There is no partitioning model that fits all situations. The choice of model to be used depends on one's appreciation of the sorption-desorption processes that attend the situation at hand. There are some simple rules that can be followed, and some simple experiments that can be conducted would provide one with an appreciation of the kinds of partitioning processes that might be operative in the transport of metal contaminants in a clay. The basic question that needs to be answered is, Is the time taken for equilibrium partitioning in batch equilibrium experiments achieved well within the time frame for transport of the contaminant leachate through the soil mass under consideration? If the answer is *Yes*, then it is a matter of selection of an appropriate equilibrium adsorption isotherm.

7. For nonlinear kinetic relationships, one has a choice of k_{des}, k_{sorb}, c, and c^* to describe the following: (a) nonlinear desorption and linear sorption, (b) nonlinear desorption and nonlinear sorption, and (c) linear desorption and nonlinear sorption. Options available include bilinear and multilinear desorption and sorption models. These should be tested with the results obtained from batch experiments conducted to obtain sorption and desorption isotherms, with proper registration of the time frames. The procedure is tedious but essential if one seeks to apply nonlinear kinetic desorption-sorption models to describe partitioning of contaminants during transport through a clay buffer/barrier system.

References

Benjamin, M.M., and Leckie, J.O., 1982, Effects of complexation by Cl, SO$_4$, S$_2$O$_4$, on adsorption behaviour of Cd on oxide surfaces, *Environ. Sci. Technol.*, 16:152–170.

Bohn, H.L., 1979, *Soil Chemistry*, John Wiley & Sons, New York, 329 pp.

Bollag, J.-M., 1992, Decontaminating soil with enzymes. *Environ. Sci. Technol.*, 16(10): 1876–1881.

Bresler, E., McNeal, B.L., and Carter, D.L., 1982, Saline and sodic soils, in *Principles— Dynamics—Modeling*, Springer-Verlag, Berlin, 85–101.

Coles, C.A., Rao, S.R., and Yong, R.N., 2000, Lead and cadmium interactions with Mackinawite: retention mechanisms and role of pH, *Environ. Sci. Technol.*, 34:996–1000.

Doner, H.E., 1978, Chloride as a factor in mobilities of Ni (II), Cu(II), and Cd(II) in soil, *Soil Sci. Soc. Amer. J.*, 42:882–885.

Einstein, A., 1905, Uber die von der molekularkinetischen theorie der warme geforderte bewegung von in ruhenden flussigkeiten suspendierten teilchen, *Ann. Physick*, 4:549–660.

Elliott, H.A., Liberati, M.R., and Huang, C.P., 1986, Competitive adsorption of heavy metals by soils, *J. Environ. Qual.*, 15:214–219.

El-Rahman, K.M.A., El-Sourougy, M.R., Abdel-Monem, N.M., and Ismail, I.M., 2006, Modelling the sorption kinetics of cesium and strontium ions on zeolite A, *J. Nucl. Radiochem. Sci.*, 7(2):21–27.

Elzahabi, M., and Yong, R.N., 1997, Vadose zone transport of heavy metals, in R.N. Yong and H.R. Thomas (Eds.), *Geoenvironmental Engineering—Contaminated Ground: Fate of Pollutants and Remediation*, Thomas Telford Publ. Ltd., London, 73–180.

Evans, J.L., 1989, Chemistry of metal retention by soils, *Environ. Sci. Technol.*, 23:1046–1056.

Frenkel, M., 1974, Surface acidity of montmorillonites, *Clay. Clay Miner.*, 22:435–441.

Garcia-Miragaya, J., and Page, A.L., 1976, Influence of ionic strength and inorganic complex formation on the sorption of trace amounts of Cd by montmorillonite, *Soil Sci. Soc. Amer. J.*, 40:658–663.

Hester, R.E., and Plane, R.A., 1964, A Raman spectrophotometric comparison of inter-ionic association in aqueous solutions of metal nitrates, sulphates and perchlo-rates, *Inorg. Chem.*, 3:769–770.

Jeffers, P.M., Ward, L.M., Woytowitch, L.M., and Wolfe, N.L., 1989, Homogeneous hydrolysis rate constants for selected chlorinated methanes, ethanes, ethenes, and propanes. *Environ. Sci. Technol.*, 23:965–969.

Jones, G., and Dole, M., 1929, The viscosity of aqueous solutions of strong electrolytes with special reference to barium chloride, *J. Amer. Chem. Soc.*, 52:29–50.

Jost, W., 1960, *Diffusion in Solids, Liquids, Gases*, Academic Press, New York.

Klanberg, F., Hunt, J.P., and Dodgen, H.W., 1963, Nuclear magnetic resonance studies on the resistance of perchlorate acid, *Inorg. Chem.*, 2:139–142.

Larson, R.A., and Weber, E.J., 1994, *Reaction Mechanisms in Environmental Organic Chemistry*, Lewis Publishers, Boca Raton, FL, 433 pp.

Lerman, A., 1979, *Geochemical Processes: Water and Sediment Environments*, John Wiley & Sons, New York, 481, pp.

Li, Y.H., and Gregory, S., 1974, Diffusion of ions in sea water and in deep-sea sedi-ments, *Geochem. Cosmochim. Acta*, 38:603–714.

MacDonald, E., 1994, Aspects of competitive adsorption and precipitation of heavy metals by a clay soil, M. Eng. thesis, McGill University.

Mannheim, F.T., and Waterman, L.S., 1974, Diffusimetry (diffusion constant esti-mation) on sediment cores by resistivity probe, Initial Report of the Deep Sea Drilling Project, Vol. 22, U.S. Govt. Printing Office, 663–670.

McCarty, P.L., and Semprini, L., 1994, Ground-water treatment for chlorinated solvents, in *Bioremediation of Ground Water and Geologic Material: A Review of In-Situ Technologies*, Government Institutes, Inc. MD, Section 5.

Mortland, M.M., 1970, Clay-organic complexes and interactions, *Adv. Agron.*, 22:75–117.

Nernst, W., 1888, Zur knetik der in losung befinlichen korper, *Z. Phys. Chem.*, 2:613–637.

Onsager, L., 1931, Reciprocal relation in irreversible processes, II, *Phys. Rev.*, 38:2265–2279.

Paissioura, J.B., 1971, Hydrodynamic dispersion in aggregated media: I. Theory, *Soil Sci.*, 111:339–344.

Perkins, T.K., and Johnston, O.C., 1963, A Review of diffusion and dispersion in pourous media. *J. Soc. of Petroleum Engr.*, 17:70–84.

Phadungchewit, Y., 1990, The role of pH and soil buffer capacity in heavy metal retention in clay soils, Ph.D. thesis, McGill University.

Plastourgou, M., and Hoffmann, M.R., 1984, Transformation and fate of organic esters in layered-flow systems: the role of trace metal catalysis, *Environ. Sci. Technol.*, 18:756–764.

Puls, R.W., and Bohn, H.L., 1988, Sorption of cadmium, nickel, and zinc by kaolinite and montmorillonite suspensions, *Soil. Sci. Soc. Amer. J.*, 52:1289–1292.

Renkin, E.M., 1954, Filtration, diffusion, and molecular sieving through porous cellulose membranes, *J. Gen. Physiol.*, 38:225–243.

Robinson, R.A., and Stokes, R.H., 1959, *Electrolyte Solutions*, 2nd ed., Butterworths, London.

Schwarzenbach, R.P., Gschwend, P.M., and Imboden, D.M., 1993, *Environmental Organic Chemistry*, John Wiley & Sons, New York, 681 pp.

Solomon, D.H., and Murray, H.H., 1972, Acid-base interactions and properties of kaolinite in non-aqueous media, *Clay. Clay Miner.*, 20:135–141.

Sposito, G., and Mattigold, S.V., 1979, GEOCHEM: A computer programme for calculation of chemical equilibria in soil solutions and other natural water systems, University of California, Riverside, CA.

Stone, A.T., 1989, Enhanced rates of monophenyl terephthalate hydrolysis in aluminium oxide suspension, *J. Colloid Interface Sci.*, 127:429–441.

U.S. Environmental Protection Agency (USEPA), 1998, Oxygenates in water: Critical information and research needs EPA/600/R-98/048, Office of Research and Development, Washington, DC.

Vogel, T.M., Criddle, C.S., and McCarty, P.L., 1987, Transformation of halogenated aliphatic compounds, *Environ. Sci. Technol.*, 21:722–736.

Wagenet, R.J., 1983, Principles of salt movement in soils, in *Chemical Mobility and Reactivity in Soil Systems*, SSSA spec. Publ. 11:123–140.

Xu, D., Zhou, X., and Wang, X., 2008, Adsorption and desorption of Ni^{2+} on Na-montmorillonite: Effect of pH, ionic strength, fulvic acid, humic acid and addition sequences, *J. Appl. Clay Sci.*, 39:133–141.

Yong, R.N., and Samani, H.M.V., 1987, Modelling of contaminant transport in clay via irreversible thermodynamics, *Proc. Geotechnical Practice for Waste Disposal '87*, ASCE, Ann Arbor, pp. 846–860.

Yong, R.N., and Xu, D.M., 1988, An identification technique for evaluation of phenomenological coefficients in unsaturated flow in soils, *Int. J. Num. Anal. Method Geomech.*, 12:283–299.

Yong, R.N., and Sheremata, T.W., 1991, Effect of chloride ions on adsorption of cadmium from a landfill leachate, *Can. Geotech. J.*, 28:378–387.

Yong, R.N., and Phadungchewit, Y., 1993, pH influence on selectivity and retention of heavy metals in some clay soils, *Can. Geotech. J.*, 30:821–833.

Yong, R.N., Desjardins, S., Farant, J.P., and Simon, P., 1997, Influence of pH and exchangeable cation on oxidation of methylphenols by a montmorillonite clay, *Appl. Clay Sci.*, 12:93–110.

Yong, R.N., and MacDonald, E.M., 1998, Influence of pH, metal concentration, soil component removal on retention of Pb and Cu by an illitic soil, in E.A. Jenne (Ed.), *Adsorption of Metals by Geomedia*, Academic Press, San Diego, chap. 10, 230–254.

Zachara, J.M., 1986, Quinoline sorption to materials: role of pH and retention of the organic cation, *Environ. Sci. Technol.*, 20:620–627.

10

Environmental Soil Behaviour

10.1 Introduction

Soils are living dynamic systems that will (a) from time of soil formation, continue to mature and age as a result of *natural ageing processes*, and (b) from time of interaction with chemical contaminants, continue to change their physical, chemical, and biological soil attributes. The changes of soil attributes occur slowly or quickly in response to the impacts, depending on the circumstances. We consider *environmental soil behaviour* to be the time-related changes of soil attributes in response to the impacts from natural and anthropogenic stressors. The changes are evidences of *soil evolution* in the long process of soil maturation and ageing. The nature of a soil will change as it ages, and changes in the soil attributes that characterize a soil at any instance of time will be the result of the impacts from all kinds of stressors, such as the soil environment and anthropogenic *stressors shown in* Table 5.1 (Chapter 5). In short, evolution of a soil mass is a function of actions from (a) anthropogenic activities and events, and (b) regional natural environment processes. Accordingly, *soil evolution* is defined as the result of impacts of stressors on a soil from all sources (anthropogenic and natural ageing), from initial formation to time-present, and is manifested in terms of corresponding changes in its nature and attributes.

Soil evolution is directly related to such factors as (a) origin and composition, and (b) types and intensity of stressors acting on the soil mass. The evolving nature of soil has a direct impact on the short-term and long-term functional service life of soil barriers, embankments, foundations, and agricultural land use. In this chapter, the discussions will focus on some of the more significant multiprocesses and kinetics involved in the evolution of soils, that is, factors such as alteration and dissolution of soil minerals, attenuation of contaminants, biogeochemical transformation of soil composition, and seasonal freeze–thaw kinetics. The kinetics induced in soils from the various stressors, and their effects on soil attributes, such as the soil properties shown in Table 5.1 in Chapter 5, raise many pertinent issues. These include

- One's knowledge (or lack thereof) of all the kinds of time-related activities or actions that enter into the ageing process of the particular soil mass and, hence, the evolution of the soil, and how these affect soil properties and behaviour.

- The fact that design or planned soil functionality generally uses information on "present-day" soil properties, characteristics, and performance obtained on soil samples or in situ testing on a soil mass that has yet to undergo the various processes that attend soil evolution. This means that the changes in soil functionality that occur will not be fully realized in the planning process.

- The necessary requirements to prescribe the sources and nature of stressors (environment and anthropogenic) and hence the types of stressor impacts on the soil mass in relation to its planned functional service life.

- The need to fully and accurately predict future performance of the soil mass, requiring the development of predictive tools such as analytical-computer models that can successfully simulate the various interactions, reactions, and processes in the soil–water system as it responds to the stressors over the course of its planned soil functional service life.

10.2 Soil Evolution and Ageing Processes

10.2.1 Natural Ageing Processes

Natural soil ageing processes can be defined as those processes associated with geochemical reactions and biogeochemical activities and reactions in soils. These soil ageing processes occur without benefit of anthropogenic activities and events. It is not easy to distinguish or differentiate between the results obtained from geochemical reactions and those obtained from biogeochemical activities and reactions. Theoretically, we can consider geochemical processes and reactions separately. However, in real-life field situations where natural microorganisms exist in the ground, the activities of these microorganisms affect the pH and Eh of the soil–water system, which in turn will influence geochemical reactions. In practice, it is difficult to determine from observations or test results whether these are due to geochemical processes/reactions or biogeochemical activities/reactions. Accordingly, it is more expedient to consider all of these activities and reactions as geochemical processes with a microbiological overlay.

To understand soil evolution due to soil ageing processes, one needs to have a clear insight into the interrelationships and interactions between microorganisms and inorganic/organic fractions in a soil, and their influence on the various transformations and alterations of soil fractions. Some of the more prominent geochemical reactions that require study include dissolution–precipitation of carbonates, sulphates, and silica; oxidation-reduction reactions and protonation–deprotonation. These are considered to be influential in changing the nature of a soil, the result of which will impact on its design or planned functional role. In studies on abiotic- and biotically mediated transformation of clay minerals, Moodie and Ingledew (1990) show that some transformations are preferentially biotically mediated and others are abiotically mediated—for example, oxidation of S^0 into SO_4^{2-} with reduction of Fe^{3+} into Fe^{2+} for the former (biotically mediated) and oxidation of S^{2-} into S^0 with reduction of Fe^{3+} into Fe^{2+} for the latter (abiotically mediated). These reactions could result in (a) transformation of the clay mineral fraction in a soil and hence the composition of the soil, (b) degradation of chemical and physical buffering potential, and (c) changes in the hydraulic, mechanical, physical, and chemical conditions and properties of a soil.

10.2.2 Soil Fractions and Natural Ageing Processes

The soil fractions that are most susceptible to natural soil ageing processes include clay minerals, soil organics, carbonates, sulphates, and amorphous materials such as the oxides and hydrous oxides. This is not to say that the larger-sized particles (silts and sands) do not undergo age-related changes. By and large, if ageing processes would occur for the silts and sands, these would be changes in the nature and extent of coatings that tend to surround these particles.

Since any change in the clay fractions of a soil will mean a change in the nature of the soil itself, for the purpose of discussions in this book, we consider these to refer to changes in soil attributes. Specifically, this would mean (a) physical and chemical changes in a soil such as changes in density, porosity, restructuring of microstructural units, soil functional groups, and so forth, and (b) changes in properties and characteristics of a soil such as hydraulic conductance, strength characteristics, compressibility, swelling characteristics, accumulative properties, buffering capacity, and water uptake. Chemical transformation of clay fractions in a soil includes both abiotic and biologically mediated transformation reactions. Abiotic transformations occur without mediation of microorganisms and include chemical reactions such as hydrolysis, oxidation-reduction, and dissolution; and the transformation products are generally other kinds of clay minerals, organic compounds, and carbonates. Biologically mediated (microbially mediated) transformation reactions, also known as biotic transformation processes, include associated chemical reactions such as oxidation-reduction, cation exchange reactions, dissolution–precipitation of silica and carbonates, and so forth. Microbial activity contributes to these reactions by acting as catalytic agents or as precursor processes, especially in respect to the spatial structuring of redox dynamics. According to Kurek (2002), microbial activities can increase weathering of clay minerals by a million times. Kurek (2002), Ehrlich (2002a), and Huang et al. (2005) state that the modes of microbial attack leading to dissolution of clay minerals by microbes under both aerobic and anaerobic conditions include

- Direct attack—(a) direct enzymatic oxidation of a reduced mineral component and (b) direct enzymatic reduction of an oxidized mineral component
- Indirect attack—(a) with a metabolically produced redox agent or inorganic and organic acids, (b) with a metabolically produced alkali, (c) with a metabolically produced ligand forming a highly soluble product with a mineral component, and (d) by biopolymers

10.2.3 Transformation of Clay Minerals

10.2.3.1 Weathering Processes

Transformation of clay minerals occurs naturally in soils as a result of chemical and biologically mediated weathering processes. The schematic chart in Figure 10.1 shows how the proportions of leached magnesium (Mg) and potassium (K) of basic and acid igneous rocks are attained in the weathered products. Weathering of basic igneous rocks containing significant amounts of Mg can lead to the formation of montmorillonite or kaolinite, depending on the Mg content of the weathered product. If Mg is allowed to remain in the weathered material upon breakdown release from the parent mineral in basic igneous rocks with high Mg content, montmorillonite will be obtained as the weathered product (Grim, 1953). On the other hand, if Mg is immediately or very quickly transported away from the weathered material, the product will be kaolinite. The same result is obtained if Mg is rapidly leached from the weathered product montmorillonite—that is, kaolinite will be obtained from montmorillonite with rapid leaching of Mg.

In acid leaching environments, weathering processes involved in the removal of Mg will cause breakdown of the minerals. This leads to the release and transport of Al and Fe, and the retention Si in the weathered product. If the leaching environment is neutral or mildly alkaline, Fe and Al will remain with the weathered product whilst the released Si will be transported away. This demonstrates the importance of the pH of the leaching environment, that is, (a) acid leaching environments will most likely result in the removal of Fe and Al and concentration of Si in the weathered product, and (b) neutral or mildly alkaline leaching environments will most likely result in the concentration of Al and Fe in the weathered product.

Weathering of the parent mineral in acid igneous rocks containing significant proportions of Mg and K will produce alteration mineral products (i.e., weathered products) that could be montmorillonite or illite, depending on the proportion of Mg or K present in the weathered product. Montmorillonite is obtained when the Mg content is the dominant proportion, whereas illite will

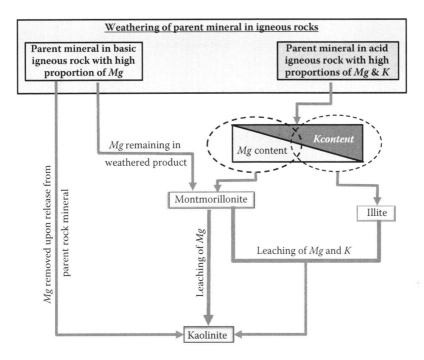

FIGURE 10.1
Effect of leaching of *Mg* and *K* in parent minerals of basic and acid igneous rocks. The right-hand side of the diagram shows that higher proportions of *Mg* content in the weathered product (from acid igneous rock) will yield montmorillonite, whereas higher *K* content will yield illite.

be obtained if the *K* content is the dominant proportion. For the in-between case, a mixed-layer mineral will be obtained. As shown in Figure 10.1, subsequent leaching of *Mg* and *K* will yield kaolinite as the alteration mineral product. The dissolution of some minerals and the reprecipitation of others contribute to the variety of transformed minerals. Temperature and moisture are key elements in the weathering process. The intensity of weathering increases with increasing temperature, and the degree of weathering is greater in most soils. Different minerals have different degrees of robustness—that is, more resistant to weathering. Depending on the type of clay mineral, chemical reactions that promote dissolution and precipitation are greatest at higher temperatures and higher moisture contents.

10.2.3.2 Illitization from Smectite

Transformation of clay minerals to other kinds of minerals occurs because of reactions with the chemistry of the porewater. Take the case of the transformation of smectite as an example. The transformation of smectite to different clay minerals such as illite or saponite involves a process that is

also dependent to some extent on the solid:liquid ratio. The following relationships show transformational reactions:

Smectite + Feldspar + Mica + Others → Illite + Quartz + Chlorite and others

The general chemical reaction of smectite is assumed to be

$$\text{Smectite} + K^+ + Al^{3+} + H_2O \rightarrow \text{Illite} + Si^{4+} + \text{other ions} \tag{10.1}$$

$$\text{Montmorillonite} + Mg^{2+} + Fe^{2+} + H_2O \rightarrow \text{Saponite} + Si^{4+} + \text{other ions} \tag{10.2}$$

In the model proposed by Takase and Benbow (Grindrod and Takase, 1994)
for dissolution of smectite and formation of reaction products, a smectite-to-
illite as the basic reaction is not preset. Taking $O_{10}(OH)_2$ as a basic unit, and X
as the interlamellar adsorbed cation (Na for Na-montmorillonite), the general
formulae for smectite and illite are obtained as

$$X_{0.35}\, Mg_{0.33}\, Al_{1.65}\, Si_4O_{10}(OH)_2 \text{ and } K_{0.5\text{-}0.75}\, Al_{2.5\text{-}2.75}\, Si_{3.25\text{-}3.5}\, O_{10}(OH)_2 \tag{10.3}$$

The reactions in the illitization process are

$$Na_{0.33}\, Mg_{0.33}\, Al_{1.67}\, Si_4O_{10}(OH)_2 + 6H^+ = 0.33Na + 0.33Mg^{2+}$$
$$+ 1.67Al^{3+}\, 4SiO_{2(aq)} + 4H_2O \tag{10.4}$$

causing precipitation of illite and siliceous compounds:

$$K\, Al_3Si_3\, O_{10}\, (OH)_2 + 10H^+ = K^+ + 3Al^{3+} + 3\, SiO_2(aq) + 6H_2O \tag{10.5}$$

The two reactions proposed for conversion of smectite to illites are (a) successive intercalation of illite lamellae in smectite stacks forming mixed-layer
minerals (Weaver, 1979; Nadeau and Bain, 1986) and (b) neoformation of illite
by co-ordination of *Si, Al, K,* and hydroxyls to an extent controlled by access
to any of them.

Assuming that the rate of change of smectite:illite ratio is a function of
(a) temperature, (b) potassium content in the porewater, and (c) activation
energy, the reaction rate can be theoretically computed using the Pytte
model (Pytte, 1982). This model, which is based on the Arrhenius-type theory, uses an activation energy that is commonly taken to be 104.6–125.52 kJ/
mol. The rate equation, which allows one to determine the rate of illitization
from smectite, is given as

$$-\frac{dS}{dt} = \left[Ae^{-\frac{U}{RT(t)}} \right]\left[\frac{K^+}{Na^+} mS^n \right] \tag{10.6}$$

where S represents the mole fraction of smectite in illite to smectite assemblages, R is the universal gas constant, T is the absolute temperature, t represents time, and m and n are coefficients. The results of calculations using

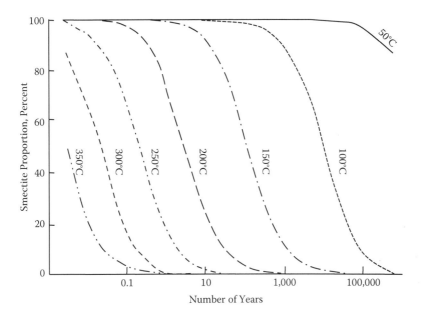

FIGURE 10.2
Conversion of smectite to illite in relation to temperature, for an activation energy of 100.42 kJ/mol, as calculated, using Pytte's model. At 50°C, the original 100% smectite drops to 97% in 100,000 years. However, at 100°C, the original 100% smectite drops in proportion to 50% in 1000 years. (Adapted from Pusch, R., 1994, Waste disposal in rock, *Developments in Geotechnical Engineering*, 76, Elsevier, Amsterdam.)

activation energy of 100.42 kJ/mol are shown in Figure 10.2. The calculated results show that heating to 50°C causes only an insignificant loss of smectite in 10^6 years, whereas about 100% of the original smectite turns into illite in this same period of time at 100°C.

10.2.4 Biotransformation of Clay Minerals

In addition to transformation processes that are chemical in origin, transformation processes can also be driven by microbial activities. These are called *biotransformation processes*. Biotransformation of clay minerals occurs as a result of microbial activities under both aerobic and anaerobic conditions involving (a) dissolution of the original minerals, followed by (b) release of the various structural ions, and (c) formation of new minerals, as shown in Figure 10.3. There are some commonly used terms that attend the various processes and sequences that result in the final transformed products. The term *bioweathering*, meaning biologically weathering, is defined as processes associated with biological modification of rates and mechanisms of chemical and physical weathering of minerals. The terms *alteration* and *degradation* are used to refer to biologically mediated oxidation-reduction transformation of

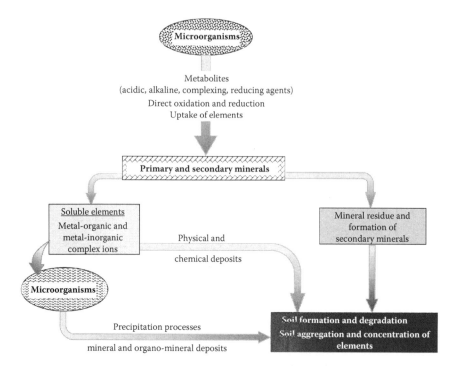

FIGURE 10.3
Schematic flow chart showing bioweathering of soil minerals. (Adapted from Robert and Berthelin, 1986, and Huang and Schnitzer, 1986).

minerals. The distinction between alteration and degradation is partly one of a qualitative nature and one of judgement—meaning that one needs to decide whether the changes observed in the transformed product are an altered form of the original or a lesser (degraded) form. With the preceding definition of terms, we can define *biotransformation* as the bioweathering and alteration/degradation of clay minerals.

Dissolution–precipitation and formation of new minerals are some of the direct consequences of the change in the geochemical environment, a process that is sometimes classed as a form of biomineralization. *Biomineralization,* which is broadly defined as minerals produced by processes attributable to living organisms, has application in several disciplines. However, the genetic basis for biomineralization is not well understood. Strictly speaking, in respect to soils, we can broadly define *biomineralization* to mean the biologically mediated process by which one obtains amorphous and crystalline materials from aqueous ions. The two general paths are (a) biologically induced mineralization, and (b) boundary organized mineralization.

The rates and extent of microbially mediated transformations will be a function of the types of clay minerals and microorganisms. The reaction of microorganisms with clay particles resulting in the production of

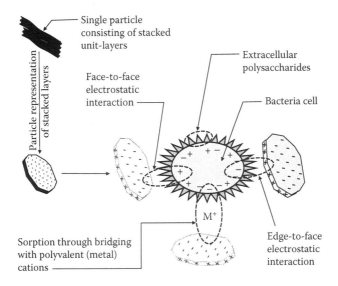

FIGURE 10.4

Bacterial cell with coating of extracellular polysaccharides in interaction with charged clay mineral particle, showing the various sorption mechanisms between the bacterial cell and clay mineral particles. As with the clay mineral particles, the bacterial cell will generally have a net negative surface charge at circum neutral pH conditions. (Adapted from Huang, P.M., et al., 2005, *Pedobiologia*, 49:609–635.)

extracellular polysaccharides emanating from their surfaces can promote the kinds of interactions shown in Figure 10.4. These include (a) coating of the surfaces of particles with extracellular polysaccharides, (b) rearrangement of the distribution and packing of particles, and (c) changes in the nature of the microstructural units in a soil, in effect changing macrostructure of the soil.

10.2.4.1 Structural Iron Reduction

Microorganisms such as facultative anaerobic bacteria, fungi, and even anaerobes can be active in reducing iron. Porewater in soils provides the opportunity for both acid-base and oxidation-reduction reactions, with the latter being abiotic and/or biotic. The reduction of the structural Fe^{3+} in the octahedral and tetrahedral sheets of a clay mineral to Fe^{2+} will significantly alter the short-range forces between the layers in the layer-lattice structure of the clay minerals and could result in (a) a lower specific surface area, (b) a high degree of layer collapse, (c) decreased water-holding capacity, (d) reduced swelling capability, and (e) stacking of the layers. Reduction of octahedral Fe^{3+} to Fe^{2+} in smectites affects (a) the interaction of H_2O molecules with the oxygen ions on the basal surfaces of the mineral particles, and (b) the vibrational energies of the Si-O groups in the tetrahedral sheets. Since swelling of smectites involve two types of interlayers—fully expanding and partially/fully collapsed interlayers—reduction of octahedral Fe^{3+} results in more of the clay layers to collapse in comparison to the oxidized state. The end result is a diminished water uptake and water-holding capability.

10.2.4.2 Microbial Dissolution of Clay Minerals

The attack of bacteria on the clay fractions of soils results in dissolution of the clay minerals, thereby releasing the bonding energy established between atoms that make up the clay minerals, and resulting in the release of these atoms. We can estimate the rate and extent of dissolution by taking into account (a) mass conservation and (b) energy balance or conservation law. In cases where the estimation or measurement of mass can be properly determined for dissolved atoms that are recomposed into various phases—as determined, for example, from batch experiments—the mass conservation law can be usefully applied to determine the rate and extent of dissolution. In nature, however, it is impossible to precisely track the atoms released from clay minerals since some ionized atoms are discharged into the environment. Under these circumstances, an alternative method is to use the energy balance or conservation law to estimate dissolution of clay minerals by bacteria. In essence, this requires determination of the balance between energy consumed by bacteria and energy released from clay minerals.

The energy within the clay minerals, that is, the bonding energy between atoms, can be utilized for growth and life sustenance for bacteria, and can be chemically and physically transformed into the bonding energy for resultant precipitation, metabolites, and gas in vivo and in vitro. A part of the bonding energy will be dissipated as heat. Bacteria would not utilize the whole energy released by dissolution of clay minerals. Clay minerals exposed on the wall of pores or on the surface of compacted clays would not be thoroughly dissolved. A part of them would remain undissolved. Taking all of these into account, when the bio-films are formed, the energy balance established between clay minerals and bacteria can be described as (Nakano and Kawamura, 2008)

$$\varepsilon_{cm} \frac{\partial(\xi\rho_{cm}L)}{\partial t} = E_\mu \frac{\partial}{\partial t}\left(\int_0^\chi \rho_\mu dx\right) + E_l \int_0^\chi \rho_m dx$$

$$+ E_{met} \frac{d}{dt}\left(\int_0^\chi \rho_{met} dx\right) + (1-\varsigma)\varepsilon_{cm} \frac{\partial(\xi\rho_{cm}L)}{\partial t}$$

(10.7)

where L is the corrosion depth of clays; ρ_{cm} is the density of clay minerals; and ρ_μ, ρ_m, and ρ_{met} are the density of bacteria increasing by the net growth, the density of living bacteria, and the density of metabolites, respectively. ε_{cm} is Gibbs free energy of clay minerals, and E_μ, E_l, and E_{met} are Gibbs free energy required for the net growth, living maintenance, and the production of metabolites, respectively. ξ is an erosion factor $(0 < \xi \le 1)$ by which the extent of erosion is defined when clays of L in thickness are partly dissolved by bacteria, and ζ is the ratio of the energy used per unit time by bacteria to the Gibbs free energy released from clays of thickness L $(0 < \zeta \le 1)$; it denotes the energy consumption efficiency of microorganisms. χ is the thickness of bio-films, x is a distance from the surface of clays, and t is time. The term on the left-hand side expresses the energy released from clays of thickness L. The first, second, and third terms on the right-hand side are the energy used per unit time for the net growth, life maintenance, and metabolism, respectively.

Rearranging Equation (10.7), we can obtain the following equation expressing the surface corrosion depth L of a clay mass as

$$L = \frac{\rho_0}{\xi \rho_{cm}} \cdot \frac{\varepsilon_\mu}{\zeta \varepsilon_{cm}} \int_0^t \left\{ \left(\frac{\varepsilon_l}{\varepsilon_\mu} + \mu \right) \chi \left(\frac{\rho_m}{\rho_0} \right) \right\} dt \tag{10.8}$$

where ε_μ is the apparent energy required for the growth that would be approximately equivalent to the energy measured in batch experiments, and ε_l is the apparent energy required for life maintenance. Since it is difficult to choose the appropriate values for ε_l, one could hypothesize that $\varepsilon_l = \alpha \varepsilon_\mu$, where the coefficient α is assumed to be $0.1 < \alpha < 2$.

Since the growth kinetics of bacteria are expressed by a generalized logistic curve using Monod's equation: $(d\rho_m/dt) = (\mu - \kappa)\rho_m$ (Monod, 1949), and the intrinsic growth and decay rate have been previously expressed as Equations (2.4) and (2.5) in Section 2.8.4, Chapter 2, the density of bacteria ρ_m will be given as

$$\rho_m = \rho_0 \cdot \exp \left[\int_0^t \left\{ (\mu - \kappa) - \frac{\partial}{\partial t} (\ln \chi) \right\} dt \right] \tag{10.9}$$

where ρ_m refers to the mean density of bacteria in bio-films, ρ_0 is the initial density of bacteria, and χ is the thickness of bio-films. Figure 10.5 shows several growth curves of living bacteria calculated from Equation (10.9).

As bacteria grow in their habitat, the population of living bacteria gradually increases during a certain characteristic time in the initial stage, and thereafter reaches the saturation state at the final stage. At this juncture, the thickness of bio-films reaches a certain maximum thickness. Accordingly, the thickness of bio-films can be expressed as follows:

$$\chi = \chi_{max} \left(1.1 - a_\chi \cdot \exp(-b_\chi t^2) \right) \tag{10.10}$$

where χ_{max} is the maximum value of χ and a and b are constants.

Since $\mu \to \mu_{max}$, $\kappa \to \kappa_{max}$, $\chi \to \chi_{max}$, and $\rho_m/\rho_0 \to (\rho_m/\rho_0)_{max}$, when t becomes extremely large, that is, larger than about 20 days as shown Figure 10.5, we can conveniently approximate Equation (10.8) with the following simple

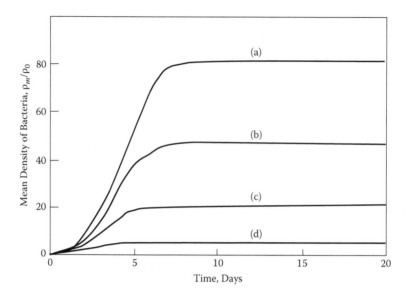

FIGURE 10.5
Logistic curves with different biological parameters μ, κ, and χ ($b_\kappa = b_\chi = 0.06$ (a), 0.07 (b), 0.09 (c), and 0.15 (d)) when $\mu_{max} = \kappa_{max} = 2.0$, $a_\mu = 0.5$, $a_\kappa = a_\chi = 1.0$, and $b_\mu = 4.0$, in Equations (2.4), (2.5), and (10.10). (Data by Nakano, M., and Kawamura, K., 2010.)

expression, although the equation will slightly overestimate the corrosion depth:

$$L = \frac{\rho_0}{\xi \rho_{cm}} \cdot \frac{\varepsilon_\mu}{\varsigma \varepsilon_{cm}} \cdot \left\{ \frac{\varepsilon_l}{\varepsilon_\mu} + \mu_{max} \right\} \cdot \chi_{max} \cdot \left(\frac{\rho_m}{\rho_0} \right)_{max} \cdot t \tag{10.11}$$

The subscript *max* expresses the maximum of each microbial parameter in the stable state.

10.3 Time-Related Changes and Soil Evolution

In situations where soils are associated with human activities/events, the set of processes generated by these activities/events will interact with the set of natural ageing processes because natural ageing processes exist from initial soil formation. There are interrelationships and interdependent reactions formed between the two sets of processes, as the driving forces for both sets will be active throughout the service life (and longer) of the soil mass under consideration. The various kinds of activities, reactions, and phenomena generated

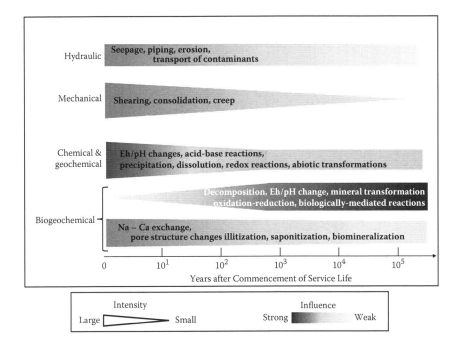

FIGURE 10.6
Major stressor forms and time-related changes and/or phenomena in a soil mass. The top two stressor forms (hydraulic and mechanical) are generally associated with time-related events and the two bottom stressor forms are often considered as soil ageing processes leading to soil evolution.

by the two sets of processes are shown in the sketch in Figure 10.6. It is postulated that whilst some processes from each set might remain independent, the greater proportion of the processes will be integrated. In Figure 10.6, the processes leading to consolidation and shearing of a soil due to *mechanical* stressors might be considered as separate processes over the service life of a structure or facility, so long as chemical and geochemical processes are minimal or absent. This is because by comparison, the service lives of facilities or structures are short in comparison to the time spans involved in transformations of clay fractions from geochemical and biogeochemical reactions.

There are at least two reasons why one might want to differentiate between (a) time-related changes in soil attributes due to human activities, and (b) soil evolution due to natural soil ageing processes. One needs to (a) determine the magnitude or severity of changes (that is, impact) in soil attributes for planning of contemplated anthropogenic activities and projects, and (b) predict service life performance of planned projects at various elapsed time periods. For such purposes, analytical-computer performance models need to be developed, the validity and success of which will depend on one's ability to fully comprehend and represent the various types of activities and phenomena that will occur in the soil mass over the time period of interest.

In general, however, it is sufficient to consider all time-related changes in soil attributes and behaviour as being due to soil evolution.

10.3.1 Stressors and Time-Related Activities

To illustrate the complex interactions and integration of the two different sets of processes discussed in the previous section, consider the case of the major stressor form represented by "Hydraulic" in Figure 10.6. The short time-scale phenomena resulting from application of hydraulic pressures are (a) porewater pressures leading to disruption of soil structure if these become excessive, (b) hydraulic flow and transport of solutes, and (c) piping and/or erosion. In the longer term, one needs to consider the reactions of the solutes transported from outside the soil mass under consideration. These solutes are by definition contaminants since they are "foreign" to the soil mass. Contaminant-soil interactions occurring will now occupy time spans ranging from short to long term, depending on the types and concentrations of the contaminants brought into the soil mass from hydraulic transport. These interactions and reactions fall under the category of "Chemical and geochemical" in Figure 10.6, and as the diagram shows, they will continue to exist for long periods of time. During this longer time period, porewater pressure changes could also occur because of the interactions, together with seepage and piping phenomena. A cycle of events, activities, and phenomena will result from all these interactions/reactions and changes, albeit at a diminishing pace. One needs to pay attention to these situations as hydraulic–mechanical–chemical interactions.

10.3.2 Transported Contaminants and Soil Permeability

The transport of contaminants into a soil mass of interest brings some critical concerns regarding not only partitioning of contaminants—discussed previously in Chapter 8—but also in regard to the hydraulic properties of the soil. The differences or modifications in soil permeability in relation to the chemistry of the permeating fluid are the result of such actions and factors as

- Extraction of lattice aluminium ions from the octahedral sheets of clay minerals
- Ion exchange on the surface of the clay minerals due to replacement of the naturally sorbed cations of lower valence by the extracted Al^{3+} ions and hence a reduction in the thickness of the diffuse ion layer
- Increase in the effective pore sizes and a decrease in the tortuosity factor, resulting in a higher permeability to water
- Disruption of the soil microstructure and its micropores to an extent that may increase the microstructure permeability

- Formation of quasi-crystals between exchangeable calcium ions and a pair of opposing siloxane ditrigonal cavities (Sposito, 1984)
- Hydrolysis of Fe^{3+} ions to form iron hydroxy species that form coatings around the quasi-crystals, thereby affecting their aggregation by electrostatic bonding
- Higher sorption of K^+ ions in the Stern layer
- Partial fixation of cations in the hexagonal oxygen holes in the surface silicate layers of clay minerals

The results of experiments conducted on a Na-smectite (unbuffered and buffered with sodium carbonate) contaminated with different concentrations of $Cu(NO_3)_2$ are shown in Figure 10.7. The permeability results between the different samples illustrate some of the described actions and factors. The permeabilities of the buffered smectite samples are slightly affected by $Cu(NO_3)_2$ contamination. This is because the addition of $Na_2(CO_3)$ increased the pH of the smectite to more than one pH unit (from pH 9.65 to 10.9), and so long as the pH of the samples remained above the precipitation pH of Cu^{2+}, there would be little replacement of the interlayer cations because of precipitation of Cu^{2+} ions. Replacement of Na^+ ions by Cu^{2+} reduces the pH of the soil, consistent with the hydrolysis relationship (Elliott et al., 1986) which shows that

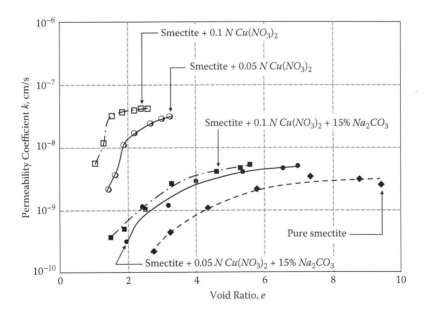

FIGURE 10.7
Variation in permeability coefficient k in buffered (with Na_2CO_3) and unbuffered smectite contaminated with different concentrations of $Cu(NO_3)_2$. (Data reported by Yong, R.N., et al., 2009, *ASCE J. Geotech. Geoenvironmental Eng.*, pp. 1981–1985.)

the presence of metal cations in the pore fluid will result in a decrease in the pH because of the increase in hydrogen ion concentration in the pore fluid due to the following reaction:

$$M^{2+}(aq.) + nH_2O \leftrightarrow M(OH)_n^{2-n} + nH^+ \qquad (10.12)$$

The Darcy permeability coefficients k shown in Figure 10.7 vary in relation to (a) the level of contamination by Cu^{2+} ions, and (b) whether the buffering capacity of the smectite is enhanced by sodium carbonate ($Na_2(CO_3)$).

In regard to the chemistry of the transporting fluid, inorganic acids may solubilise clay minerals containing alumina and iron. In the case of organic chemicals as part of the transporting fluid, changes in interlayer spacing of mineral particles could occur. For organic chemicals with dielectric constants lower than that of water, contraction of individual particles results in producing thinner interlayer spacings and reorientation of particles and microstructural units. The evidence shows that there is a good correlation between the permeability of a soil permeated with an organic chemical and its dielectric constant. The greater the dielectric constant, the greater the value of the permeability coefficient. This is not surprising since the dielectric constant serves as a measure of a liquid's hydrophobicity or hydrophilicity.

Organic molecules permeating a soil will move by diffusion through the micropores and advection and diffusion through the macropores. As a general rule, as the molecular weight of an organic chemical increases, the permeability of the soil decreases, since more water will be displaced at the greater contact points of soil particles by the large organic molecules with the active surface of the soil particles. The greater the molecular weight, the higher the tendency of the organic chemical to be hydrophobic. In consequence, movement of such molecules through the aqueous channels in a soil will be slower.

Weakly sorbed molecules will move more quickly through the aqueous channels, whereas hydrophobic chemicals such as heptane, xylene, and aniline, which are highly partitioned onto soil particle surfaces, would develop soil–heptane, soil–xylene, and soil–aniline permeabilities that would be lower than those of soil–water permeability. The partitioning effect seen in relation to partitioning of organic chemicals in soils suggest that there may be a relationship between the coefficient of permeability of the soil with the particular chemical and the octanol–water partition coefficient k_{ow} (Figure 10.8). The greater the value of k_{ow}, the lesser the value of the permeability coefficient, meaning that the more hydrophilic the organic chemical, the more rapidly it will move through the soil (Yong et al., 1992).

10.3.3 Transported Contaminants and Soil Compression

Transport of contaminants into a soil can have consequences far beyond those encountered in permeability modification. The results of interactions

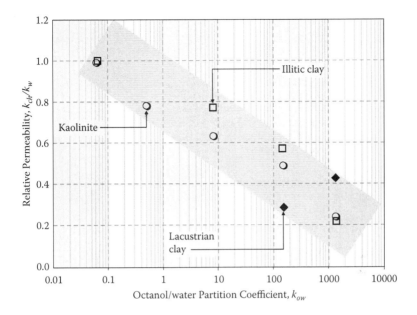

FIGURE 10.8
Variation of relative permeability ratio kch/kw with octanol/water partition coefficient k_{ow}. k_{ch} and k_w refer to permeability coefficient with respect to permeation with organic chemical and with respect to water, respectively (results reported by Yong et al., 1992).

and reactions due to introduction of transported solutes into a soil are shown in Figure 10.9 for a lacustrian clay undergoing creep because of an applied constant load. The leaching fluids used for the tests include sodium silicate ($0.025\ N\ Na_2SiO_3{\cdot}9H_2O$), sodium acetate ($1\ N\ CH_3COONa$), and distilled water. The results demonstrate the importance of recognizing the effect of hydraulic stressors, and interactions and reactions, initiated by chemical stressors. Depending on the type of solutes introduced as leachates into the sample, the mechanical resistance of the soil to applied loads will be more or less degraded, as shown by the increased axial strains. Simple leaching with distilled water also reduced the creep resistance strength of the soil (curve B in Figure 10.9). One deduces that natural solutes in the soil, which were responsible for providing some of the mechanical resistance of the soil through interactions with the soil particles, were replaced or removed by the leaching fluids, leading to disruption of the internal forces acting on and within the microstructural units and macrostructure.

The cations replaced or removed from the original porewater chemical composition for the tests conducted and reported in Figure 10.9 are shown in the right-hand side of Figure 10.10 for the case of the sample leached with sodium silicate ($0.025\ N\ Na_2SiO_3{\cdot}9H_2O$). The graph shows the percentage of cations Mg^{2+}, Ca^{2+}, K^+, and Na^+ removed or gained by the soil sample as a result of the leaching process under the creep load. The negative ordinate in

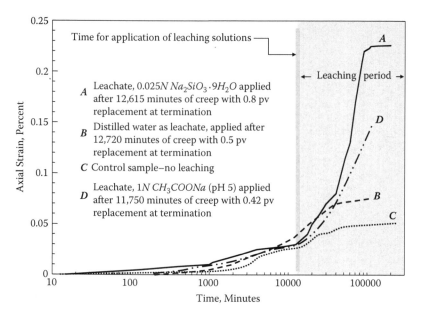

FIGURE 10.9

Effect of exposure of a natural lacustrian clay consisting of illite, chlorite, kaolinite, and other nonclay minerals, under a creep load (4.8 kPa) to various leachates after a certain period of elapsed time; pv refers to pore volume. (Adapted from data presented by Yong, R.N., et al., 1985, *J. Eng. Geol.,* 21:279–299.)

the graph means that ions were gained from the exchange process in leaching. In this case, Na^+ was added to the soil, whilst Mg^{2+}, Ca^{2+}, and K^+ were removed. Leaching with distilled water does not remove as large an amount of the various cations, as can be seen in the graph shown in the left-hand side of Figure 10.10.

10.4 Biodegradation and Biotransformation of Contaminants

In time-related processes and soil ageing, interactions between geochemical and microbial processes occur throughout a soil at any given time, whether or not external stressors are applied to the soil. The coupling of these process-based biogeochemical interactions provides the soil with biogeochemical tools for management of soil functionality. The primary agents responsible for biogeochemical reactions are the diverse range of microorganisms that inhabit the soil–water system such as the chemoorganotrophs, heterotrophs, and chemolithotrophs, to name a few.

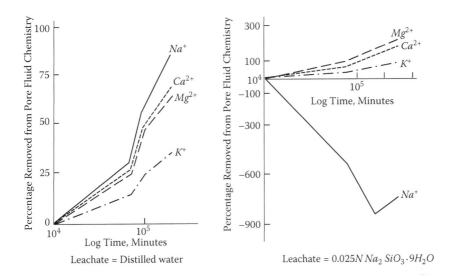

FIGURE 10.10
Percentage of cations removed (or added) to chemical composition of porewater due to introduction of leachate. Positive ordinate indicates percentage of cations removed, whilst negative ordinate indicates percentage of cations added. Left side results are for distilled water as the leachate, and right side results used sodium silicate (0.025N $Na_2SiO_3 \bullet 9H_2O$) as the leachate.

Microorganisms in a soil–water system modify the chemical and physical nature of their surrounding environment through their utilization of substrates and nutrients as energy sources, and production of biomass. Because bioavailable organic matter constitutes one of the main sources of energy for microorganisms (for example, chemoorganotrophs and heterotrophs), it is natural to expect that organic chemical contaminants that find their way into a soil could also serve as energy sources, meaning that microorganisms would have the capability for degradation of organic chemical compounds. Inorganic compounds can also serve as an energy source for microorganisms (e.g., chemolithotrophs). The biodegradative capability of microorganisms is a significant factor in the management of soil attributes and also in the determination of soil functionality.

Chemical contaminants serve as source material for metabolic activities by microorganism, particularly when they in a free form in a soil–water system. Production of microbial surfactants in an aquifer enhances desorption of hydrocarbons from the soils. Substances called *chelants*, which are produced by microorganisms, can bind insoluble metals which in turn can be solubilised and mobilized. Organic chemicals can penetrate the micropores of the microstructural units of a soil, and since these micropores are less than one micron in diameter, bacteria cannot easily grow and neither can their enzymes penetrate.

The process of degradation of substances that are normally biodegradable can be slowed or stopped by the presence of a toxic component (Section 2.8.4 in Chapter 2). The mechanisms involved include inhibition of a single enzyme in a metabolic pathway. This will adversely affect microbial growth or substrate utilization. An example of this is sulfonylurea, known to specifically inhibit acetolacetate synthase, which is required for the synthesis of amino acids that are the basis of proteins (LaRossa and Schloss, 1984). Binding of mercuric, cupric, or uranyl metal ions to the cell surface and breakage of the cell membrane can nonspecifically inhibit the functioning of the cell membrane, thereby causing difficulties in maintaining concentration gradients across the cell (Khovrychev et al., 1974). Inhibition of cell energy-generating processes occurs when the haem iron in cytochromes (hemoproteins), required for many electron transport reactions, are bound by contaminants such as cyanide and azide (Brock et al., 1984), which can also be toxic, depending on their concentrations.

Chemical compounds considered toxic to many kinds of microorganisms include organic acids, ethers, aldehydes, phenol, chlorophenol, cyanide, ammonia compounds, antibiotics, and dyes, depending on their concentration and time span of interaction (contact time). By and large, lower concentrations at short contact times are less toxic, as illustrated by 2,4-D which will degrade much faster at 1 to 100 ppm than at higher concentrations where it can remain undegraded for years. The biodegradability of a substance is dependent on the degree of saturation or branching, and the nature and amount of substitution. Water solubility is important since components enter the cell by water transportation. Increasing the saturation in a compound tends to decrease its solubility.

10.4.1 Biodegradation of Organic Contaminants

The composition of the various compounds that constitute petroleum, pesticides, industrial solvents, and so forth vary widely, ranging from varying degrees of branching, chain lengths, molecular sizes, and substitution with nitrogen, oxygen, or sulphur atoms. Accordingly, their susceptibility to biodegradation, involving a variety of microbial species, would also vary, from highly biodegradable to not at all (recalcitrant).

10.4.1.1 Alkanes and Cycloalkanes

Components of petroleum hydrocarbons include alkanes, cycloalkanes, aromatics, polycyclic aromatic hydrocarbons, asphaltenes, and resins. Of the many isomers for alkanes (C_nH_{2n+2}), low molecular weight alkanes are the most easily degraded by microorganisms, and as the chain length increases from C_{20} to C_{40}, hydrophobicity will increase; solubility and biodegradation rates will decrease. Conversion of alkanes leads to the formation of an alcohol, followed by oxidation to an aldehyde and then to a fatty acid (Pitter and

Chudoba, 1990). Further oxidation of the fatty acid, called β-oxidation, yields products less volatile than the original contaminants.

Because of their cyclic structure, cycloalkanes (C_nH_{2n}, $n > 2$) are not as degradable as alkanes, and with an increasing number of rings, their biodegradability will decrease, partly because of decreasing solubility (Pitter and Chudoba, 1990). Oxidation of the cycloalkanes with the oxidase enzyme leads to production of a cyclic alcohol followed by a ketone that forms to a lactone by insertion of an oxygen in the ring (Bartha, 1986).

10.4.1.2 Gasoline Components BTEX and Methyl Tertiary-Butyl Ether (MTBE)

Degradation of benzene, toluene, ethylbenzene, and xylene (*BTEX*), which are volatile water-soluble hazardous components of gasoline, is catalysed by oxygenases from fungi and other eukaryotes to form transdihydrodiols. Conversion of the dihydrodiol is by dehydrogenation to catechol, which is then either converted by an ortho pathway to produce muconic acid (splitting of the ring between carbons with the hydroxyl groups) or the meta pathway to produce 2-hydroxymuconic semialdehyde (splitting of the ring at the carbon with a hydroxyl group) (Cerniglia, 1984). Aerobic degradation of all components of *BTEX* occurs rapidly when oxygen is present at a concentration of 2 mg/L in water or 5% in gaseous vadose zone, and will continue until oxygen is depleted (Salanitro, 1993; Brown et al., 1995). Under anaerobic conditions, bacterial metabolism proceeds through a series of steps, depending on availability of electron acceptors. Degradation is less assured and is slower than under aerobic conditions. Denitrification occurs until nitrate or the carbon source is depleted, followed by $Mn(IV)$ reduction until manganese oxide concentrations are limiting. The next reaction is iron ($Fe(III)$) reduction followed by sulphate reduction until the carbon or sulphate becomes limited.

Methyl tert-butyl ether (*MTBE*), which is an additive to gasoline in concentrations as high as 15% by volume, is highly resistant to biodegradation since it is reactive with microbial membranes. Some believe that it is slowly biodegraded (Borden et al., 1997) whilst others believe that it partially degrades to tert-butyl alcohol (Landmeyer et al., 1998).

10.4.1.3 Halogenated Aliphatic and Aromatic Compounds

Pesticides such as ethylene dibromide (*DBR*) or $CHCl_3$, $CHCl_2Br$ and industrial solvents such as methylene chloride (CH_2Cl_2) and trichloroethylene (C_2HCl_3) are halogenated aliphatic compounds. The lower energy and higher oxidation state of the compounds, due to the presence of halogen, makes it more difficult to achieve aerobic degradation. Anaerobic biodegradation is a more viable alternative. Methylene chloride, chlorophenol (C_6H_5ClO), and chlorobenzoate (for example, $C_9H_9ClO_2$, $C_8H_7ClO_2$) are the most aerobically biodegradable. Because of the decreasing electronegativity, the ease of replacement of the halogens in decreasing order is as follows: iodine, bromine, chlorine, and fluorine. For short carbon chain lengths, removal of the halogen and replacement by a hydroxide group is often the first step of the degradation process, as demonstrated by methylene chloride (CH_2Cl_2) where formaldehyde (CH_2O), 2-chloroethanol ($HOCH_2CH_2Cl$), and 1,2-ethanediol ($C_2H_6O_2$) are intermediates and carbon dioxide is the final product (Pitter and Chudoba, 1990). For long chains, the methyl group at the end of the molecule is oxidized to a halogenated alcohol since the halogen has less influence on the reaction. Anaerobic reductive dehalogenation involves replacement of the halogen with hydrogen or formation of a double bond when two adjacent halogens are removed (dihalo-elimination), as demonstrated in Figure 9.15 (Chapter 9).

Halogenated aromatic compounds include pesticides such as DDT ($C_{14}H_9Cl_5$), 2,4-D ($C_8H_6Cl_2O_3$), and 2,4,5-T ($C_8H_5Cl_3O_3$), and plasticizers such as pentachlorophenol (C_6HCl_5O) and polychlorinated biphenyls ($C_{12}H_{10-n}Cl_n$, $n =$ 1–10). Although PCBs have been banned since the 1970s, they are still found in aqueous and sediment systems. Mechanisms of conversion include hydrolysis (replacement of halogen with hydroxyl group), reductive dehalogenation (replacement of halogen with hydrogen), and oxidation (introduction of oxygen into the ring causing removal of halogen). Reductive dehalogenation is more likely as the number of halogens increases.

10.4.2 Biotransformation of Inorganic Contaminants

Microbial cells accumulate heavy metals through ion exchange, precipitation, and complexation on and within the cell surface containing hydroxyl, carboxyl, and phosphate groups. Bacterial oxidation/reduction can alter the mobility of the metals, as demonstrated in the reduction of $Cr(VI)$ in chromate (CrO_4^{2-}) and dichromate ($Cr_2O_7^{2-}$) by bacteria to $Cr(III)$. Not only is $Cr(III)$ less toxic, but it is less mobile in environments where pH values are above its precipitation pH (that is, pH > 5) (Bader et al., 1996). Metabolism of mercury occurs through aerobic and anaerobic mechanisms through uptake, converting $Hg(II)$ to $Hg(0)$, methyl and dimethylmercury or to insoluble $Hg(II)$ sulphide precipitates that will be immobile if sufficient levels of sulphate and electron donors are available.

Forms of arsenic found in the environment include As_2S_3, elemental As, arsenate (AsO_4^{3-}), arsenite (AsO_2^-), and organic forms that include trimethyl arsine ($(CH_3)_3As$) and methylated arsenates such as monosodium methyl arsenate (CH_4AsNaO_3). Whilst all forms of arsenic are toxic, the anionic forms, which are mobile, are highly toxic. Mechanisms or processes involved in microbial transformation of arsenic include methylation, oxidation, and reduction under anaerobic or aerobic conditions. Aerobic microbial transformation produces energy through oxidation of arsenite for living microorganisms.

Radionuclides of interest include uranium and plutonium and neptunium, with the most common forms of uranium being the insoluble

$U(IV)O_2$ and soluble $U(VI)O_2^{2+}$ (Lovley, 1995). The reduction of plutonium by sulphate-reducing bacteria from the insoluble $Pu(IV)$ to the soluble $Pu(III)$ makes this contaminant more mobile. Neptunium can be immobilized by microorganisms by reduction to $Tc(IV)$ and $Tc(V)$ oxides from $Tc(VII)$.

10.4.2.1 Bacterial Metabolism of Nitrogen and Sulphur

Some significant portions of the nitrogen and sulphur cycles are directly related to the activities of microorganisms in the soil–water system. Although nitrogen N can exist in valence forms that range from a valence of +5 (represented by the nitrate ion NO_3^-) to a valence of –3 (represented by the ammonium cation, azanium, NH_4^+), it is the reduced form that is most relevant to the problem at hand. Ammonification of biomass, which relates directly to the microbial decomposition of the biomass, provides the most reduced forms of nitrogen, ammonia (NH_3), or azanium (NH_4^+). Oxidation of NH_4^+ to NO_3^- by chemolithotrophic prokaryotes, known otherwise as nitrifying bacteria, occurs under oxic conditions. This process, which is generally known as *nitrification*, is sensitive to redox and temperature. Redox values lower than 200 mV are considered to be inhibitory to nitrification.

Denitrification, which is dissimulative nitrate reduction—that is, reduction of nitrate (NO_3^-), to nitrite (NO_2^-), then nitric oxide (NO), then nitrous oxide (N_2O), and finally dinitrogen (N_2)—is a naturally occurring process that removes nitrogen from the soil and releases it into the atmosphere. It is a process that can only be performed by bacteria. Oxidized nitrogen compounds that result from dissimilatory nitrate reduction can serve as terminal electron acceptors by other microorganisms under anoxic conditions. In assimilative nitrate reduction, performed by bacteria, fungi, and plants, ammonia and then amino acids and proteins are produced for growth. Assimilative nitrate reduction can occur in the presence of oxygen whereas dissimulative nitrate reduction cannot.

Sulphur S can exist also in valence forms that range from a valence of +6 (represented by sulphate, SO_4^{2-}) to a valence of –2 (represented by S^{2-}). Amino acids such as sulphydryl thiol groups ($R-SH$), which are sulphur analogues of alcohol, can be desulphydrated by both eukaryotes and prokaryotes, releasing S^{2-}. Abiotic and biotic oxidation of the S^{2-} result in the production of sulphate (SO_4^{2-}), the reduction of which provides for formation of iron sulphides and pyrite (FeS_2). The sulphides have a tendency to form on metal surfaces, thereby creating problems for electrodes used in electrokinetics. Acid production and transport due to the oxidation of FeS_2 to H_2SO_4 constitute one of the major problems of metal mining industries.

Sulphur bacteria are facultative autotrophs that use sulphur, hydrogen sulphide, thiosulphate, and organic sulphides or organic materials for energy. Typical reactions are

$$H_2S + 2O_2 \rightarrow SO_4^{2-} + 2H^+$$

$$S^0 + H_2O + \frac{3}{2}O_2 \rightarrow SO_4^{2-} + 2H^+ \tag{10.13}$$

$$S_2O_3^{2-} + H_2O + 2O_2 \rightarrow 2SO_4^{2-} + 2H^+$$

The conversion of hydrogen sulphide to sulphate is a multistep process in which sulphur is produced. Microorganisms using the very insoluble sulphur must grow on the sulphur solids, slowly oxidizing the gradually soluble sulphur. In all of these reactions, sulphuric acid is produced, lowering the pH to 1 in high concentrations. Sulphate-reducing bacteria (*SRB*) oxidize organic compounds with sulphate ions and form metal (*Me*) sulphides that are insoluble, as shown by the following reactions:

$$CH_3COOH + SO_4^{2-} \rightarrow 2HCO_3 + HS^- + H^+$$

$$H_2S + Me^{2+} \rightarrow MeS + 2H^+ \tag{10.14}$$

10.4.3 Natural Attenuation of Contaminants

The term *natural attenuation* refers to natural processes occurring in the soil that reduce the toxicity and/or the concentration of the contaminants in the soil. This term (*natural attenuation, NA*) will always be italicized when used in the context of processes. The common definition of natural attenuation describes it as a process that "involves the biodegradation, dispersion, dilution, sorption, volatilization of contaminants, together with chemical and biochemical reactions and transformations of the contaminants to reduce contaminant toxicity, volume, mass, and concentrations to levels considered as nonthreatening to biotic receptors and the environment."

The primary mechanisms involved in the natural attenuation of contaminants include:

- *Organic chemical compounds:* volatilization, biodegradation, biotransformation, abiotic transformation, and sorption
- *Inorganic contaminants:* biotransformation, abiotic transformation, sorption, precipitation, dilution, and dispersion
- *Radionuclides:* dilution, dispersion, and radioactive decay

10.4.3.1 Monitored Natural Attenuation

Monitored natural attenuation (*MNA*) has been defined in USEPA-SAB (2001) to mean "a remediation approach based on understanding and quantitatively documenting naturally occurring processes at a contaminated site that protect humans and ecological receptors from unacceptable risks of exposure to hazardous contaminants." The report further indicates that this is "a knowledge-based remedy because instead of imposing active controls, as in engineered remedies, scientific and engineering knowledge is used to understand and document naturally occurring processes to clearly establish a causal link." Monitored natural attenuation (*MNA*), because of its adherence to "remedy by natural processes," necessitates a proper understanding of the many principles involved in the natural processes of contaminant attenuation by soils that contribute to the end result.

Monitored natural attenuation (*MNA*), as a remediation tool, is very dependent on information obtained, that is, it is a knowledge-based method of management of the transport and fate of contaminants in the soil. Monitoring of various strategic positions in a contaminated site is a key element of the use of *MNA*. It (*MNA*) is a site-specific tool that utilizes the hydrogeologic setting and accumulative properties of the soil in the site to attenuate contaminant concentrations and toxicity. Much work is required to provide the necessary assessments that will allow this type of passive remediation tool to be applied (Figure 10.11).

The data and information inputs shown on the left-hand side of Figure 10.11 tell us what is required to satisfy site-specific conditions, and

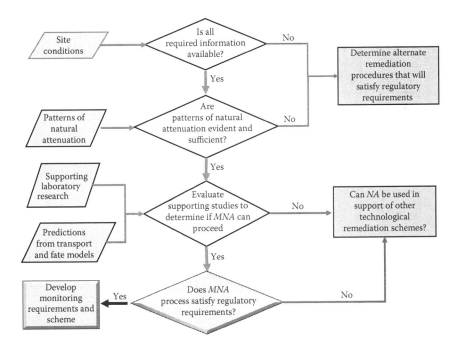

FIGURE 10.11
General protocol that might be used when considering monitored natural attenuation (*MNA*) as a remediation tool. *NA* refers to natural attenuation.

whether the indicators for evidence of natural bioremediation are sufficient to proceed with further assessment to determine whether *MNA* is a viable treatment tool. A "no" response from the first two decision steps will automatically trigger a technological and/or engineered solution to the remediation problem. The laboratory research input in the third step refers to the microcosm studies and the various laboratory procedures to determine partitioning. These, together with transport and fate predictions from the models developed for the specific situation, constitute the "supporting studies" shown in the third decision box. Analysis of the results from the supporting studies (laboratory research and predictions) should inform one about the ability of the site materials and conditions to attenuate the pollutants. A "no" response from the third decision step will allow one to incorporate natural attenuation processes as part of a technological remediation solution.

Monitored natural attenuation (*MNA*) as a remediation process has been called, at times, passive remediation and/or intrinsic remediation. The implicit requirement for application of the natural processes in the soil to remediate contamination by natural attenuation is a strict monitoring programme that would supply essential information according to the criteria set forward in the *lines of evidence*. Detailed protocols and guidelines issued by the regulatory or oversight agencies can be site-specific or generic, and are designed to cover the essential requirements, procedures, standards, criteria, and so forth that need to be used in determining the *lines of evidence* and the effectiveness of the *MNA* process. Figure 10.12 shows the suggested protocols for establishing evidence of success in application of *MNA* as a remediation tool. *Lines of evidence* consists of several parts: (a) corroborating laboratory and bench-type treatability tests, (b) predictions from transport/fate models, (c) information on the hydrogeology of the setting, (d) performance of contaminants, and (e) composition and properties of contaminants at various locations and time.

10.4.3.2 *Attenuation of Organic Chemical Compounds*

Organic chemical compounds, generally classified as xenobiotic compounds, in the land environment have origins in various chemical industrial processes and as commercial substances for use in various forms. These include chemical products such as organic solvents, paints, pesticides, oils, gasoline, creosotes, and greases. The more common organic chemicals found in contaminated sites can be grouped into three convenient groups as follows:

- Hydrocarbons—including the PHCs (petroleum hydrocarbons), the various alkanes, alkenes, and aromatic hydrocarbons, MAHs (multicyclic aromatic hydrocarbons), and PAHs (polycyclic aromatic hydrocarbons);

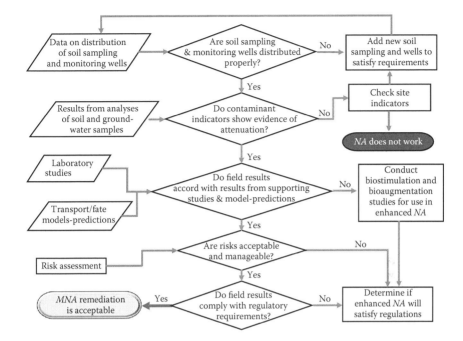

FIGURE 10.12
Protocols for establishing *evidence of success* for monitored natural attenuation (*MNA*) using monitoring information and supporting laboratory research and predictions from fate and transport models. *NA* refers to natural attenuation.

- Organohalide compounds—including the chlorinated hydrocarbons, TCE (trichloroethylene), carbon tetrachloride, vinyl chloride, hexachlorobutadiene, PCBs (polychlorinated biphenyls), and PBBs (polybrominated biphenyls);
- Miscellaneous compounds—including oxygen-containing organic compounds such as phenol and methanol, and nitrogen-containing organic compounds (for example, TNT, trinitrotoluene).

The chemical properties of the functional groups of the soil fractions that contribute appreciably to the acidity of the soil particles are significant properties, since surface acidity is very important in the adsorption of ionizable organic molecules. Surface acidity plays a major role in clay adsorption of amines, s-triazines, amides, and substituted urea due to protonation on the carbonyl group. Examples of this are the hydroxyl groups in organic chemical compounds consisting of alcohols (ethyl, methyl, isopropyl, etc.), phenols (monohydric and polyhydric), and the two types of compound functional groups, that is, those having a *C-O* bond (carboxyl, carbonyl, methoxyl, etc.) and the nitrogen-bonding group (amine and nitrile). Amine, alcohol, and other organic chemicals possessing dominant carbonyl groups that are positively charged by protonation can be readily sorbed by clays. The NH_2 functional group of amines can protonate in soil, resulting in replacement of the inorganic cations from the clay complex by ion exchange. The extent of sorption of these kinds of organic molecules depends on (a) the CEC of the clay minerals, (b) the composition of the clay soil (soil organics and amorphous materials present in the soil), (c) the amount of reactive surfaces, and (d) the molecular weight of the organic cations.

Large organic cations are adsorbed more strongly than inorganic cations due to their length and higher molecular weights. Polymeric hydroxyl cations are adsorbed in preference to monomeric species because of the lower hydration energies, higher positive charges, and stronger interactive electrostatic forces. Since carbonyl compounds have dipole moments due to the unsymmetrically shared electrons in the double bond, they can sorb onto clay minerals by hydrogen bonding between the OH group of the adsorbent and the carbonyl group of the ketone or through a water bridge. For the carbonyl group of organic acids such as benzoic and acetic acids, sorption onto clays occurs directly with the interlayer of cation or by formation of hydrogen bonds with the water molecules (water bridging) coordinated to the exchangeable cation of the clay complex.

The outcome of the application of MNA as a remediation process is the evidence of occurrence of biodegradation and transformation of the target organic chemicals. For abiotic and biotic transformation processes, the indicators used to satisfy *lines of evidence* requirements involve determination of both the decrease in concentration of the organic chemical compounds and their transformations. As has been noted in previous discussions, the major difference between the transformation products from abiotic and biotic processes is that abiotic transformation products are generally other kinds of organic chemical compounds, whereas transformation products resulting from biotic processes are mostly seen as stages (intermediate products) towards mineralization of organic chemical compounds.

10.4.3.3 Attenuation of Heavy Metals

For sites contaminated by heavy metals (HMs), environmental mobility of these metals is dependent upon whether they are in the pore water as free and complexed ions or sorbed onto the soil particles. Mobility of free ions and complexed ions in the pore water will be governed by advection and diffusion mechanisms, and so long as the full assimilative potential of the soil for HMs is not reached, attenuation of the HMs will continue. Metals that are sorbed onto the soil particles are held by different sets of forces, determined to a large extent by the soil fractions and the pH of the soil–water system. Although precipitation of HMs is not strictly speaking a sorption phenomenon, precipitation of HMs as hydroxides, sulphides, and carbonates generally is classified as part of the assimilative mechanism of soils because the precipitates form distinct solid material species.

They are classified under the category of *partitioning,* and thus are most often considered as part of the attenuation process. Either as attached to soil particles or as void pluggers, precipitates of *HMs* can contribute significantly to attenuation of *HMs* in contaminant plumes. Hydroxide precipitation is favoured in alkaline conditions as, for example, when $Ca(OH)_2$ is in the groundwater in abundance. With available sulphur and in reducing conditions, sulphide precipitates can be obtained. Sulphide precipitates can also be obtained as a result of microbial activity, except that this will not be a direct route. Sulphate reduction by anaerobic bacteria will produce H_2S and HCO_3^-, thus producing the conditions for formation of metal sulphides.

10.4.4 Enhanced and Engineered Natural Attenuation of Contaminants

Enhanced natural attenuation (ENA) refers to the situation where, for example, (a) nutrient packages are added to the soil system to permit enhanced biodegradation to occur, (b) catalysts are added to the soil to permit chemical reactions to occur more effectively, and (c) techniques such as biostimulation and/or bioaugmentation are used. Engineered aids such as barriers and others can be used to help in attenuating contaminants as they are transported in the soil. Procedures adopted with engineering aids lead to attenuation that is called *engineered natural attenuation (EngNA)*.

10.4.4.1 Biostimulation

The simplest procedure for improving the intrinsic bioremediation capability of a soil is to provide a stimulus to the microorganisms that is already in the soil. This procedure, which is called *biostimulation*—adding nutrients and other growth substrates, together with electron donors and acceptors—seeks to promote increased microbial activity with the set of stimuli to better degrade the organic chemical pollutants in the soil. The addition of nitrates and other stimuli such as *Fe (III)* oxides, *Mn (IV)* oxides, sulphates, and CO_2 allow anaerobic degradation to proceed. This technique, which is best used for sites contaminated with organic chemical compounds as contaminants, is perhaps one of the least intrusive of the methods for enhancement of natural attenuation.

10.4.4.2 Bioaugmentation

When the native or indigenous microorganisms are not capable of degrading the organic chemicals in the soil because of contaminant concentrations, inappropriate consortia, and so forth, other microorganisms (called exogenous microorganisms) can be introduced into the soil. The purpose of these exogenous microorganisms is to augment the indigenous microbial population such that effective degradative capability can be obtained. Biostimulation, as a technique, can also be used in conjunction with bioaugmentation to further increase the likelihood of effective degradative capability. The risks involved relate to unknown results obtained from interactions between the genetically engineered microorganisms and the various chemicals in the contaminated ground. It is possible that the interaction of chemicals with microorganisms may result in mutations in the microorganisms themselves, and/or microbial adaptations. There is also a possibility that the use of microorganisms grown in uncharacterized consortia that include bacteria, fungi, and viruses can produce toxic metabolites (Strauss, 1991).

10.4.4.3 Geochemical and Biogeochemical Intervention

Geochemical and/or biogeochemical interventions such as pH and pE or Eh manipulation are well suited for treatment of inorganic contaminants in the soil. Changes in toxicity for inorganic contaminants such as *Cr* and *As* can be obtained by changes in their oxidation state. *Cr(III)*, which is an essential nutrient that helps the body use sugar, protein, and fat, can be oxidized to *Cr(VI)* by dissolved oxygen and manganese dioxides. However, since this is an undesirable situation, geochemical and/or biogeochemical means can be used to create a reducing environment through such means as depletion of oxygen, thereby disallowing oxidation of *Cr(III)* in the subsurface. The opposite intervention technique is necessary for the case of *As(III)*, which is more toxic than *As(V)*. In this instance, the intervention technique should provide an oxygen source to ensure oxidation of the *As(III)* in the ground.

pH manipulation, as an intervention technique, can precipitate of heavy metals in the porewater or dissolve precipitated heavy metals. Changes in pH will also result in changes in the sign of surface electrostatic charges for those soil fractions with amphoteric surfaces, that is, surfaces that have pH-dependent charges. When this happens, bonding of metals or release of heavy metals from disruption of bond disruptions will occur. Both pH and Eh changes will have considerable effect on the various acid-base reactions and on abiotic and biotic electron transfer mechanisms. Transformations and degradation of organic chemical pollutants resulting from acid-base and oxidation-reduction reactions are less significant than those obtained in relation to biotic processes. Reaction kinetics for such processes, and those initiated by the catalytic action of soils resulting in abiotic transformation, are considered to be relatively slow.

10.4.4.4 Permeable Reactive Barriers (PRBs)

The intent of a permeable reactive barrier (*PRB*), also known as *treatment walls*, is to intercept a contaminant plume in transport in the subsoil (see Figure 10.13). The choice of soil material in the barrier, its thickness, and size are all dependent on the size, position, and nature of contaminants (in the contaminant plume) to be intercepted. The intent is to have the contaminant plume pass through the reactive barrier such that contaminant attenuation

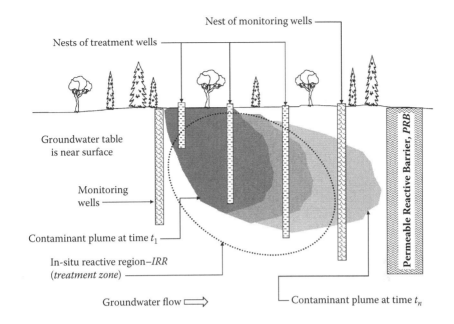

FIGURE 10.13
Enhancement of natural attenuation using treatment wells. Treatments for enhancement can be any or all of the following: geochemical intervention, biostimulation, and bioaugmentation. Treatment occurs in the pollutant plume and downgradient from the plume.

by the soil material in the *PRB* can occur, ending with a "clean" plume that exits the *PRB*. The *PRB* needs to be strategically located downgradient to intercept the contaminant plume, and if needed, the pollutant plume can be channelled to flow through the *PRB* using several kinds of channelling techniques. A *funnel-gate* technique is one of the more common ones used to channel the pollutant plume to flow through the reactive barrier. This funnel, which is constructed with impermeable confining boundaries, is placed in the contaminated ground in such a way as to funnel or direct the contaminant plume to the *PRB*.

The range of soil materials in the *PRBs* can include a variety of oxidants and reductants, chelating agents, catalysts, microorganisms, zero-valent metals, zeolite, reactive clays, ferrous hydroxides, carbonates and sulphates, ferric oxides and oxyhydroxides, activated carbon and alumina, nutrients, phosphates, and soil organic materials. The choice of any of these treatment materials is made on the basis of site-specific knowledge of the interaction processes between the target contaminants and material in the *PRB*. Laboratory tests and treatability studies are essential elements in the design of PRBs. When designed properly, a PRB provides the capability for assimilation of the contaminants in the contaminant plume as it moves through the barrier. The in-situ *reactive region* (IRR) shown in Figure 10.13 is a treatment zone, whose purpose is to provide not only pretreatment or preconditioning in support of another treatment procedure, but also as a posttreatment process for sites previously remediated by other technological procedures.

10.5 Freeze–Thaw Seasonal Impact

In regions where winter temperatures fall significantly below the freezing point of water and soil freezing occurs, several issues relating to soil properties and behaviour—that is, time-related changes or soil evolution—need consideration. These are (a) freezing phenomena in coarse- and fine-grained soils, (b) frost heaving of soils, (c) thawing of soils with ice lenses, and (d) phenomenon and relevance of unfrozen water content in frozen soils. The nature of a geothermal profile—variation of soil temperature with depth—at a particular location and time is a function of the net transfer of heat between the ground surface and air, and the thermal properties of the subsurface soil. The heat transfer mechanisms include radiation, convection, and conduction. The main factors and variables involved in radiation heat transfer mechanisms include (a) location and view factor, (b) albedo, which governs resultant absorbtivity to radiation, (c) the greenhouse effect, and (d) cloud cover and atmospheric vapour pressure—factors that determine equivalent sunshine and solar absorption and scattering. For convection and conductive heat transfer mechanisms, the following factors and variables are important: (a) ground surface features and inclination (vegetative cover, snow cover, surface roughness, trees, and other features that affect the characteristics of wind and/or air flow at the surface), (b) prevailing wind velocity, and evaporation and condensation.

10.5.1 Freezing and Frost Penetration

10.5.1.1 Freezing Index

The temperature at the ground surface at any time can be graphically depicted to show the intensity of temperature in terms of duration of existing temperature (for example, for one year), as demonstrated with the idealized typical ambient air temperature variation with time shown in Figure 10.14. The freezing temperature intensity is the shaded area contained by the curve below freezing, and is defined as the *freezing index F*. If T_f represents the mean surface temperature during the freezing period t, then F is $T_f \cdot t$. This refers specifically to the number of degree-days below freezing for the particular location where the temperatures were recorded. When coupled with the magnitude of the surface freezing temperature, F is seen to contribute directly to the penetration of the freezing front—also known as the frost front—in the soil at that location.

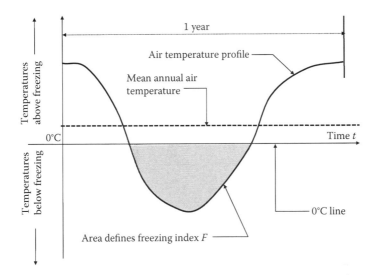

FIGURE 10.14
Idealized annual air temperature profile. Shaded area under the 0°C line is the freezing index *F*.

10.5.1.2 Geothermal Profiles

The idealized geothermal profiles for winter and summer are shown in Figure 10.15. The active layer shown in the figure is the layer in the subsurface that freezes in winter and thaws in summer. Its thickness depends on several factors, but chiefly on the penetration of the frost front into the subsurface. In the left-hand sketch, the summer temperatures in the soil do not fall below the freezing point of water. However, in the right-hand sketch, the geothermal profile shows a layer of soil where the temperatures remain lower than the freezing point of water, meaning that this layer is permanently frozen. This layer is generally identified as the permafrost layer. The thickness of this layer depends on several factors and conditions that include (a) ground surface cover, (b) soil properties and available water, and (c) duration, intensity, and penetration of the frost front (freezing index). Since about one-fifth of the land area of the world is underlain by permafrost, the behaviour of soils in the environment that produces the conditions favourable for ground freezing and permafrost requires attention, particularly in respect to (a) mechanics and effects of frost penetration, and (b) thawing of frozen soils.

In permafrost regions, the thicknesses of both the active layer and the permafrost layer are established and controlled by surficial environmental factors, temperature regimes, and substrate conditions. There exists a very delicate dynamic thermal equilibrium between these layers, and in particular, with the processes that are responsible for the development and maintenance of the layers. This equilibrium is more stable at lower temperatures, but becomes more unstable and precarious at higher temperatures—especially

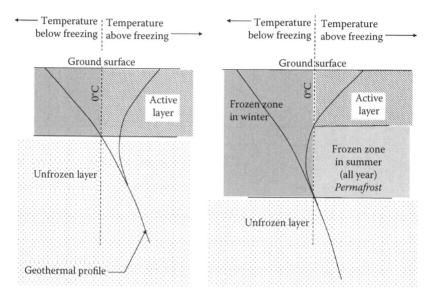

FIGURE 10.15
Typical geothermal profiles for winter and summer. The active layer in both diagrams is the layer that freezes in winter and thaws in summer. The frozen zone in the summer, shown in the right-hand drawing, is generally defined as the permafrost (permanently frozen) layer.

at the southern limit of the permafrost belt. Changes in the surficial cover (at the ground surface) can have a very significant effect on the thickness of a previously established permafrost layer—to an extent that might, with time, eliminate (destroy) the permafrost layer.

The idealized geothermal profiles shown in Figure 10.16 illustrate the changes in the profiles as one reduces the effectiveness of the thermal blanket on the ground surface. Because of the delicate balance of energy established between the above and below ground regimes, any change of the interface properties between the two regimes would contribute to the establishment of thermal erosion effects. Decreases in the effectiveness of the surface cover—for example, removal of surficial vegetative cover and topsoil—would increase penetration of the active layer. Bare ground surfaces provide conditions for optimal penetration of the active layer. The increased penetration of the active layer and thawing of the permafrost can lead to (a) development of thermokarsts and other similar features, (b) slope instability and failure, and (c) loss of ground support for overlying structures and facilities.

10.5.1.3 Frost Penetration

Frost penetration in the active layer is a problem of note for soils that have a tendency to heave in winter because of the formation of ice lenses in the soil that literally push up the soil, thus causing ground heave. Advance of the

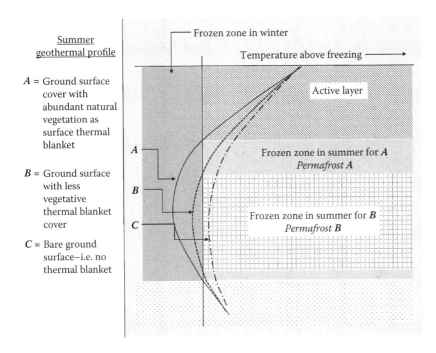

Summer
geothermal profile

A = Ground surface
cover with
abundant natural
vegetation as
surface thermal
blanket

B = Ground surface
with less
vegetative
thermal blanket
cover

C = Bare ground
surface–i.e. no
thermal blanket

Frozen zone in winter

Temperature above freezing ⟶

Active layer

Frozen zone in summer for *A*
Permafrost A

Frozen zone in summer for *B*
Permafrost B

FIGURE 10.16
Schematic illustration of penetration of the active layer as a result of changes in the effectiveness of the surficial thermal blanket to the point where the permafrost layer is totally eliminated.

frost front into the subsoil is a function of such factors as (a) freezing index and associated temperature and climatic factors, (b) ground surface cover and topography, (c) soil type and properties, (d) chemistry of porewater in subsoil and groundwater, and (e) thermal properties of the soil–water system—specifically, the heat of soil solids, volumetric heat of the system, latent heat of porewater, and thermal conductivity of the soil.

The depth of frost penetration can be determined using an idealized geothermal profile such as that shown in Figure 10.17. Whilst this highly simplistic representation of the geothermal profile will lead to an overestimation of the depth of frost penetration, it is however thought to be useful for scoping calculations. Proceeding with this simple view of the temperature profile, we note that at point Θ, the continuity condition that needs to be satisfied states that the latent heat released as the porewater freezes to a depth Δx in time Δt must be equal to the rate of heat removed and conducted to the ground surface. Thus, for small values of Δx and Δt, we have

$$k_f \frac{T_f}{x} = L \frac{dx}{dt} \tag{10.15}$$

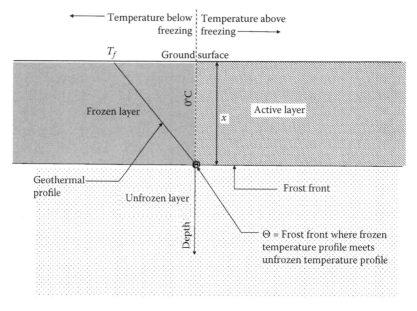

FIGURE 10.17
Idealized geothermal profile used for calculation of frost penetration (freezing front) depth.

where T_f is the mean surface temperature below the freezing point, k_f is the thermal conductivity of the soil with temperatures below the freezing point, L is the latent heat of the porewater, x is the depth of front penetration, and t is time. The solution to Equation (10.15) should give us

$$x = \sqrt{\frac{2k_f \int T_f dt}{L}} , \tag{10.16}$$

where $\int T_f dt$, which is reported in degree-hours, is the <u>freezing</u> index F in degree-days. Making the necessary substitutions: $x = \sqrt{48k_f F/L}$.

For a more realistic determination of the depth of frost penetration, we can use the diffusion equation to characterize the geothermal profile as follows:

$$\frac{\partial T_f}{\partial t} = a_f \frac{\partial^2 T_f}{\partial x^2} \qquad \textit{for frozen layer}$$

$$\frac{\partial T_u}{\partial t} = a_u \frac{\partial^2 T_u}{\partial x^2} \qquad \textit{for unfrozen layer} \tag{10.17}$$

where x is the spatial coordinate, T is the temperature, a is the diffusivity coefficient, and the associated subscripts f and u represent *frozen* and *unfrozen*, respectively. The thermal diffusivity coefficient a is defined as the ratio of the thermal conductivity k of the soil to its volumetric heat capacity C, that is, $a = k/C$ (see Section 7.2.3 in Chapter 7).

The continuity condition at the point Θ requires that the net rate of heat flow from the frost (freezing) front must be equal to the latent heat supplied by the subsurface soil water as it freezes to a depth Δx and Δt. For infinitesimal values of Δx and Δt, we will have

$$L\frac{dx}{dt} = \Delta q \tag{10.18}$$

where Δq is the net rate of heat flow at the freezing front. It follows that

$$k_f\frac{\partial T_f}{\partial x} - k_u\frac{\partial T_u}{\partial x} = L\frac{dx}{dt} \tag{10.19}$$

As with Equation (10.16) the solution for x in Equation (10.19), the depth of frost penetration, will be obtained as $x = \lambda\sqrt{48k_f F/L}$, with a correction coefficient λ that is dependent on the thermal properties of both the frozen and unfrozen soils.

10.5.1.4 Frost Heave and Heaving Pressures

Frost heaving and formation of ice lenses occur in soils if they are frost susceptible and if water for ice lens growth is available. The Casagrande classification of frost susceptible soils considers (a) well-graded soils where 3% or more of the soil particles are 0.02 mm in particle size, and (b) uniformly graded soils where 10% or more of the soil particles are 0.02 mm in particle size, to be highly susceptible to frost heave and ice lens formation, provided that water is available for lens formation and that the rate of freezing must be relatively slow to allow for growth of ice lenses. This means that the rate of freezing must be consistent with the rate of water movement into a growing ice crystal.

In ice lens growth, one begins with ice nucleation in the "free" water in the pore spaces of a frost-susceptible soil—that is, water in pore spaces where the matric potential ψ_m is relatively high, in comparison to water within the Stern layer (adjacent to soil particle surfaces), which has a low matric potential ψ_m. It is recalled that since the quantitative value of the matric potential ψ_m is negative, a low matric potential ψ_m signifies a high soil suction, whereas a high matric potential ψ_m signifies a low soil suction. After ice nucleation and subsequent crystallization in the pore spaces, growth of the ice will continue so long as heat loss exceeds heat gain (from water attracted to the bud of crystallization). For this to develop into a mature ice lens, a delicate balance between heat loss and heat gain must be obtained. If heat gain from water attracted to the bud is too great in comparison to heat loss from the bud, melting would occur. If heat gain is equal to heat loss, equilibrium is obtained and no growth would result. As the ice in the pore spaces grows and water from the region immediately adjacent to the ice crystals is being depleted, more water will need to be drawn in from regions further from the growing crystals. The temperature must continue to decrease at these locations—meaning that the freezing front must penetrate further into the ground. If the ice crystals in the pore spaces cannot grow into and through the pore channels connecting contiguous pore spaces, there will be a tendency for ice segregation to occur. If this continues, this will begin the process of ice lens growth. Too rapid a freezing rate will limit the growth of the ice lens, whereas too slow a freezing rate will only create a frozen soil layer, that is, frozen pore water in the void spaces.

The extent of local consolidation or compression of the unfrozen soil that occurs in front of the freezing front is a function of (a) water available for ice lens growth, and (b) degree of heave experienced in the layer above the ice lens. With the right freezing conditions and water availability, and with compatible frost susceptible soils, ice lenses will form in different portions and different positions in the soil as the freezing front penetrates into the soil. The continuity condition that must be satisfied at the freezing front says that the rate of change in thermal energy must be balanced by the spatial rate of change of the heat flux. In this case, the heat flux includes the heat brought in as mass transfer. In all instances, there is an infinitesimal layer of unfrozen water separating the ice lens from the particles and microstructural units. This water layer is the hydrate layer (or layers, depending upon the activity of the soil particle surfaces) and is often called the adsorbed water layer. Numerous studies have reported on this unfrozen water phenomenon in freezing soils.

The frost heaving pressures created at the ice lens interface with the overlying soil layer can be theoretically calculated. This requires one to determine the frost heaving pressures as a function of the equivalent radii of representative pore (void) and particle or microstructural unit, beginning with the Maxwell thermodynamic relations written for a curved ice–water interface in a pore (Figure 10.18). For small incremental changes:

$$\Delta T = -\frac{2\sigma_{iw}T}{\rho_i L r} \tag{10.20}$$

where ΔT is the freezing point change at the curved surface, σ_{iw} is the ice–water interfacial energy, T is the temperature, L is the latent heat of fusion, ρ_i is the density of ice, and r is the radius of the curved surface. Since the free energy of the ice is equal to that of the water at equilibrium—for the geometric configuration shown in Figure 10.18—the reversed curvature of the ice front between the pore space and microstructural units (*msu*) must be accounted for. In addition, since the minimum potential energy configuration for the ice front must be a flat plane, both the ice radii in the pore and around the microstructural unit need to be considered. Assuming a cubicle packing of microstructural units:

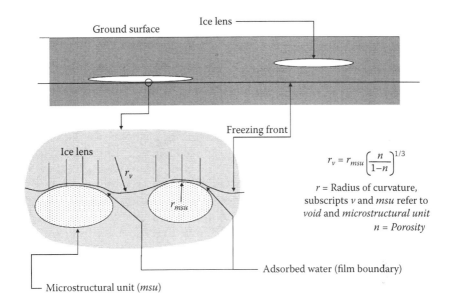

Ice lens

Ground surface

$$r_v = r_{msu}\left(\frac{n}{1-n}\right)^{1/3}$$

Freezing front

r = Radius of curvature, subscripts v and msu refer to *void* and *microstructural unit*

n = Porosity

Ice lens

r_v

r_{msu}

Adsorbed water (film boundary)

Microstructural unit (*msu*)

FIGURE 10.18
Schematic illustration of the local regime at the freezing front showing the curvatures at front surface of an ice lens.

$$r_v = r_{msu}\left(\frac{n}{1-n}\right)^{\frac{1}{3}}$$

$$\Delta T = -\frac{2\sigma_{iw}T}{\rho_i r_v L} \qquad \text{in the pore} \tag{10.21}$$

$$\Delta T = \frac{2\sigma_{iw}T}{\rho_i r_{msu} L} \qquad \text{for the microstructural units}$$

where r_v signifies the radius of curvature of the ice in the pore, and r_{msu} identifies the radius of curvature around the microstructural unit (see Figure 10.18).

In the region that encompasses a microstructural unit and a pore space, the net ΔT will be given as

$$\Delta T = -\frac{2\sigma_{iw}T}{\rho_i L}\left[\frac{1}{r_{msu}} + \frac{1}{r_v}\right] \tag{10.22}$$

The pressure difference generated between the ice and soil particles or microstructural units, that is, frost heaving pressure, ΔP, can be expressed by the Clapeyron relation as

$$\Delta P = \frac{L\Delta T}{T(V_w - V_i)} \qquad (10.23)$$

where V_w and V_i refer to the volume of water and ice, respectively.

Representative radii of curvature for the voids and for the microstructural units can be obtained from a study of electron micrographs. Whilst it would be normal to see a distribution of sizes of microstructural units and macro-voids (voids between microstructural units), it should not be difficult to make a judicious choice of a representative microstructural unit. Experimental frost heaving tests reported by Yong (1967) on fine inorganic silts show that the predictions accord well with measured values of heaving pressures. It is noted that if volume change under frost heaving pressures is allowed, stress relief occurs, resulting thereby in a lower measured heaving pressure. This means that the theoretically calculated values, which are based on a no-volume change condition, are the conservative maximum values.

10.5.1.5 Cyclical Effect of Freeze–Thaw

Cyclical freezing and thawing of clays with high water contents can result in significant changes in the microstructure of the clays. This is because of the various forces associated with movement of water to ice nuclei and resultant growth of pore ice—creating volumetric expansion forces. Figure 10.19 shows the change in microstructural units from the first freeze–thaw period to the 32nd freeze–thaw cycle of a natural, high water content clay from northern Québec, Canada. Note the scale for all three pictures is the same, as shown by the 1 nm scale-bar (black bar) near the centre-bottom for the pictures. Restructuring of the microstructural units occurs in response to the forces exerted by the growing ice crystals during the freezing period. The result of restructuring is more stable microstructural units as the number of freeze–thaw cycles is increased, for both open and closed systems, closed or open to the ingress of water. The microstructural units obtained in the open system are less compact than those obtained in the closed system—that is, the micro-porosities of the open system microstructural units are larger than those units resulting from the closed system. These impact directly on the production of larger macropores, resulting in greater hydraulic transmission characteristics.

Observations reported by Yong et al. (1984, 1986) for studies on cyclic freeze–thaw effects on soil microstructure and properties of clays include the following:

- The formation of numerous ice lenses complicates the prediction of depth of frost penetration and frost heaving pressures.
- Repetitive freezing and thawing causes significant variation in the thermal properties of the clays. The production of greater macro-pores as a result of the cyclical freeze–thaw reorganization of the microstructures seems to be one of the contributing factors.

Mineralogy in order of abundance.

Quartz
Illite/mica
Feldspar
Kaolinite
Chlorite
Amphibole

1 cycle - Open

32 cycles - Open

32 cycles - Closed

FIGURE 10.19
Scanning electron microscope (SEM) pictures showing change in size and shape of microstructural units in high water content clay (92% natural water content) following one cycle and 32 cycles of freeze–thaw in open and closed systems. The open system allows water to be drawn into the clay or to escape from the clay as it freezes and thaws. The closed system does not allow any water to enter or escape from the clay during freezing and thawing. Note the scale of both pictures is the same, as shown by the scale-bar at the bottom-right corner for all three pictures.

- The thermal properties of the clays depend substantially on the boundary conditions imposed, particularly on the amount of constraint applied against frost heaving.
- Loss of support and potential instability of slopes are likely results of restructuring of the microstructural units.

10.5.2 Unfrozen Water in Frozen Soils

The water layers adjacent to surfaces of soil particles do not necessarily freeze at the freezing point of bulk water because of the bonding mechanisms established between water and the soil particles, and also because of the presence of dissolved solutes in the Stern layer. The surface forces and the various mechanisms that define the nature and amount of water that surrounds soil particles and microstructural units have been described in the earlier chapters (Chapter 3). The quantity of unfrozen water content, as a proportion of the total (initial) water content of a soil, varies with (a) freezing

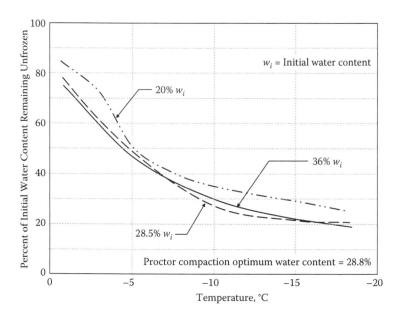

FIGURE 10.20
Percentage of water content remaining unfrozen in laboratory-prepared medium "plastic" clay (liquid limit = 67%, plastic limit = 28.8%); w_i refers to the initial water content of the samples at the start of the freezing test.

temperature and freezing index F, (b) soil composition, (c) soil structure and proportion of active clay minerals in the soil, (d) electrolyte concentration, (e) degree of water saturation, (f) charge density of the soil particles, (g) freezing history, and (h) original water content. Figure 10.20 shows the percentage of water content remaining unfrozen in clay samples prepared in the laboratory at three different initial water contents. The freezing point depression of the water immediately next to the soil particle surfaces, which is the difference between the freezing point of bulk water and the freezing point of the water next to the particle surfaces, may be as high as 15°C.

10.5.3 Water Movement in Frozen Soils

The ability for a portion of the water content in soils to remain unfrozen in subfreezing temperatures creates the situation where water movement in soils under subfreezing temperatures is possible. When a soil freezes, temperature differences between adjacent soil elements will create temperature gradients, resulting in movement of water in the unfrozen water layers adjacent to the soil particle surfaces in response to the gradients. This is generally identified as part of the phase boundary transport phenomenon. The overall phenomenon of phase boundary transport includes the transport of salt in the unfrozen water layers in response to temperature and concentration gradients. The magnitude and rate of water movement is a function of the magnitude of the temperature gradient and on the thermal and hydraulic properties of the soil.

Gradual freezing of water in soil pores, beginning with ice nucleation and subsequent crystal growth, will generally "push" the salts out from the porewater. This process, which is often called *brine exclusion* or *salt exclusion*, produces a film of brine surrounding ice crystals within a pore space. Since pore spaces in a soil are not regular or uniform, it follows that concentrations of salt will vary from point to point in a soil. Phase boundary transport will provide the means for ionic diffusion in response to increased or decreased activities resulting from temperature and concentration differences.

The flow of soil water and salt in an unsaturated frozen soil can be expressed as (Cary and Mayland, 1972):

$$J_w = -n_w k \frac{d\psi_m}{dz} - \beta \frac{DpH}{R^2T^3} \frac{dT}{dz}$$
$$J_s = -n_s k \frac{d\psi_m}{dz} - \bar{\gamma}\tau D_s \frac{dn_s}{dz}$$

(10.24)

where J is the flux at depth z; subscripts w and s refer to water and salt, respectively; $\bar{\gamma}$ refers the fraction of unfrozen water at depth z; τ is the tortuosity; D_s is the diffusion coefficient of salt in water; n_w and n_s are the mole fraction of water and salt, respectively; β is a dimensionless constant, generally taken to be about 2.5; D is the diffusion coefficient of water vapour in air; p is the pressure of ice; k is the hydraulic conductivity; ψ_m is the matric potential; R is the gas constant; T is the temperature; and H is the latent of vaporization.

10.5.4 Salt Concentration and Unfrozen Water Content—Swelling Clays

The amount of water remaining unfrozen in swelling clays is higher than in most other types of clays because of the water-holding characteristics of the clays. The influence of salt concentrations on the freezing phenomenon in such soils and their effect on the amount of unfrozen water have been studied in relation to problems arising from freezing of landfill multibarrier systems consisting of engineered clay liners.

Models predicting unfrozen water content in frozen swelling clays need to combine surface forces and dissolved solute influences. The migration of interlayer water to growing ice crystals in pore spaces during freezing will result in a decrease in interlayer spacing and an increase in the swelling pressure of the clay. Further reduction in the temperature will cause more water to move to the growing crystals and will also further reduce interlayer spacing and increase swelling pressure. With these mechanisms in mind, it follows that one can use the swelling pressure as a parameter to index the freezing point depression. The partial molar free energy difference $(\bar{F}_l - F_l^o)$ with respect to the standard state can be related to the swelling pressure Π by the relationship:

$$(\bar{F}_l - F_l^o) = -\bar{v}_l \Pi \tag{10.25}$$

where $\bar{F}_l$ and F_l^o are the partial molar free energy of water in porewater and at standard state, respectively, and $\bar{v}_l$ is the partial molar volume of the water in the soil.

The difference in partial molar free energy can be thermodynamically expressed as:

$$(\bar{F}_l - F_l^o) = RT \ln a_l \tag{10.26}$$

where R is the gas constant, T is the temperature, and a_l is the activity of liquid water.

Assuming the partial molar free energy of the porewater adjacent to a soil particle surface to be equal to the free energy of ice (solid water) at the equilibrium state, that is, $\bar{F}_l = \bar{F}_s^o$, where the subscripts s and l and the superscript o refer to solid, liquid, and standard states, respectively. With the appropriate substitutions and differentiations, the following relationship can be obtained:

$$\frac{d \ln a_l}{dT} = -\frac{1}{R} \frac{d(\Delta \bar{F}_f^o / T)}{dT} = \frac{\Delta \bar{H}_f^o}{RT^2}, \tag{10.27}$$

where $\Delta \bar{F}_f^o = F_l^o - \bar{F}_s^o$, $d(\Delta \bar{F}_f^o / T)/dT = \Delta \bar{H}_f^o / T^2$, and ΔH_f refers to the heat of fusion of ice transformation. Assuming $\Delta \bar{H}_f^o$ to be piecewise constant over a small temperature change from T_f to the freezing temperature T_f^*, and defining the freezing point depression as δ, that is, $\delta = T_f - T_f^*$, integration of Equation (10.27) from T_f to T_f^* leads to

$$\ln a_l = -\frac{\Delta \bar{H}_f^o \delta}{RT_f T_f^*} \tag{10.28}$$

Taking note that at the freezing temperature T_f^*, $(\bar{F}_l - F_l^o) = RT_f^* \ln a_l$, and making the necessary substitution in Equation (10.26), one obtains

$$(\bar{F}_l - F_l^o) = -\frac{\Delta H_f^o \delta}{T_f} \tag{10.29}$$

With the preceding, and assuming pure ice formation and a flat surface for the ice, and the specific density of the porewater to be the same as bulk water, Equation (10.25) can be solved to obtain

$$\delta = \frac{\bar{v}_l \Pi T_f}{\Delta \bar{H}_f^o},$$

(10.30)

where the swelling pressure Π is at the temperature T_f^*.

For swelling pressures, the effect of salt or brine exclusion during ice formation in the freezing process can be calculated by the following relationship:

$$\Pi = 2RT_f^* c_o (\cosh y_m - 1),$$

(10.31)

where c_o is the electrolyte concentration in mol/l remaining in the porewater (soil water) when ice forms, and y_m is the dimensionless potential in the plane midway between two parallel plates (see Equation [3.16] in Chapter 3). The calculations reported by Yong et al. (1979) for the unfrozen water contents for Na-montmorillonite with different salt concentrations and at various subfreezing temperatures, which can be seen in Figure 10.21, are compared with

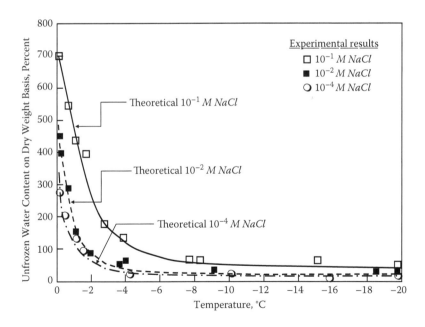

FIGURE 10.21
Comparison of theoretically calculated and experimental results of unfrozen water content of Na-montmorillonite with three different sodium chloride concentrations. Initial unfrozen water content of the Na- montmorillonite was 700%, with specific surface area S = 770 m²/g, and half spacing $d = 100\omega/S$, where ω is the weight water content of the clay. (Data reported by Yong, R.N., et al., 1979, *J. Eng. Geol.*, 13:137–155.)

actual measurements obtained in laboratory tests using a differential scanning calorimetry technique for generation of multiple endotherms of frozen test samples.

10.6 Concluding Remarks

The subject of *environmental soil behaviour* covers a wide range of topics, the details and extent of which are related to one's discipline and interests. Of the many different topics that comprise environmental soil behaviour, we have chosen to focus on two particular encompassing aspects that have considerable impact on soil behaviour and functionality.

Soil ageing is a continuing process, and the nature of a soil at any one instance of time is a reflection of the history of interactions and processes. Laboratory and/or field determinations of soil properties and characteristics—that is, soil attributes—do nothing more than provide one with a snapshot of those attributes. One needs to pay attention to the evolving nature of those attributes, and consider how their evolution will affect the planned functionality of the soil under consideration. It can be argued that almost all of the ageing processes are site specific, and that since changes in soil attributes are miniscule in relation to the life-span of most anthropogenic projects, changes in soil attributes due to ageing processes are not significant issues. Whilst this may be true, there is one overarching concern that must be addressed—changes in soil attributes due to anthropogenic events and activities, that is, soil evolution (as distinct from soil ageing, see Section 10.1).

The result of one of the most significant sets of anthropogenic events/activities is the presence of contaminants (inorganic and organic chemical compounds) in the ground. These contaminants are the stressors that have considerable impact on soil attributes. The discussions in the first part of this chapter have focused on three aspects of the soil contamination problem: (a) processes involved in the transport and fate of contaminants in soils, (b) effects of contaminant presence on soil attributes, and (c) the role of soil, and especially the microorganisms in soil, in management of soil pollution through its bioremediation capabilities.

It is manifestly clear that contaminants in soils will have significant effects on the properties and characteristics of soils. When soils are required to use their mechanical, hydraulic, and transmission capabilities for their planned (project) functionality, degradation of such capabilities through contaminant presence and interactions will have a negative impact on their functionality. In short, soil evolution needs serious consideration. The other side of the coin shows that soil attributes developed from the natural presence of microorganisms and their activities can be used advantageously to manage adverse effects of contaminant presence in soils. The discussions in this chapter have shown that natural attenuation capability of soils is a powerful tool that can be used not only in a passive capacity, but also aggressively as a control barrier system.

Seasonal freeze–thaw performance of soils is an environmental soil behaviour phenomenon, as demonstrated by frost heave and subsequent thaw-settlement problems. What is not fully appreciated in the study of soil freezing and related effects/problems is the fact that unfrozen water in subfreezing soils can pose significant problems through the phenomenon of phase boundary transport. In situations where the presence of contaminants in soils is a fact, the transport of these contaminants in the unfrozen boundary layers of soil particles can pose significant health threats to biotic receptors. There has not been sufficient attention and research paid to such phenomena in the past, perhaps because of the lack of awareness of the level of significance transport in the unfrozen boundary layer. The discussions mounted in this chapter in this area of concern are meant to highlight this phenomenon, with the hope that the problem will be given its proper attention in the near future.

References

Bader, J.L., Gonzales, G., Goodell, P.C., Pilliand, S.D., and Ali, A.S., 1996, Bioreduction of hexavalent chromium in batch cultures using indigenous soil microorganisms, HSRC/WERC, Joint Conference on the Environment, Albuquerque, NM, April 22–24.

Bartha, R., 1986, Biotechnology of petroleum biodegradation, *Microbial Ecol.,* 12:155–172.

Borden, R.C., Daniel, R.A., LeBrun, L.E., and Davis, C.W., 1997, Intrinsic biodegradation of MTBE and BTEX in a gasoline-contaminated aquifer, *Water Resources Res.,* 33:1105–1115.

Brock, T.D., Smith, D.W., and Madigan, M.T., 1984, *Biology of Microorganisms,* 4th ed., Prentice-Hall, Englewood Cliffs.

Brown, R.A., Hicks, P.M., Hicks, R.J., and Leahy, M.C., 1995, Postremediation bioremediation, in R.E. Hinchee, J.T. Wilson, and D.C. Downey (Eds.), *"Intrinsic Bioremediation,* Batelle Press, Columbus, OH.

Cary, J.W., and Mayland, H.F., 1972, Salt and water movement in unsaturated frozen soil, *Soil Sci. Soc. Am. Proc.,* 36:549–555.

Cerniglia, C.E., 1984, Microbial metabolism of polycyclic aromatic hydrocarbons, *Adv. Appl. Microbiol.,* 30:31–39.

Ehrlich, H.L., 2002a, *Geomicrobiology.* 3rd ed., Marcel Dekker, New York.

Elliott, H.A., Liberati, M.R., and Huang, C.P., 1986, Competitive adsorption of heavy metals by soils, *J. Environ. Qual.,* 15:214–219.

Grim, R.E., 1953, *Clay Mineralogy,* McGraw Hill Book Co. New York, 384p.

Grindrod, P., and Takase, H., 1994, Reactive chemical transport within engineered barriers. In *Proc. 4th Int. Conf. on the Chemistry and Migration Behaviour of Actinides and Fission Products in the Geosphere*, Oldenburg Verlag, pp. 773–779.

Huang, P.M., Wang, M-K., and Chiu, C-Y., 2005, Soil mineral-organic matter-microbe interactions: impacts on biogeochemical processes and biodiversity in soils, *Pedobiologia*, 49:609–635.

Khovrychev, M.P., Ivanova, I.I., and Taptykova, S.D., 1974, Chemical composition and physiological properties of *Candida utilis* in the presence of inhibition of growth by copper ions, *Microbiology*, 43:405–409.

Kurek, E., 2002, Microbial mobilization of metals from soil minerals under aerobic conditions, in P.M. Huang, J.M. Bollag, and N. Senesi (Eds.), *Interactions between Soil Particles and Microorganisms: Impact on the Terrestrial Ecosystem*, Vol. 8, IUPAC Series on Analytical and Physical Chemistry of Environmental Systems, Wiley, Chichester, U.K., pp. 189–225.

Landmeyer, J.E., Chapelle, F.H., Bradley, P.M., Pankow, J.F., Church, C.D., and Tratnyek, P.G., 1998, Fate of MTBE relative to benzene in a gasoline-contaminated aquifer (1993–1995), *Ground Water Monit. Rev.*, 18(4):93–102.

LaRossa, R.A., and Schloss, J.V., 1984, The sulfonyl berbicie sulfometuron methyl is an extremely potent and selective inhibitor of acetolactate synthase in *Salmonella typhimurium*, *J. Biol. Chem.*, 259:8753–8757.

Lovely, D.R., 1995, Bioremediation of and metal contaminants with dissimilatory metal reduction. *J. Industrial Microbiol.*, 14:85–93.

Monod, J., 1949, The growth of bacterial culture, *Annu. Rev. Microbiol.*, 3, 371.

Moodie, A.D., and Ingledew J., 1990, Microbial anaerobic respiration, *Adv. Microbiol. Physiol.*, 31:225–269.

Nadeau, P.H., and Bain, D.C. 1996, Composition of some smectites and diagenetic illicit clays and implications for their origin, *Clays and Clay Minerals*, 14:455–464.

Nakano, M., and Kawamura, K., 2010, Estimating the corrosion of compacted bentonite by a conceptual model based on microbial growth dynamics, *Appl. Clay Sci.*, 47:43–50.

Pitter, P., and Chudoba, J., 1990, *Biodegradability of Organic Substances in the Aquatic Environment*, CRC Press, Boca Raton, FL.

Pusch, R., 1994, Waste disposal in rock, in *Developments in Geotechnical Engineering*, 76, Elsevier, Amsterdam.

Pytte, A.M., 1982, The kinetics of smectite to illite reactions in contact metamorphic shales, M.A. Thesis, Dartmouth College, New Hampshire.

Reynolds, T.D., and Richards, P.A., 1996, *Unit Operations and Processes in Environmental Engineering*, PWS Publishing Company, Boston.

Salinitro, J.P., 1993, The role of bioattenuation in the management of aromatic hydrocarbon plumes in aquifers. *Ground Water Monit. Remed.*, 13:150–161.

Sposito, G., 1984, The surface chemistry of soils. Oxford University Press, New York, 234p.

Strauss, H., 1991, Final report: An overview of potential health concerns of bioremediation, Env. Health Directorate, Health Canada, Ottawa, 54 pp.

USEPA-SAB, 2001, Monitored natural attenuation: USEPA research programme—an EPA science advisory board review, EPA-SAB-EEC-01-004.

Weaver, C.E., 1979, Geothermal alteration of clay materials and shales: Dianesis, Tech. Report, ONWI-21, ET-76-C-06-1830. Battelle Office on Nuclear Waste Isolation, Paris.

Yong, R.N., 1965, Soil suction effects on partial soil freezing, *Highway Research Board*, 68:31–42.

Yong, R.N., 1967, On the relationship between partial soil freezing and surface forces, in H. Oura (Ed.), *Physics of Snow and Ice*, Bunyeido Printing Co., Sapporo, Japan, pp. 1375–1386.

Yong, R.N., Ccheung, C.H., and Sheeran, D.E., 1979, Prediction of salt influence on unfrozen water content in frozen soils, *J. Eng. Geol.*, 13:137–155.

Yong, R.N., Boonsinsuk, P., and Tucker, A.E., 1984, A study of frost-heave mechanics of high clay content soils, *Trans. ASME*, 106:502–508.

Yong, R.N., Elmonayeri, D.S., and Chong, T.S., 1985, The effect of leaching on the integrity of a natural clay, *J. Eng. Geol.*, 21:279–299.

Yong, R.N., Boonsinsuk, P., and Tucker, A.E., 1986, Cyclic freeze-thaw influence on frost heaving pressures and thermal conductivities of high water content clays, *Proc. Fifth Int. Offshore Mechanics and Arctic Engineering*, 4:277–284.

Yong, R.N., Mohamed, A.M.O., and Warkentin, B.P., 1992, *Principles of Contaminant Transport in Soils*, Elsevier, Amsterdam, 327 pp.

Yong, R.N., and Phadungchewit, Y., 1993, pH influence on selectivity and retention of heavy metals in some clay soils, *Can. Geotech. J.*, 30, 821–833.

Index